AF837029

Kai Borre

Mathematical Foundation of Geodesy

Kai Borre
(Editor)

Mathematical Foundation of Geodesy

Selected Papers of Torben Krarup

With 15 Figures

Springer

PROF. DR. KAI BORRE
Aalborg University
Danish GPS Center
Niels Jernes Vej 14
9220 Aalborg
Denmark

e-mail:
borre@gps.aau.dk

Library of Congress Control Number: 2006923696

ISBN-10 3-540-33765-2 Springer Berlin Heidelberg New York
ISBN-13 978-3-540-33765-2 Springer Berlin Heidelberg New York

This work is subject to copyright. All rights are reserved, whether the whole or part of the material is concerned, specifically the rights of translation, reprinting, reuse of illustrations, recitation, broadcasting, reproduction on microfilm or in any other way, and storage in data banks. Duplication of this publication or parts thereof is permitted only under the provisions of the German Copyright Law of September 9, 1965, in its current version, and permission for use must always be obtained from Springer-Verlag. Violations are liable to prosecution under the German Copyright Law.

Springer is a part of Springer Science+Business Media
springer.com
© Springer-Verlag Berlin Heidelberg 2006
Printed in The Netherlands

The use of general descriptive names, registered names, trademarks, etc. in this publication does not imply, even in the absence of a specific statement, that such names are exempt from the relevant protective laws and regulations and therefore free for general use.

Cover design: Erich Kirchner
Typesetting: camera-ready by the editor
Production: Christine Jacobi
Printing: Krips bv, Meppel
Binding: Stürtz AG, Würzburg

Printed on acid-free paper 30/2133/cj 5 4 3 2 1 0

Torben Krarup

∗ March 2, 1919 † November 24, 2005

Laudatio for Torben Krarup
Levallois Medal

by

Helmut Moritz

A committee consisting of all IAG Past Presidents unanimously recommended to award at this General Assembly the Levallois Medal to Dr.h.c. Torben Krarup. I have the honor to present the laudatio.

This is an easy and pleasant task. Easy because Krarup, since about 1969, is generally recognized as the authority on physical geodesy. The name of "least-squares collocation" is inseparably connected with Torben. I am proud that this idea goes back to the time when I was the chairman of an IAG Study Group on Mathematical Methods in Physical Geodesy of which he was the most active and inspiring member.

His "Letters on Molodensky's Problem" sent to the members of this Study Group became an influential instrument (the influence reached as far as to the famous Swedish mathematician Lars Hörmander) although he could not be persuaded to publish them. In this he followed Carl Friedrich Gauss ("Pauca sed matura"). He shows that one can influence the history of geodesy even without participating in the current paper industry.

Born March 2, 1919 in Odder (Denmark), he studied first mathematics and physics and then geodesy in Copenhagen, finishing 1952. At the Danish Geodetic Institute he was instrumental in geodetic computations, having actively participated in the construction of the electronic computer GIER built at the Geodetic Institute around 1960. So he was a pioneer also in this field.

Torben Krarup is a wonderful person: kind, gentle, helpful, unselfish and, above all, inpeccably honest. He likes to spread his ideas in discussions with friends and in letters in a generous way.

I do not know a person more worthy of the Levallois Medal. In the name of the international geodetic community, I congratulate you on this honor and thank you, Torben, for your great contributions to geodesy and for your equally great friendship.

I have the great pleasure and honour to hand over now the Levallois Medal and the related certificate to you, Torben Krarup. The certificate reads:

The International Association of Geodesy
awards the Levallois Medal
to
Torben Krarup
In recognition of distinguished service to the Association and
the science of geodesy in general.

XXII IUGG/IAG general Assembly
Birmingham, UK, July, 1999
Signed. Klaus-Peter Schwarz, IAG president

Preface

This book contains contributions from one of the outstanding geodesists of our time. From start to finish the reader will find wonderful insights and be stimulated by the manner in which the interdependence of various subjects emerge from these writings. The influence of Torben Krarup's work has been tremendous, and by publishing this volume his achievements are made available to everyone.

Most of the material in the book has been hard to obtain. The idea of collecting it dates back to the 1980's, when Helmut Moritz asked me if it were possible to publish Torben Krarup's so-called Molodensky letters. At that time the difficulties of getting this project started prevented me from trying it.

With the coming of TeX and my closer relationship with the publishing world I saw new possibilities. So I decided to create pdf-versions of all papers that never found their way into a periodical. Often I scanned the typed papers, and optical character recognition software secured a surprisingly correct result. Formulas and tables had to be entered in LaTeX and the figures were created by means of the powerful MetaPost. This last step took place on my travels abroad where I spent hundreds of hours in hotel rooms in Finland, France, Greece, Italy, Lithuania, Norway, Sweden, and the USA working on this manuscript.

The next step was to scan all printed and published material. That job was easier, but also more comprehensive because of the large number of pages. Finally all the material was compiled into a book. The project was undertaken because I felt strongly that having all of Torben's writings accessible in one volume would be of great benefit to the geodetic community.

The volume contains a variety of mathematical techniques useful for geodesy. One of Torben's hallmarks is his broad knowledge of mathematics combined with a rare innovative ability to put *relevant* topics and concepts together. Modern students of geodesy can learn a lot from his selection of mathematical tools for solving actual problems. He himself somewhere in the book says: *...Most people seem to be more interested in presenting new ideas than in* *transforming good ideas into reality.* Torben's writings are mathematically well founded and professionally relevant. It is rare nowadays to find scientific papers that fulfill both requirements.

Torben Krarup was very reluctant to publish. He found it difficult to put his ideas on paper. Typically, papers were re-written more than once, until a formulation was found that satisfied him. Sometimes this did not happen and papers were eventually put away. This painstaking process of finding the right expression often led to compact formulations which some readers find difficult to digest. To some extent he writes for the chosen few. He wants the reader to be actively involved, and therefore he puts high demands on him. So it may appear contradictory that he became disappointed when only a few people understood his main ideas after the writing had been so laborious. *It is more important to be understood than to understand by yourself*—he once remarked on this theme. It took a few years before someone understood his collocation theory. It was seen as a framework for theoretical investigations, not as a procedure for creating numbers.

Once he told me that you never should publish more than a third or a half of what you know. The papers that I have followed closely really live up to this principle. His paper on Helmert Geometry underwent at least three different derivations, all fundamentally different. Only very creative mathematicians can do this.

I have co-authored papers with Torben that never passed the final test. In my opinion they were excellent and I remember my disappointment when he did not agree to publication. Because of this, Torben has left numerous unfinished manuscripts. They are not included in this volume because the papers published here cover all relevant subjects to which he has contributed.

As a colleague, he was very generous about sharing ideas. I think he did this with a twofold aim: To educate students and colleagues to use the best possible mathematical tools, and also to start a professional discussion on the subject matter. Never in my life have I experienced more fatiguing moments than those together with him. Most often they were tremendously fruitful!

I believe Torben's finest mathematical skill is his intuition. Often he knows the result before it has been derived and proved. My most recent memory is of the prolonged process of establishing the Helmert Geometry. Many colleagues would ask for the result in the form of algorithms. He denied to give them away as he was looking for a geometric description of the problem rather than a set of algorithms. His curiosity was the driving force, the result a valuable supplementary gain.

Torben Krarup's interest in geodetic problems can be compared to a sculptor's attention to each work. His interest declines when the solution emerges. The artist's interest lies in the problem, not in the complete work. In this respect Torben Krarup is an artist.

Faced with the life work of a great scientist one asks if there is an overall structure in the topics and problems treated. Let me try to answer this question by looking at some of his major contributions.

In the late 1950s Torben worked in a team dealing with computational problems in geodesy. As an outcome of this work the possibility of building a transistorized computer arose. Krarup contributed much to its logic. It was called GIER and became an important tool for the study of the propagation of rounding errors in geodetic networks. This project also led to a better understanding of the organization and structure of geodetic network problems.

With the publication of *A Contribution to the Mathematical Foundation of Physical Geodesy* (1969) Torben created a well founded mathematical frame for a description of the reality of physical geodesy. The booklet introduced a new concept, namely *collocation.* It took a considerable time before the message was understood by a number of the leading geodesists. It actually happened when Helmut Moritz became the promotor. A number of geodesists considered it as an algorithm for performing comprehensive computations; however, it was meant as a common reference frame for description of geodetic observations, so that they could be corrected properly for the influence of the gravity field of the Earth.

It was not unexpected that a paper dealing with details on how to treat geodetic observation equations followed. A final version of *Integrated Geodesy* was published in 1978. The core concept is the definition of a local frame which is based on the reference potential. The classical geodetic observation equations are linearized in this frame and are used in an iterative collocation process. The iteration stops when a minimum condition is fulfilled.

In parallel to this development Krarup was also occupied with a rigorous formulation of the problem of physical geodesy: Given the potential field of the Earth and some discrete geodetic observations like distances, angles, absolute gravity, height differences, deflections of the vertical etc., determine the surface of the physical Earth. By means of partial differential equations Krarup succeeded in a stringent formulation of a free value boundary problem with oblique derivatives. The problem description was a reformulation of Molodensky's problem of 1945. In 1973, after ten years of thinking, he wrote the celebrated Molodensky letters which were distributed to the members of a study group on the topic. Perhaps this was his most important contribution to our science.

Models for combining discrete geodetic measurements and the continuous gravity field of the Earth were very much on Torben's mind. Therefore it was a natural continuation of his activities when he formulated the geodetic elasticity theory. In 1974 he published *Foundation of a Theory of Elasticity For Geodetic Networks* which is based on the fundamental duality between applied and numerical mathematics.

During the later years Torben Krarup and I worked on certain generalizations of the Helmert transformation. The draft paper *Helmert Geometry—or the Importance of Symmetry* looks at different aspects of the linear part, i.e. the translation, and the non-linear part, i.e. the rotation and change of scale. It turns out that singular value decomposition provides better insight into the problem than the classical solution.

Over more than a generation Torben Krarup has continuously been intrigued by central problems in geodesy. He succeeded in formulating them in a relevant mathematical framework. But equally important, he succeeded in solving the mathematical problems so formulated.

Allow me to add a comment on publishing conditions then and now. As a state geodesist at the Geodetic Institute, Torben Krarup did not suffer from the contemporary pressure on publishing. When you have genuine and new results you publish. Cutting and pasting papers was not a part of his world. If he republished, it happened in order to correct earlier mistakes. He was very honest to himself and to the scientific community in which he worked.

When he decided to publish it was most often in non-refereed journals and reports. So when circumstances made it a necessity to go through the referee system he was often annoyed and angry. In many cases the result was not to publish.

During the editing of the material contained in this book, I have come across situations where another editor certainly would have changed the presentation. Only unpublished material has been properly copy-edited. It has led to minor changes, but always with due respect to Torben's definite style.

Acknowledgements

A substantial financial support from Department of Communication Technology, Aalborg University is gratefully acknowledged.

My colleagues Petr Holota and Klaus-Peter Schwarz helped in various ways during the work. I highly appreciate this help.

I'm grateful that Springer has agreed to publish the present volume. Throughout the process I have appreciated the smooth collaboration with Dr. Christian Wischel.

This book is set with TeX, LaTeX 2ε, and AMS-TeX using Computer Modern fonts. The masterful page layout and numerous intricate TeX solutions are due to Frank Jensen. John D. Hobby's MetaPost should also be mentioned. Most figures were created by combining Donald Knuth's METAFONT and PostScript from Adobe. Using it is a sheer joy.

Fjellerad,
February 2006

Kai Borre
borre@gps.aau.dk

Contents

1

Linear Equations

Introduction

In the last years several methods for solving linear algebraic equations have been published. This chapter is not meant to be a more or less detailed survey of these methods, rather, its purpose is to give the reader such a knowledge of a few special methods or techniques that he will be able to make a series of programs for his computer for solving linear equations of the types he is likely to meet with in practice. We find it important to point out that one must have a series of programs for these problems and not merely one standard program because the problems which occur in practice vary considerably with respect to dimensions, numerical behaviour, symmetrical pattern etc. The cardinal point in our investigation has been the numerical strength of the methods, considering that large systems of equations are often ill-conditioned, and it is of little consolation to use a method that uses the central and the peripheral storages of a computer in an elegant way if the numerical result of the computation is drowned in rounding errors.

Direct Methods: The General Case

The form of the linear equations considered is supposed to be the following:

$$AX = B \tag{1}$$

where A is a given non singular $n \times n$ matrix, B a given $n \times m$ matrix, and X is the unknown $n \times m$ matrix. If B is the $n \times n$ unit matrix then X will be the inverse A^{-1} of A and the problem of matrix inversion is thus included in our problem. We suppose the numbers to be real numbers. The technique we shall use is the classical one of reducing the system of equations to successive systems of equations with triangular matrices of coefficients, that is matrices with all the coefficients below the diagonal (or above the diagonal) being zero.

We try to decompose the system (1) thus

$$\begin{aligned} LY &= B \\ UX &= Y \end{aligned} \tag{2}$$

where U is an upper triangular matrix having zeros below the diagonal, L is a lower triangular matrix having zeros above the diagonal, and

$$LU = A. \tag{3}$$

The same connection can also be expressed in another way. Let us consider the compound matrix $[A \;\; B]$ consisting of the columns of A followed by the columns of B and let us try to express this matrix as a product of a lower triangular matrix L_A and an upper triangular matrix $[U_A \;\; U_B]$

$$L_A[U_A \;\; U_B] = [A \;\; B],$$

then the system (1) is reduced to the triangular system

$$U_A X = U_B.$$

The two descriptions are only formally different, the following equations being fulfilled

$$\begin{aligned} L_A &= L \\ U_A &= U \\ U_B &= Y \end{aligned}$$

and the calculations being the same. It is however often convenient to use the latter description.

If the triangular decomposition (3) is possible at all it is possible in an infinity of ways. If LU is one triangular decomposition of A then $(LD^{-1})(DU)$, where D is any non singular diagonal matrix, is another decomposition. In the following we shall suppose that the decomposition is defined by the condition that all the diagonal elements of the matrix L be unity.

Equation (3) or, more explicitly,

$$\begin{bmatrix} 1 & & & \\ l_{21} & 1 & & \\ l_{31} & l_{32} & 1 & \\ & \cdots & & \ddots \end{bmatrix} \begin{bmatrix} u_{11} & u_{12} & u_{13} & \cdots \\ & u_{22} & u_{23} & \cdots \\ & & u_{33} & \cdots \\ & & & \ddots \end{bmatrix} = \begin{bmatrix} a_{11} & a_{12} & a_{13} & \cdots \\ a_{21} & a_{22} & a_{23} & \cdots \\ a_{31} & a_{32} & a_{33} & \cdots \\ \cdots & \cdots & \cdots & \ddots \end{bmatrix} \tag{4}$$

is equivalent to the following series of triangular systems of equations:

$$
\begin{aligned}
&& u_{11} &= a_{11};\\
l_{21}u_{11} &= a_{21}; & u_{12} &= a_{12};\\
&& l_{21}u_{12} + u_{22} &= a_{22};\\
l_{31}u_{11} &= a_{31}; & u_{13} &= a_{13};\\
l_{31}u_{12} + l_{32}u_{22} &= a_{32}; & l_{21}u_{13} + u_{23} &= a_{23};\\
&& l_{31}u_{13} + l_{32}u_{23} + u_{33} &= a_{33};\\
&& \text{etc.} &
\end{aligned}
\tag{5}
$$

The u's are found by solving the right hand systems and the l's are found by solving the left hand systems, and it can be done in the following order: First the first right hand system, then the first left hand system and alternately the next right hand and the next left hand system. When the l's are found the right hand system can always be solved, but the condition for the solvability of the left hand systems is that $u_{ii} \neq 0$ for $i = 1, 2, \ldots, n$. (It should be pointed out that the equations (5) can be solved also in other orders, as for instance the algorithm (13) below shows.)

Considering the determinants of the matrices in (4) one obtains

$$u_{11}u_{22}\cdots u_{nn} = \det(A),$$

but considering the submatrices consisting of the first p columns and the first p rows of each matrix, the formula (4) is still valid, and from the determinants one obtains

$$u_{11}u_{22}\cdots u_{pp} = \det(A_p)$$

where $\det(A_p)$ is the pth principal subdeterminant of A. It follows that if $\det(A_p) \neq 0$ for $p = 1, 2, 3, \ldots, q$ but $\det(A_q) = 0$ then

$$u_{pp} \neq 0 \quad \text{for} \quad p = 1, 2, 3, \ldots, q-1 \quad \text{but} \quad u_{qq} = 0.$$

Since it is always possible to rearrange the columns and rows of a non singular matrix so that the principal subdeterminants are non zero, it follows from the above that it is always possible—at least, after a rearrangement of columns and rows—to effectuate the triangular decomposition (3) of a non singular matrix A. In practice it is necessary to make a rearrangement not only if one of the diagonal elements of U would otherwise be zero but also if it would be relatively small.

Some special properties of the matrix A can be used to make the decomposition simpler.

If the matrix A is symmetric: $A^{\mathrm{T}} = A$, and the decomposition (3) is possible, then the decomposition can be effectuated in a symmetrical way thus:

$$A = LDL^{\mathrm{T}} \tag{6}$$

where D is a diagonal matrix the diagonal elements of which are ± 1. (The diagonal elements of L are here in general different from unity). It is, however, not always possible during the triangular decomposition to restrict the rearrangements of columns and rows to those leaving the coefficient matrix symmetric. In Section 1.2 we shall see a very important case where the symmetric decomposition is possible.

Skew symmetric matrices (matrices A for which $A^{\mathrm{T}} = -A$) have all the diagonal elements equal to zero, hence they can not be decomposed in the normal way as skew symmetric matrices. Since systems of linear equations with skew symmetric coefficient matrices should be relatively common in problems relating to partial differential equations it is surprising that the solving of such systems is dealt with so seldom, that the authors have no reference to papers treating it. Since symmetric matrices can generally be decomposed according to the formula (4), a skew symmetric matrix of even order (a skew symmetric matrix of odd order is always singular) can generally be decomposed thus:

$$A = LJL^{\mathrm{T}} \tag{7}$$

where J is a matrix for which

$$j_{rs} = \begin{cases} 1 & \text{for} \quad r, s = 2p-1,\, 2p \quad \text{and} \quad p = 1,\, 2,\, \ldots,\, n/2 \\ -1 & \text{for} \quad r, s = 2p,\, 2p-1 \quad \text{and} \quad p = 1,\, 2,\, \ldots,\, n/2 \\ 0 & \text{otherwise} \end{cases}$$

and L is a lower triangular matrix such that

$$\begin{aligned} l_{2p,2p-1} &= 0 \\ l_{2p,2p} &= l_{2p-1,2p-1} \quad \text{for} \quad p = 1,\, 2,\, \ldots,\, n/2. \end{aligned}$$

The proof and the necessary formulae are found very simply by writing (7) explicitly as a system of equations.

Many attempts have been made to make use of the fact that the coefficient matrices met with in practice very often consist of relatively few non zero elements, but with only very slight success. Virtually the only thing to say about this is the following almost obvious theorem:

> If the first p elements (p is any integer) of a column of A equal zero, then the corresponding elements of U are zero too; if the first p elements of a row of A equal zero, then the corresponding elements of L are zero too.

A general theorem about matrices that deserves to be commonly known shall be exposed here:

> For any two square matrices A and B of orders m and n respectively we define the *direct product* $[a_{rs}] \times [b_{rs}]$ as the square matrix $[a_{rs}b_{r's'}]$ of order mn in which the pair of numbers r, r' define the row, and s, s' the column. Any convention may be adopted for the ordering of the rows in the direct product but the same convention must be adopted for the columns. (A natural convention is to let (r_1, r_1') precede (r_2, r_2') for $r_1 < r_2$ and for $r_1 = r_2$ if $r_1' < r_2'$; the matrix $A \times B$ is then the

compound matrix made from A by replacing for each element a_{rs} the matrix $a_{rs}B$.)

If A and B are two matrices of order m and C and D are two matrices of order n, then we have

$$(A \times C)(B \times D) = (AB) \times (CD)$$

and if A and C moreover are non singular, we have

$$(A \times C)^{-1} = A^{-1} \times C^{-1}.$$

The reader will with a pencil and paper be able to prove this.

Before we go into details concerning the practical solution of linear equations we shall investigate how well the unknowns are defined by such equations with rounded coefficients, that is to say, how much the unknowns can alter when the coefficients undergo small alterations.

For the n-dimensional vectors we shall use the Euclidian metric, the *norm* $\|V\|$ of a vector V being defined by

$$\|V\| = \left(\sum_{i=1}^{n} V_i^2 \right)^{1/2}$$

and for matrices we define the *norm* by

$$\|A\| = \max_{x \neq 0} \frac{\|Ax\|}{\|x\|}.$$

The exact solution of the system of equations

$$Ax = b, \qquad A \text{ non singular and } b \text{ a column vector,}$$

is

$$x = A^{-1}b.$$

If b is given an increment Δb and if the new solution is denoted by $x + \Delta x$, we have

$$\begin{aligned} x + \Delta x &= A^{-1}(b + \Delta b) = A^{-1}b + A^{-1}\Delta b \\ \Delta x &= A^{-1}\Delta b \end{aligned}$$

and

$$\|\Delta x\| \leq \|A^{-1}\| \cdot \|\Delta b\|, \tag{8}$$

using the obvious inequality $\|PQ\| \leq \|P\| \cdot \|Q\|$ valid for any matrices P and Q.

The effect of an increment ΔA of the matrix of coefficients is a little more complicated:

$$
\begin{aligned}
(A + \Delta A)(x + \Delta x) &= b, \\
Ax + (A + \Delta A)\Delta x + \Delta A \cdot x &= b, \\
\Delta x &= -(A + \Delta A)^{-1}\Delta A \cdot x
\end{aligned}
$$

if $(A + \Delta A)$ is non singular. Under this assumption we have

$$\|\Delta x\| \leq \left\|(A + \Delta A)^{-1}\right\| \cdot \|\Delta A\| \cdot \|x\|. \tag{9}$$

A known theorem concerning norms says that if $\|M\| < 1$ then $I + M$ is non singular and

$$\left\|(I + M)^{-1}\right\| \leq \frac{1}{1 - \|M\|}$$

see e.g. [2], pp. 290–291.

Therefore (9) can be written

$$\|\Delta x\| \leq \left\|A^{-1}(I + A^{-1}\Delta A)^{-1}\right\| \cdot \|\Delta A\| \cdot \|x\|$$

or

$$\|\Delta x\| \leq \frac{\|A^{-1}\| \cdot \|\Delta A\| \cdot \|x\|}{1 - \|A^{-1}\Delta A\|}$$

provided that

$$\|A^{-1}\Delta A\| < 1.$$

We can now write

$$\|\Delta x\| \leq \frac{\|A^{-1}\| \cdot \|\Delta A\| \cdot \|x\|}{1 - \|\Delta A\| \cdot \|A^{-1}\|} \tag{10}$$

provided that

$$\|\Delta A\| < \frac{1}{\|A^{-1}\|}.$$

We see from (6) and (8) that if the significance of the last digit retained in the calculations is of the same order of magnitude as

$$\frac{1}{\|A^{-1}\|},$$

then the system of equations is very ill-conditioned and the rounding errors in the original coefficients and those arising during the calculations are likely to spoil the results totally.

In investigating the errors arising from the triangular decomposition we use the ingenious method of J. H. Wilkinson and it is warmly recommended that interested readers read the two papers [1] and [2] for further details.

Supposing that the matrix $[\,A \;\; B\,]$ is so scaled that the triangular decomposition can be effectuated in fixed point arithmetic in the interval $-1 \leq x \leq 1$, we shall find the matrix $[\,A' \;\; B'\,]$ which really corresponds to the computed triangular matrices. We suppose further that the accumulations of the product sums are made with double precision, and that the divisions are made as divisions of double precision numbers by single precision numbers, the quotients being correctly rounded as single precision numbers. Then we see from the equations (5) that if we denote by a_{rs} the coefficients of the matrix $[\,A \quad B\,]$ and by a'_{rs} those of the matrix $[\,A' \quad B'\,]$ we have

$$|a'_{rs} - a_{rs}| \le \begin{cases} 2^{-t-1} & \text{for } s \ge r \\ 2^{-t-1}|u_{rr}| \le 2^{-t-1} & \text{for } s < r \end{cases} \tag{11}$$

where t is the number of binary digits in the single precision numbers.

We have seen above that it can be necessary to change the orders of rows and columns to prevent that some u_{ss} should be zero. If some u_{ss} is numerically extraordinary small then some l_{rs} with $r > s$ is likely to be extraordinary large and this will probably make some later computed numbers large etc. In fact it has been possible to give an upper bound for the maximum number appearing in the calculation only under the supposition that the "complete pivoting procedure" is used, that is to say, that before every computation of a diagonal element u_{ss} the rows and columns with numbers not smaller than s are rearranged so as to give the absolutely largest u_{ss}. Unfortunately this process is not a simple one because it is necessary to compute which a_{rs} would give the absolutely largest u_{ss}, but examples show that this pivoting procedure is generally necessary. In consideration of this we find that J. H. Wilkinson's proposal ([2]) is the most recommendable. This proposal is first to reduce the matrix (or solve the equations) according to the original Gaussian method using floating point calculations and the complete pivoting procedure which in this case is less complicated, and secondly, after scaling of the rows and rearrangement of the rows and columns of the matrix using results from the first calculation, to effectuate the triangular decomposition in fixed point arithmetic. (It is possible too to multiply the columns by appropriate scaling factors and then, when the solutions are found, to divide these by the same factors. If these scaling factors are chosen so that the solutions of the scaled equations are, to a higher degree than the solutions of the original equations, of the same order of magnitude, then this method can perhaps give more accurate results than the results calculated from the unscaled equations, but the question of what strategy to use is still open.)

By multiplying the rows by appropriate constants it is possible to transform the matrix so that the largest modulus in each row is of the same order of magnitude for all the rows, and we shall suppose this to be done before the calculations are started.

The process of solution will now be described more detailed, first the Gaussian reduction (in floating point arithmetic) and secondly the decomposition in fixed point arithmetic.

The Gaussian reduction with complete pivoting consists of $n-1$ successive steps, after the pth of which the system consists of one equation with n unknowns, one with $n-1$ unknowns, ..., one with $n-p+1$ unknowns, and $n-p$ equations with $n-p$ unknowns. The pth step can be described thus:

$$\begin{array}{l} \textbf{for } i := p+1 \textbf{ step } 1 \textbf{ until } n \textbf{ do} \\ \textbf{begin } m := a[i,p]/a[p,p];\ a[i,p] := 0; \\ \quad \textbf{for } j := p+1 \textbf{ step } 1 \textbf{ until } n \textbf{ do} \\ \qquad a[i,j] := a[i,j] - m \times a[p,j]; \\ \textbf{end}; \end{array} \tag{12}$$

But before the pth step the last $n-p$ rows and the last $n-p$ columns of the (left side) coefficient matrix must be rearranged so that the absolutely largest a_{ij} for $p < i \leq n$, $p < j \leq n$ replaces $a_{p+1,p+1}$ and the permutation of rows and columns must be recorded in a catalogue, by means of which one can permute the original rows and columns before the fixed point calculation. Besides this permutation the only result of interest of the first calculation is the largest absolute value a_{ss} (which after the rearrangement must occur in the diagonal).

After rearranging rows and columns of the original matrix and multiplication of its elements by a scaling factor $c > 0$ such that $\max |a_{ss}| \times c$ is sufficiently less than unity, the triangular decomposition in fixed point arithmetic can begin. This can be described in an ALGOL-like language thus:

$$\begin{array}{l} \textbf{for } p := 1 \textbf{ step } 1 \textbf{ until } n \textbf{ do} \\ \textbf{begin for } r := p \textbf{ step } 1 \textbf{ until } n \textbf{ do} \\ \qquad a[p,r] = a[p,r] - \sum_{i=1}^{p-1} a[p,i] \times a[i,r]; \\ \quad \textbf{for } s := p+1 \textbf{ step } 1 \textbf{ until } m \textbf{ do} \\ \qquad a[s,p] := \left(a[s,p] - \sum_{i=1}^{p-1} a[s,i] \times a[i,p]\right) / a[p,p]; \\ \textbf{end}; \end{array} \tag{13}$$

Here the sums of the type $\sum_{i=1}^{p-1}$ are interpreted as zero for $p = 1$, and corresponding interpretations are made in the following algorithms (14), (16), (26), (33), and (34).

The back substitution is effectuated by the algorithm

$$\begin{array}{l} \textbf{for } s := n+1 \textbf{ step } 1 \textbf{ until } m \textbf{ do} \\ \quad \textbf{for } p := n \textbf{ step } -1 \textbf{ until } 1 \textbf{ do} \\ \qquad a[p,s] = \left(a[p,s] - \sum_{i=p+1}^{n} a[p,i] \times a[i,s]\right) / a[p,p]; \end{array} \tag{14}$$

The computations are first carried out in floating point arithmetic and then, if one wants more precise results, in fixed point arithmetic with double precision accumulation etc. using scaling factors determined by the floating point calculation. The back substitution procedure is generally less critical than the triangular reduction, and hence the fixed point back substitution is in most cases unnecessary.

The formula (11) shows that for small u_{rr} (and for ill-conditioned systems some u_{rr} is small) the values $|a'_{rs} - a_{rs}|$ are generally smaller below the small diagonal element than to the right of it. Since it is very important that $|a'_{rs} - a_{rs}|$ is small when u_{rr} is small it would probably give a better precision if the formulae were made more symmetrical using a square root method inspired by the Cholesky method for symmetric equations (see below). In place of the algorithm (13) we will then use:

$$
\begin{array}{l}
\textbf{for } p := 1 \textbf{ step } k \textbf{ until } n \textbf{ do} \\
\textbf{begin } a[p,p] := a[p,p] - \sum_{i=1}^{p-1} a[p,i] \times a[i,p]; \\
\quad a[p,p] := sign(a[p,p]) \times sqrt\big(abs(a[p,p])\big); \\
\quad \textbf{for } s := p+1 \textbf{ step } 1 \textbf{ until } m \textbf{ do} \\
\qquad a[s,p] := \big(a[s,p] - \sum_{i=1}^{p-1} a[s,i] \times a[i,p]\big) \,/\, a[p,p]; \\
\quad \textbf{for } r := p+1 \textbf{ step } 1 \textbf{ until } n \textbf{ do} \\
\qquad a[p,r] := \big(a[p,r] - \sum_{i=1}^{p-1} a[p,i] \times a[i,r]\big) \,/\, abs(a[p,p]); \\
\textbf{end};
\end{array}
\tag{15}
$$

Since this algorithm is more complicated and slower than (14) it shall be recommended only for ill-conditioned systems.

It will often be possible to increase the precision of a solution of a system of linear equations by a simple iterative method. Let us consider the case where the right hand side of the system is a column vector b:

$$Ax = b. \tag{16}$$

The methods described above will theoretically give the calculated solution x_0 as a linear function of b, say D,

$$x_0 = Db. \tag{17}$$

We define an iteration process thus

$$x_n = x_{n-1} + D(b - Ax_{n-l}), \tag{18}$$

where the meaning is that by means of an approximate solution x_{n-1} we calculate x_n as the sum of x_{n-1} and the calculated solution of (16) where b is exchanged with the column vector of discrepancies of the equations (16) when x_{n-1} is inserted. Using the exact solution x we can write (18) in the following way

$$
\begin{aligned}
x_n - x &= x_{n-1} - x + D\big(b - A(x_{n-1} - x) - b\big) \\
x_n - x &= x_{n-1} - x - DA(x_{n-1} - x) \\
x_n - x &= (I - DA)(x_{n-1} - x) \\
\|x_n - x\| &\le \|I - DA\| \cdot \|x_{n-1} - x\|;
\end{aligned}
$$

that is to say, that if the norm λ of $I - DA$ is smaller than 1, then the iteration process will converge linearly with the convergence factor λ.

It was tempting to think, that when it is so easy to get a better approximation to the solution, then is it not important to use fixed point decomposition before using the iteration, but unfortunately it will often occur that it is difficult to get $\|I - DA\|$ smaller than unity.

We shall conclude this section with two cases where the pivoting procedure is much more simple.

The first example is that of a matrix A with a dominating diagonal:

$$|a_{rr}| \geq \sum_{s \neq r} |a_{rs}| \quad \text{for} \quad 1 \leq r \leq n. \tag{19}$$

Here the pivoting procedure is entirely unnecessary as it will probably be also if the inequality in (19) is only "approximately" true.

The second example is that of a Hessenberg matrix, that is, a matrix which below the diagonal has non zero elements only immediately below the diagonal:

$$a_{rs} = 0 \qquad \text{for} \quad r > s + 1.$$

A special case of the Hessenberg matrix is the tridiagonal matrix for which

$$a_{rs} = 0 \quad \begin{cases} \text{for } r > s + 1; \\ \text{for } s > r + 1. \end{cases}$$

For a Hessenberg matrix the following very simple pivoting procedure suffices: After the calculation of a row $u_{r,r}, u_{r,r+1}, \ldots, u_{r,n}$ of the upper triangular matrix, examine whether

$$|a_{r+1,r}| > |u_{rr}|;$$

if so permute the rows r and $r + 1$, otherwise continue.

Positive Definite Symmetric Matrix of Coefficients

Theorem *A positive definite symmetric matrix A can always be decomposed thus:*

$$A = LL^{\mathrm{T}}, \tag{20}$$

where L is a lower triangular matrix.

Proof We assume the theorem to be true for positive definite symmetric matrices of order $n - 1$. The matrix A of order n can be partitioned

$$A = \begin{bmatrix} A_{n-1} & a \\ a^{\mathrm{T}} & a_{nn} \end{bmatrix}$$

where A_{n-1} is a matrix of order $n - 1$, a is a vector of order $n - 1$, and a_{nn} is a scalar. That A is positive definite means that

$$x^{\mathrm{T}} A x > 0 \qquad \text{for all vectors } x \neq 0.$$

We consider all vectors x of the form

$$\left[\frac{x'}{0}\right]$$

where x' denotes an arbitrary vector of dimension $n-1$. The above inequality then reduces to the corresponding inequality for the matrix A_{n-1}

$$(x')^{\mathrm{T}} A_{n-1} x' > 0 \qquad \text{for all } x' \neq 0$$

which shows that A_{n-1} is positive definite. By hypothesis there exists an L_{n-1} satisfying

$$L_{n-1} L_{n-1}^{\mathrm{T}} = A_{n-1}.$$

Since $\det(L_{n-1}) = \det(L_{n-1}^{\mathrm{T}}) = \sqrt{\det(A_{n-1})} > 0$ it is possible to find a vector l of order $n-1$ such that

$$L_{n-1} l = a.$$

Then

$$\begin{bmatrix} L_{n-1} & 0 \\ l^{\mathrm{T}} & l_{nn} \end{bmatrix} \begin{bmatrix} L_{n-1}^{\mathrm{T}} & l \\ 0 & l_{nn} \end{bmatrix} = \begin{bmatrix} A_{n-1} & a \\ a^{\mathrm{T}} & a_{nn} \end{bmatrix} = A \tag{21}$$

if

$$l^{\mathrm{T}} l + l_{nn}^2 = a_{nn}. \tag{22}$$

Considering the determinants of (21) one obtains

$$\det(L_{n-1}) \det(l_{nn}) \det(L_{n-1}^{\mathrm{T}}) \det(l_{nn}) = \det(A) > 0$$
$$\left(\det(L_{n-1})\right)^2 \left(\det(l_{nn})\right)^2 > 0.$$

Hence l_{nn}^2 is positive and l_{nn} can be chosen positive (which it is assumed to be in the following).

The theorem is obviously correct for $n = 1$ and is thus proved by induction.

(22) expresses that the square sum of the elements in a row of L is equal to the diagonal element of the corresponding row of A. If the diagonal elements of A (and thus the modulus of all the elements of A) are less than unity, then all the elements of L will be less than unity too.

If L is a non singular lower triangular matrix then

$$B = LL^{\mathrm{T}}$$

must be symmetric:

$$B^{\mathrm{T}} = \left(LL^{\mathrm{T}}\right)^{\mathrm{T}} = LL^{\mathrm{T}} = B,$$

and it will be positive definite too:

$$x^{\mathrm{T}} B x = x^{\mathrm{T}} L L^{\mathrm{T}} x = \left(L^{\mathrm{T}} x\right)^{\mathrm{T}} \left(L^{\mathrm{T}} x\right) > 0 \qquad \text{for } x \neq 0$$

as

$$L^{\mathrm{T}} x \neq 0 \qquad \text{for } x \neq 0.$$

Since the determinant of L is the product of the diagonal elements the product of the squares of the diagonal elements of L is equal to the determinant of A.

We will prove that

$$m(A) \leq l_{pp}^2 \leq M(A) \qquad \text{for all } p, \tag{23}$$

where $M(A)$ and $m(A)$ are the maximal, respectively the minimal value of

$$\frac{\|Ax\|}{\|x\|} \qquad \text{for } x \neq 0,$$

$\|x\|$ being the Euclidean norm.

For the vector $x = [x_i]$, $x_i = 0$ if $i \neq p$ and $x_p = 1$, and we have

$$M(A) \geq x^{\mathrm{T}} A x = \left(L^{\mathrm{T}} x\right)^{\mathrm{T}} \left(L^{\mathrm{T}} x\right) = l_{p1}^2 + l_{p2}^2 + \cdots + l_{pp}^2. \tag{24}$$

For the vector

$$y = [y_i]$$

satisfying

$$\begin{aligned} y_p &= 1 \\ y_i &= 0 \qquad \text{for } i < p \\ l_{i1} y_1 + l_{i2} y_2 + \cdots + l_{ii} y_i &= 0 \qquad \text{for } i > p \\ l_{p1} y_1 + l_{p2} y_2 + \cdots + l_{pp} y_p &= l_{pp} \end{aligned}$$

(it is evidently possible to have these conditions fulfilled) we have

$$m(A) \leq y^{\mathrm{T}} A y = \left(L^{\mathrm{T}} y\right)^{\mathrm{T}} \left(L y^{\mathrm{T}}\right) = l_{pp}^2; \tag{25}$$

$\|y\|$ being not less than unity.

The inequalities (23) then follows from (24) and (25).

We will now investigate how to use this triangular decomposition (termed the *Cholesky method* or the *square-root method*) in practice.

The above mentioned property, that if the elements of the matrix A are (fixed-point) machine numbers then the elements of L will be machine numbers too, makes the Cholesky method well suited for fixed point calculation.

Since the matrix A is symmetric it is practical only to store the diagonal and the elements above the diagonal. It is possible to compute L^{T} and store the elements of this matrix in the same memory cells as the corresponding elements of A. The computation is described thus:

for $r := 1$ **step** 1 **until** n **do**
begin for $s := 1$ **step** 1 **until** $r - 1$ **do**
$\quad a[r,s] := \left(a[r,s] - \sum_{p=1}^{s-1} a[p,r] \times a[p,s]\right)/a[s,s];$ (26)
$a[r,r] := sqrt\left(a[r,r] - \sum_{p=1}^{r-1} a[p,r]^2\right);$
end;

In common mathematical notation the two formulae would be

$$l_{sr} = \frac{a_{rs} - \sum_1^{s-1} l_{pr} l_{sp}}{l_{ss}}; \tag{27}$$

$$l_{rr} = \sqrt{a_{rr} - \sum_1^{r-1} l_{rp}^2}. \tag{28}$$

Owing to round off errors $L^{\mathrm{T}}L$ will not be exactly equal to the original matrix A, but the arithmetical operations can be executed so as to make the difference very small. To attain this it is important to accumulate the product sums with double precision, which in most computers can be done without much extra trouble for fixed point calculation. The square root extraction of a double precision number and the division of a double precision number with a single precision number both giving the correctly rounded single precision number may also be executed in a relatively simple way. It is thus for

$$E = LL^{\mathrm{T}} - A \tag{29}$$

possible to attain

$$|E_{rs}| \le \begin{cases} 2^{-t-1}|l_{rr}|, & s > r \\ 2^{-t-1}|l_{ss}|, & r > s \\ 2^{-t}|l_{rr}|, & r = s, \end{cases} \tag{30}$$

(where t is the number of binary digits in a single precision number) if the moduli of the elements of A are so much smaller than unity and A so far from being singular that the matrix LL^{T} has the same properties. A sufficient condition for this is that

$$(n+1)2^{-t}\|A^{-1}\| < 1; \qquad |A_{rs}| < (1 - 2^{-t}).$$

It is surprising that it is possible to arrive at such a satisfactory result. It shows that generally the rounding errors introduced by the triangular decomposition are smaller than the initial rounding errors of the elements of A produced by transforming these into binary form. Moreover it follows from (29) and (30) that if any l_{rr} is small then the corresponding elements of E are far smaller than 2^{-t} and this has a very favourable effect on the solution of ill-conditioned equations.

The result of the triangular decomposition is that the solution of the equations

$$AX = B$$

has been reduced to that of solving

$$LL^{\mathrm{T}}X = B$$

or

$$LY = B \tag{31}$$

$$L^{\mathrm{T}}X = Y \tag{32}$$

that is, the original system is reduced to two systems with triangular matrices.

The system (31) is solved by the algorithm

$$\begin{array}{l}\textbf{for } s := 1 \textbf{ step } 1 \textbf{ until } m \textbf{ do}\\ \quad \textbf{for } r := 1 \textbf{ step } 1 \textbf{ until } n \textbf{ do}\\ \quad\quad b[r,s] := \left(b[r,s] - \sum_{i=1}^{r-1} a[r,i] \times b[i,s]\right) / \, a[r,r];\end{array} \tag{33}$$

where y_{rs} are the resulting values of b_{rs}. Formally these operations resemble the operations (26) but generally we can not be sure that

$$|y_{rs}| < 1.$$

Since the numerically critical part of the solution of linear equations is the triangular decomposition it will generally suffice to execute in floating point arithmetic the algorithm (33) and the algorithm

$$\begin{array}{l}\textbf{for } s := 1 \textbf{ step } 1 \textbf{ until } m \textbf{ do}\\ \quad \textbf{for } r := n \textbf{ step } -1 \textbf{ until } 1 \textbf{ do}\\ \quad\quad b[r,s] := \left(b[r,s] - \sum_{i=r+1}^{n} a[r,i] \times b[i,s]\right)/a[r,r];\end{array} \tag{34}$$

where x_{rs} are the resulting values of b_{rs}. If this does not suffice one may apply these algorithms first in floating point arithmetic and then in fixed point arithmetic with scaling factors determined by the floating point calculation.

In practice there are often many zeros among the coefficients, and it is therefore desirable to have a program where those zeros that stand above all non-zero coefficients in the same column shall not be stored and because of the symmetry of the coefficient matrix only the upper triangular part of it is stored. The author has written such a program which is described below.

The machine first computes a table for two functions p_s and q_s, where s is the column number. Here q_s is the number of ignored zeros in column s. The coefficients $a[r,s]$ that have to be stored are stored columnwise and in each column in order of ascending r. The ith element in this array is called $b[i]$ and the function p_s is defined from

$$a[r,s] = b[p_s + r] \quad \text{for} \quad q_s < r \leq s.$$

The triangular decomposition is then described by

$$\begin{array}{l}\textbf{for } s := 1 \textbf{ step } 1 \textbf{ until } n \textbf{ do}\\ \quad \textbf{for } r := q_s \textbf{ step } 1 \textbf{ until } s \textbf{ do}\\ \quad \textbf{begin } z := 0;\\ \quad\quad \textbf{for } t := max(q_r, q_s) + 1 \textbf{ step } 1 \textbf{ until } r \textbf{ do}\\ \quad\quad\quad z := z - b[p_r + t] \times b[p_s + t];\\ \quad\quad b[p_s + r] := \textbf{if } r < s \textbf{ then } (b[p_s + r] + z)/b[p_r + r]\\ \quad\quad\quad\quad\quad\quad\quad\quad \textbf{else } sqrt(b[p_s + s] + z);\\ \textbf{end};\end{array} \tag{35}$$

The back substitution is described by

```
for r := n step −1 until 1 do
begin z := c[r];
  for s := n step −1 until r + 1 do
  begin if q_s < r then z := z − c[s] × b[p_s + r];
  end;
  c[r] := z/b[p_r + r]
end;
```

Here the quantities $c[r]$ are used to supply the reduced right hand constants on entry into the calculation and, on exit, they will contain the solutions. The right hand reduction is not described here but can naturally be included in (35) if the q_{n+1} is put equal to zero.

References

[1] J. H. Wilkinson. Rounding errors in algebraic processes. In *Proceedings of International Conference on Information Processing*, UNESCO, 1959.

[2] J. H. Wilkinson. Error analysis of direct methods of matrix inversion. *Journal of Association for Computing Machinery*, 8:281–330, 1961.

2

The Adjustment Procedure in Tensor Form

The situation when applying the adjustment procedure is the following: Measurements of m parameters exist for the determination of a physical quantity A, which depends on n parameters $(m > n)$. According to the mathematical model in which the wanted physical quantity may be expressed by the n parameters, the measured quantities must fulfill certain conditions which in general are not fulfilled by the observed quantities, as these are encumbered with errors. The question now is how to define the value of the physical quantity A which is to appear from the existing observations.

According to ideas, which can at least be tracked back to Galileo, we will have to choose the values of A which correspond to the values of the observed quantities and is at the same time as close as possible to the observations.

We are now in a position to make precise definitions.

The observed parameters can be described by a vector

$$x^i \qquad i = 1, 2, \ldots, m$$

in a linear vector space $\mathcal{O}$ in which the positive definite quadratic form

$$g_{ij}x^i x^j, \qquad g_{ij} = g_{ji} \quad \text{and} \quad g_{ij} \text{ are constants}, \qquad i, j = 1, 2, \ldots, m \qquad (1)$$

defines a Euclidian metric.

Those values for x^i which are compatible with the mathematical model constitute a subset $\mathcal{C}$ of $\mathcal{O}$, and we presuppose that $\mathcal{C}$ be an n-dimensional twice differentiable variety in $\mathcal{O}$. Consequently, the points on $\mathcal{C}$ are in one-to-one correspondence with the values of the wanted quantity A.

Now the adjustment procedure is defined as the mapping of $\mathcal{O}$ into $\mathcal{C}$ which to any point in $\mathcal{O}$ corresponds to that point on $\mathcal{C}$, the distance of which relative to the defined metric (1) from the point in $\mathcal{O}$ is minimum, taking it that the point on $\mathcal{C}$ is uniquely determined.

Thus the adjustment procedure is defined by the specification of (a) the mathematical model which describes $\mathcal{C}$ and (b) the quadratic form $g_{ij}x^i x^j$

which determines the metric in $\mathcal{O}$ being an expression of the statistical model. A specification of these models lies outside the scope of the present paper.

Let P_μ be a point on $\mathcal{C}$, the distance of which to a given point P_σ in $\mathcal{O}$ is minimum, then the line $P_\mu P_\sigma$ is orthogonal to the tangent space to $\mathcal{C}$ at P_μ. In general, we cannot find P_μ directly, as $\mathcal{C}$ cannot be presupposed to be a subspace of $\mathcal{O}$, but we must find P_μ by successive approximation.

$\mathcal{C}$ may be specified in various ways. For the present we suppose that $\mathcal{C}$ is given on a parameter form so that

$$y^\alpha, \qquad \alpha = 1, 2, \ldots, n,$$

constitutes a permissible coordinate system on $\mathcal{C}$ within the region of interest. The actual parameter representation is

$$x^i = a^i(y^\alpha), \tag{2}$$

where a^i are m twice differentiable functions of n independent variables. Any tangent to $\mathcal{C}$ at P has the direction

$$dx^i = \frac{\partial x^i}{\partial y^\alpha}\, dy^\alpha = a^i_\alpha\, dy^\alpha,$$

where dy^α is an arbitrary vector. If $P_\mu P_\sigma$ has to be orthogonal to this tangent, then

$$a^i_{\mu,\alpha}\, dy^\alpha\, g_{ij}(x^j_\mu - x^j_\sigma) = 0.$$

As $P_\mu P_\sigma$ must be orthogonal to all tangents at P, we have

$$a^i_{\mu,\alpha}\, g_{ij}(x^j_\mu - x^j_\sigma) = 0 \qquad \text{for} \quad \alpha = 1, 2, \ldots, n. \tag{3}$$

When determining P_μ, we start from a neighboring point on $\mathcal{C}$, e.g., in the point P_1. Equation (3) is not fulfilled for this point, but we have

$$a^i_{\mu,\alpha}\, g_{ij}(x^j_\mu - x^j_\sigma) = b_{1,\alpha}. \tag{4}$$

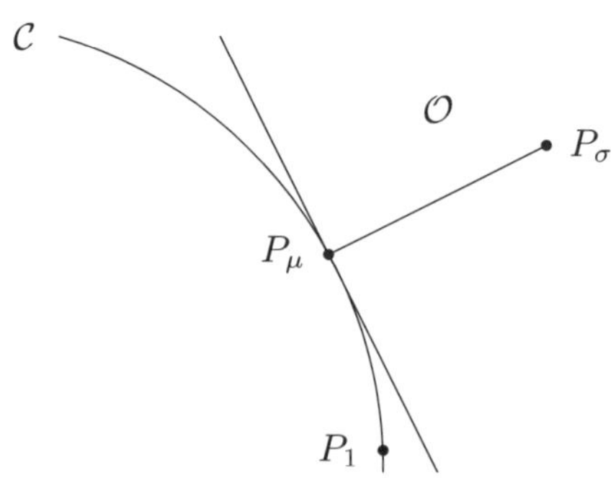

Fig. 2.1. The linear vector space $\mathcal{O}$ and the subset $\mathcal{C}$

Now it would be natural to try to solve (3) by means of Newton's method. Differentiation of (4) yields

$$db_{1,\alpha} = a^i_{1,\alpha}\, g_{ij} a^i_{1,\beta}\, dy^\beta + a^i_{1,\alpha\beta} g_{ij} (x^j_1 - x^j_\sigma) dy^\beta .$$

This means that the following point of approximation P_2 is given by the equations

$$\left(g_{ij} a^i_{1,\alpha} a^i_{1,\beta} + g_{ij} a^j_{1,\alpha\beta}(x^j_1 - x^j_\sigma)\right)(y^\beta_2 - y^\beta_1) = -b_{1,\alpha}$$

or, in a rewritten form,

$$h_{\alpha\beta} = g_{ij} a^i_\alpha a^j_\beta + g_{ij} a^i_{\alpha\beta}(x^j - x^j_\sigma), \tag{5}$$

$$h_{1,\alpha\beta}(y^\beta_2 - y^\beta_1) = -b_{1,\alpha}. \tag{6}$$

However, the matrix $h_{\alpha\beta}$ is not necessarily positive definite except in a certain region around P, so (6) will only be suitable for points P_1, which are already quite close to P. On the other hand, the first term on the right side of (5)

$$g_{\alpha\beta} = g_{ij} a^i_\alpha a^j_\beta$$

is positive definite at all points in the area where y^α is a permissible coordinate system. Experience also shows that by iterative use of

$$g_{1,\alpha\beta}(y^\beta_2 - y^\beta_1) = -b_{1,\alpha} \tag{7}$$

we obtain convergence within a larger region. By the way, (7) in connection with (4) constitute the classical normal equations.

$g_{\alpha\beta}$ is the induced metric tensor for $\mathcal{C}$ and is thus covariant to permissible transformations of the coordinates on $\mathcal{C}$. If $\mathcal{C}$ is a linear subspace of $\mathcal{O}$, then, as is well known, $g_{\mu,\alpha\beta}$ and $g^{\alpha\beta}_\mu = \left(g_{\mu,\alpha\beta}\right)^{-1}$ are respectively a covariant and a contravariant weight tensor of the coordinates y^α. In the nonlinear case they may be used as weight tensors.

In general, $h_{\alpha\beta}$ is not covariant, but $h_{\mu,\alpha\beta}$ is. The proof is as follows: Let

$$\ell = \tfrac{1}{2}\left(P_\mu P_\sigma\right)^2$$

and we get

$$\ell = \tfrac{1}{2} g_{ij}\left(x^i_\mu - x^i_\sigma\right)\left(x^j_\mu - x^j_\sigma\right).$$

Differentiating with respect to y^α therefore gives

$$\left(\frac{\partial \ell}{\partial y^\alpha}\right)_\mu = g_{ij} a^i_{\mu,\alpha}\left(x^j_\mu - x^j_\sigma\right)$$

and

$$\left(\frac{\partial^2 \ell}{\partial y^\alpha \partial y^\beta}\right)_\mu = g_{ij} a^i_{\mu,\alpha} a^j_{\mu,\beta} + g_{ij} a^i_{\mu,\alpha\beta}\left(x^j_\mu - x^j_\sigma\right) = h^{\alpha\beta}_\mu .$$

Here $(q)_\mu$ denotes q evaluated at the point P_μ. As ℓ is a scalar field on $\mathcal{C}$, $\partial\ell/\partial y^\alpha$ is a tensor. Covariant differentiation with respect to y^β gives

$$H_{\mu,\alpha\beta} = \Big(\frac{\partial^2 \ell}{\partial y^\alpha \partial y^\beta}\Big)_\mu - \big(\Gamma^\gamma_{\alpha\beta}\big)_\mu \Big(\frac{\partial \ell}{\partial y^\gamma}\Big)_\mu = h_{\mu,\alpha\beta}$$

since

$$\Big(\frac{\partial \ell}{\partial y^\alpha}\Big)_\mu = 0.$$

As the distance $P_\mu P_\sigma$ is minimum, it must also be true that $h_{\mu,\alpha\beta}$ is non-negative in a region around P_μ. These results indicate that $h_{\mu,\alpha\beta}$ may have a meaning rather similar to that of $g_{\alpha\beta}$ in the linear case.

Now let us look at the differential map of a neighborhood of P_σ into a neighborhood of P_μ. Now (3) can be written:

$$\Big(\frac{\partial x^i}{\partial y^\alpha}\Big)_\mu g_{ij}\,(x^j_\mu - x^j_\sigma) = 0. \tag{8}$$

We differentiate and get

$$\Big(\frac{\partial^2 x^i}{\partial y^\alpha \partial y^\beta}\Big)_\mu g_{ij}\,(x^j_\mu - x^j_\sigma)\,dy^\beta_\mu + \Big(\frac{\partial x^i}{\partial y^\alpha}\Big)_\mu g_{ij}\,dx^j_\mu = \Big(\frac{\partial x^i}{\partial y^\alpha}\Big)_\mu g_{ij}\,dx^j_\sigma. \tag{9}$$

In virtue of (9) and

$$dx^j_\mu = a^j_{\mu,\alpha}\,dy^\alpha_\mu$$

we get

$$h_{\mu,\alpha\beta}\,dy^\beta_\mu = g_{ij} a^i_{\mu,\alpha}\,dx^j_\sigma. \tag{10}$$

As opposed to what is valid for the linear mapping, the differential mapping in this case depends on $P_\mu P_\sigma$; this is due to the fact that the direction of projection is not constant owing to the conditions of curvature on $\mathcal{C}$. □

The very process of iteration can be described as follows:

1. Select a preliminary point P_1 on $\mathcal{C}$
2. Compute $\epsilon^i_1 = x^i_1 - x^i_\sigma = a^i\big(y^\alpha_1\big) - x^i_\sigma$
3. Compute $\Delta y^\alpha = y^\alpha_2 - y^\alpha_1$ by solving the symmetric linear equation system

$$\begin{aligned} -g_{ij}\big(x^j_2 - x^j_\sigma\big) + g_{ij} a^j_{1,\alpha} \Delta y^\alpha &= g_{ij}\epsilon^i_1 \\ a^i_{1,\alpha} g_{ij}\big(x^j_2 - x^j_\sigma\big) &= 0 \end{aligned} \tag{11}$$

4. Compute

$$y^\alpha_2 = y^\alpha_1 + \Delta y^\alpha$$

5. If P_2 is not close enough to P_μ, then put $P_1 = P_2$ and start again at 2.

Until now we have only considered *adjustment by elements*. Perhaps it would be appropriate to tell something about adjustment by correlates. Reverting to the considerations in the beginning of the paper, we remember that $\mathcal{C}$ was given in a parametrical form. But by *adjustment by correlates* we take, as a starting point, the equations which x^i must fulfil for the coordinates of a point on $\mathcal{C}$.

We suppose that $\mathcal{C}$, at least in the considered region can be described by the m $(< n)$ equations

$$b^\alpha(x^i) = 0 \tag{12}$$

where b^α are twice differentiable functions of n independent variables x^i.

The tangent space to $\mathcal{C}$ at the point $P_1 \in \mathcal{C}$ has the equations

$$b_i^\alpha (x^i - x_1^i) = 0 \tag{13}$$

where

$$b_i^\alpha = \left(\frac{\partial b^\alpha}{\partial x^i}\right)_1.$$

At the point P_μ any tangent vector to $\mathcal{C}$ is orthogonal to $P_\mu P_\sigma$, i.e.,

$$g_{ij}(x_\mu^i - x_\sigma^i) z^j = 0 \tag{14}$$

where all z^i satisfy

$$b_i^\alpha z^i = 0. \tag{15}$$

If

$$g_{ij}(x_\mu^i - x_\sigma^i) = b_j^\beta k_\beta \tag{16}$$

where k_β is an arbitrary m dimensional vector, (14) will be fulfilled for all z^i satisfying (15). A consideration of dimensions will show that (16) is necessary for a minimum, too. k_β is determined from (13) by insertion of x_μ^i as given by (16). Then we have the symmetric linear system of equations for $\Delta x^i = x^i - x_1^i$:

$$\begin{aligned} -g_{ij}\Delta x^i + b_j^\beta k_\beta &= g_{ij}(x_1^i - x_\sigma^i) \\ b_i^\alpha \Delta x^i &= 0 \end{aligned} \tag{17}$$

according to the system (11). By eliminating x^i we get the normal equations for determination of k_β:

$$g^{ij} b_i^\alpha b_j^\beta k_\beta = b_i^\alpha (x_1^i - x_\sigma^i) \tag{18}$$

according to (13). If the vector x^i we obtain by solving (17) is not equal to x_μ^i, we may use it as an approximation to this quantity and then calculate the corresponding b_i^α and continue iterating. If the first selected vector x_μ^i is sufficiently "good," the process will converge and exact computations will yield the same result as the adjustment by elements.

3

The Theory of Rounding Errors in the Adjustment by Elements of Geodetic Networks

Shortly after the appearance of the first electronic computers the investigation of the effect of rounding errors on the solution of linear equations was initiated by John von Neumann, see [2]. In the last years especially J. H. Wilkinson has contributed in this field; and his lucid exposition of the problems and the results in [3] is to day of the greatest importance for everybody who works seriously with linear algebraic equations.

For a long time it has been known that the linear equations arising from adjustment problems are often ill-conditioned. Therefore many proposals for better algorithms to solve these equations have been made, see e.g. [1].

For some years I have tried to apply these results and methods on the adjustment of geodetic networks and although these investigations have not yet been concluded, I have found that a presentation of some of the tools, I have used, could be of some interest.

A very important notion is that of the condition number. Let us consider the set of linear equations

$$Ax = b \tag{1}$$

where the matrix A is symmetric and non singular. As we know it is possible to find n eigenvalues λ_i and n eigenvectors ϕ_i, $i = 1, 2, \ldots, n$, where n is the dimension of the system so that

$$0 < |\lambda_1| \leq |\lambda_2| \leq \cdots \leq |\lambda_n| \qquad \text{and} \qquad \|\phi_i\|^2 = \phi_i^{\mathrm{T}} \phi_i = 1.$$

If the vector b is perturbed by a vector $\delta b = \epsilon\phi_i$, then x will be perturbed by the vector $\delta x = \frac{\epsilon}{\lambda_i}\phi_i$, i.e. the smaller $|\lambda_i|$ is, the greater is the perturbation. It is easy from this to show that for the relative perturbation of x, the following inequality must hold

$$\frac{\|\delta x\|}{\|x\|} \leq \frac{|\lambda_n|}{|\lambda_1|} \frac{\|\delta b\|}{\|b\|} \tag{2}$$

for an arbitrary perturbation δb of b. It is λ_n/λ_1 that is called the condition number.

For perturbations of x arising from perturbations of A we have the inequality

$$\frac{\|\delta x\|}{\|x\|} \le \frac{|\frac{\lambda_n}{\lambda_1}||\frac{\nu}{\lambda_n}|}{1 - |\frac{\lambda_n}{\lambda_1}||\frac{\nu}{\lambda_n}|} \tag{3}$$

provided $|\frac{\nu}{\lambda_1}| < 1$, where ν is the numerically greatest eigenvalue of δA.

We shall now consider the condition number of the normal equations for adjustment by elements.

The linearized observation equations can be written as

$$Az = o, \tag{4}$$

where A is an m by n matrix ($m > n$), z is the n-vector of the (differentials of the) parameters to be calculated and o is the m-vector of the (differentials of the) observations. It is possible to normalize these equations so that all the weights are unity, and we shall assume that this has been done. If v is the m-vector of the residuals we can write all the equations to be solved

$$\begin{aligned} -Iv + Az &= o, \\ A^{\mathrm{T}} v \qquad &= 0. \end{aligned} \tag{5}$$

An elimination of v yields the normal equations

$$A^{\mathrm{T}} A z = A^{\mathrm{T}} o,$$

where $A^{\mathrm{T}} A = N$ is the n by n matrix, the condition number of which we are investigating. The eigenvalue problem for this matrix is the problem to find the n-vectors ϕ_i and the numbers λ_i ($i = 1, 2, \ldots, n$) so that

$$N\phi_i = \lambda_i \phi_i, \qquad \|\phi_i\| = 1 \qquad \text{for} \quad i = 1, 2, \ldots, n \tag{6}$$

$$0 < \lambda_1 \le \lambda_2 \le \cdots \le \lambda_n. \tag{7}$$

We can assume that

$$\phi_i^{\mathrm{T}} \phi_j = 0 \qquad \text{for } i \ne j \qquad \text{also if } \lambda_i = \lambda_j.$$

Then the condition number of N is λ_n / λ_1.

Being vectors in the parameter space, the ϕ_i can be regarded as linear functions of the differentials of the parameters

$$\phi_i \approx \phi_i^{\mathrm{T}} z,$$

where the square sums of the coefficients are unity. We denote such functions unity functions.

Using the matrix $C = N^{-1}$ the equations (6) can be written:

$$C\phi_i = \lambda_i^{-1}\phi_i, \tag{8}$$

from which follows

$$\phi_i^{\mathrm{T}} C\phi_i = \lambda_i^{-1}, \qquad \text{for } j = i, \tag{9}$$

$$\phi_j^{\mathrm{T}} C\phi_i = 0, \qquad \text{for } j \neq i. \tag{10}$$

As C is proportional to the covariance matrix of the parameters, the equations (10) express that ϕ_i regarded as functions of the differentials of the parameters are independent (they are "free functions"), and the equations (9) express that the variance of the ϕ_i are inversely proportional to the λ_i. We see that the condition number λ_n/λ_1 equals the quotient the highest possible variance of a unity function by the lowest possible variance of a unity function. If we decompose z in its ϕ_i-components, $\phi_i^{\mathrm{T}} z$, we see that the better a component is defined by the measurements, the better it can be found by calculation, but while the mean square error (from the measurements) is proportional to $\sqrt{1/\lambda_i}$, then the uncertainty from rounding errors is proportional to the square of $\sqrt{1/\lambda_i}$. We can say that the more well planned the network is, the less is the condition number.

Let us now consider a very critical network. I choose a traverse connecting two points with given coordinates. We consider all the sides known to have the same length and let all the angles be measured with the same weight and we consider them to be theoretically 180° each. If the number of unknown points is n, then the eigenvalues and the eigenvector will be:

$$\lambda_i = \left(2 \sin \frac{i\pi}{2n+1}\right)^4, \tag{11}$$

$$\phi_{i,j} = K_i \sin \frac{ij}{2n+1}, \tag{12}$$

where K_i are normalizing constants. The condition number will be given by the asymptotical relation

$$\frac{\lambda_n}{\lambda_i} \approx \left(\frac{n+1}{\pi}\right)^4 \approx \left(\frac{n}{\pi}\right)^4.$$

For $n = 100$ we shall have the condition number of about 10^6 and for $n = 1000$ we get 10^{10}, numbers that give reason to some precaution.

A classical example in the literature of least-squares problems is one where the matrix A in (4) is given by

$$A = \begin{bmatrix} 1 & 1 & 1 & 1 & 1 \\ \epsilon & 0 & 0 & 0 & 0 \\ 0 & \epsilon & 0 & 0 & 0 \\ 0 & 0 & \epsilon & 0 & 0 \\ 0 & 0 & 0 & \epsilon & 0 \\ 0 & 0 & 0 & 0 & \epsilon \end{bmatrix} \tag{13}$$

where $\epsilon = 10^{-5}$ and the calculations are performed with nine decimals. The matrix of the normal equations is then

$$A^{\mathrm{T}}A = \begin{bmatrix} 1+\epsilon^2 & 1 & 1 & 1 & 1 \\ 1 & 1+\epsilon^2 & 1 & 1 & 1 \\ 1 & 1 & 1+\epsilon^2 & 1 & 1 \\ 1 & 1 & 1 & 1+\epsilon^2 & 1 \\ 1 & 1 & 1 & 1 & 1+\epsilon^2 \end{bmatrix} \tag{14}$$

and the eigenvalues are ϵ^2, ϵ^2, ϵ^2, ϵ^2, and $5+\epsilon^2$, but the matrix of the normal equations will be calculated so as to consist of unity coefficients overall and will thus be singular and it will not be possible to find the solutions, no matter how accurate a method is used for solving the normal equations. I shall try to show, how such a problem can be solved, but first I shall say something about solving the normal equations.

The most natural way to solve the normal equations is, as we all know, to use the Cholesky-method. Its essential part consists in finding an upper triangular matrix U so that

$$N = U^{\mathrm{T}}U. \tag{15}$$

From (15) follows that every element of N is the product sum of two columns of U, and from this follows again, that if N is "normalized" then every element of U and of N is numerically less than one. By normalizing the matrix of the normal equations I mean multiplying the ith row and the ith column of it by such a power of two, that the ith diagonal element n_{ii} satisfies

$$\tfrac{1}{4} \le n_{ii} < 1$$

for every i. Therefore it is possible to perform the calculations in fixed point arithmetic.

As pointed out by Wilkinson very much is gained in numerical precision by calculating the product-sums in double precision arithmetic, finding the diagonal elements of U as square roots correctly rounded to single precision of the double precision square-sums, and finding the nondiagonal elements of U as the quotients correctly rounded to single precision of the double precision product-sums by the single precision diagonal elements. This sounds complicated but in many computers it can be performed so with only very little extra cost in time and storage, and the gain in numerical precision is astonishing. For the proof and as an illustrating example I can only refer to [3]. I call this variant of the Cholesky-method the Cholesky-Wilkinson-method. I think that the normal equations should be solved by this method whenever possible.

Now we can return to the example (13). Let us look at the system (5) where A is given by (13). This is a symmetrical set of linear equations, and we shall try to use the Cholesky-Wilkinson-method for it, slightly modified so as to be used also when the diagonal elements are negative. Ignoring some normalizing problems which every computer-user should be able to solve we get by this method the matrix U of the reduced normal equations:

$$U = \begin{bmatrix} 1 & 1 & 1 & 1 & 1 \\ & \epsilon & 1 & 1 & 1 \\ & & \epsilon & 1 & 1 \\ & & & \epsilon & 1 \\ & & & & \epsilon \end{bmatrix}.$$

This matrix is not singular and the normal equations corresponding to it are:

$$N' = \begin{bmatrix} 1 & 1 & 1 & 1 & 1 \\ 1 & 1+\epsilon^2 & 1 & 1 & 1 \\ 1 & 1 & 1+\epsilon^2 & 1 & 1 \\ 1 & 1 & 1 & 1+\epsilon^2 & 1 \\ 1 & 1 & 1 & 1 & 1+\epsilon^2 \end{bmatrix}.$$

As I said before there has been proposed many methods for solving ill-conditioned least-square problems but this is the most simple and is as far as I can see numerically not less precise than any other of them. Unfortunately all these methods have the drawback that during the solution of the equations all the coefficients of the observation equations must be present in the storage or every such coefficient must be calculated several times. This fact causes a much lower limit for the dimension of the networks that can be solved on a computer with a given capacity.

Nevertheless I think that the method mentioned here can be used also on relatively small computers after an adjustment using the solution of normal equations explicitly in the following way: Express the unknown corrections to the coordinates already found as polynomials in the coordinates with coefficients to be found. Then let these coefficients be parameters in an adjustment using the numerically strong method. This will be possible if the parameters are few enough. This polynomial will give "systematical" (by "systematical" I mean here non-local) corrections to the coordinates and will probably give most of the residual corrections because the ϕ_i corresponding to low eigenvalues are likely to be approximated by polynomials of low degree.

This paper has not been intended to give final results of my investigations, I hope to be able to present them in a publication in the near future, but I hope I have pointed out that there are other problems in the geodetic computations than finding good formulas, there are numerical problems that have to be solved and these are methods to solve them.

References

[1] F. L. Bauer. Elimination with weighted row combinations for solving linear equations and least squares problems. *Numerische Mathematik*, 7:338–352, 1965.

[2] John von Neumann and H. H. Goldstine. Numerical inverting of matrices of high order. *Bulletin of American Mathematical Society*, 53:1021–1099, 1947.

[3] J. H. Wilkinson. *Rounding Errors in Algebraic Processes.* Prentice-Hall, Englewood Cliffs, 1963.

4

A Contribution to the Mathematical Foundation of Physical Geodesy

Exposition of the Ideas

Traditionally geodesists look on the problem of determination of the potential of the Earth as a boundary value problem: in principle we can measure some local property of the potential—the gravity—at the surface of the Earth and until recently only there. Seeing that today the artificial satellites make it possible to measure effects of the potential also in the outer space, and that, on the other hand, we shall never be able to make gravity measurements at more than a finite number of points, it would perhaps be more natural to look on the main problem of physical geodesy as one of interpolation. The present paper was primarily meant to be an attempt to solve some of the theoretical and computational problems connected with this changed point of view, but gradually, as my ideas took their form and I became aware of the difficulties, my impression that the mathematical foundation of physical geodesy had reached a state of crisis grew stronger and stronger.

I think that in reality the cause of this crisis is a crisis of communication, not communication between geodesists but between geodesy and mathematics, and that physical geodesy shares this difficulty with other sciences which are interested in applied mathematics. Looking around on the mathematical methods used, e.g. in physical geodesy, you will find ideas which seem to have reached us by accident from the closed land of mathematics rather than methods naturally suited for our problems.

I shall give a few examples which, I hope, will make it clear what I mean.

The methods used for determining the potential from satellite measurements and from gravity measurements on the Earth are completely different, and although they should be able to complete one another—the satellite data being best suited to give the more global information and the gravity data to give the more local information—it is nevertheless very difficult to combine these two sets of data. It is curious that geodesists who are accustomed to use and perhaps misuse adjustment methods in geometrical geodesy have not been able to find an adjustment method that can combine these two sorts of data.

Some geodesists have suspected that Molodenskiy's boundary value problem has some form of unstableness. Would it not have been natural then to try to formulate the boundary value problem as an adjustment problem, i.e. to try to obtain an improvement of the boundary values so as to make the boundary value problem uniquely solvable and so that the improvement in some least-squares sense would be as small as possible? We have, however, stuck to the traditional formulation of boundary value problems, used and perhaps relevant in other parts of physical mathematics, but—at least in my opinion—not relevant here.

This crisis of communication is well understandable. The mathematical foundation of physical geodesy has been the classical potential theory as it was, say in the late thirties, which theory is well described in a few well-known books. After that time mathematics has developed fast and has in the opinion of many people become an esoteric science. If you want to know something about modern potential theory you will first learn that nothing is written today about potential theory, but that this subject is generalized to the theory of linear elliptic partial differential equations, and that it is preferable also to read something about linear partial differential equations in general. But when you see the books you realize that in order to understand anything in them you have first to read one or two books on functional analysis (and to understand these you must first read something about topology). Well, after a few years of study you may begin to suspect what a modern potential theory would look like and to see that there really are new things of importance to physical geodesy in it. But a book on modern potential theory for geodesists could—if it existed—be read in a few weeks; only such a book does not exist. I understand why modern books on mathematics are written as they are, but as a user of mathematics I must deplore it.

As a type of problem that would be treated in a modern book on potential theory I can mention the problems concerning approximation of potentials and in this connection stability problems.

I shall here give a few examples:

We have learned in the classical potential theory that two potentials which are identical in an open set are identical in their whole domain of regularity. The new potential theory gives a valuable complement to this theorem:

Let us consider potentials regular in some domain Ω, and consider another domain $\Omega_0 \in \Omega$. For two potentials ϕ_1 and ϕ_2 regular in Ω we assume that

$$|\phi_1(P) - \phi_2(P)| < \epsilon \qquad \text{for } P \in \Omega_0 \tag{1}$$

where ϵ is some positive number.

We must now distinguish between two different cases:

1. The boundary of Ω is a part of the boundary of Ω_0. Then there exists a constant $\delta > 0$ so that

$$|\phi_1(P) - \phi_2(P)| < \delta \qquad \text{for } P \in \Omega. \tag{2}$$

2. The boundary of Ω is not a part of the boundary of Ω_0. Then there does not exist a constant δ so that (2) is satisfied.

From this important theorem many conclusions can be drawn, most of which are more or less known by geodesists, e.g.:

The upward continuation of the potential (or the gravity) of the Earth is stable and the downward continuation is unstable.

By going to the limit in (1) we see that the Dirichlet problem is stable but the Cauchy problem is unstable.

The potential at the surface of the Earth (or the figure of the Earth) cannot be found from dynamical satellite measurements alone.

These conclusions are just expressed as slogans; the exact meaning can be found only on the basis of the theorem. Thus the words stable and unstable have been given an exact content by (1) and (2).

In a parenthesis I should like to say a few words more on the stability problem in another connection. We know from the classical potential theory that both Dirichlet's and Neumann's problems have one and only one solution and that this is also the case as regards the third boundary value problem, where

$$\frac{\partial \phi}{\partial n} + h\phi \tag{3}$$

is given at the boundary, if $h \leq 0$. But for certain functions h for which $h \leq 0$ does not hold, the values of (3) at the boundary cannot be arbitrarily prescribed, but must satisfy certain linear condition(s) for the problem to have a solution, which is now not unique (Fredholm's alternative). If h is not given exactly, but for example found by measurements, it may happen that it has no meaning to say that h is or is not one of the exceptional coefficient functions. If this is the case, the boundary value problem will be unstable.

And now back to the approximation problems.

As far as I can see, the most important point of view introduced in physical geodesy since the appearance of Molodenskiy's famous articles is Bjerhammar's idea of calculating an approximation of the potential by collocation, at the points where gravity anomalies have been measured, of potentials that are regular down to a certain sphere situated inside the surface of the Earth.

From the classical theory this idea looks very venturesome, because we know that the actual potential of the Earth is not regular down to the Bjerhammar sphere. Nor does the evidence that Bjerhammar produces in support of his idea seem convincing to me at all. As we shall see later on from the point of view of the new potential theory, very much can be said in favour of the determination of the potential by interpolation methods as well as of the use of potentials that are regular outside a Bjerhammar sphere.

It is curious that from the very beginning these two things (I mean the interpolation and the Bjerhammar sphere) have appeared together, exactly as I was forced to accept the Bjerhammar sphere as soon as I drew the conclusions from Moritz' interpolation formula for the gravity anomalies.

Moritz' interpolation formula rests upon the idea of an auto-correlation (the word "covariance" is used in this sense in the geodetic literature in the same way as it will be used in this paper from the first section and on) function for the gravity anomalies which is invariant under the group of rotations around the centre of the Earth. In the first section I shall prove that from the existence of such an auto-correlation function on a sphere follows the existence of an auto-correlation function for the potential in the whole space outside the surface of the sphere and also that this auto-correlation function is invariant under the group of rotations. It appears from some of Kaula's articles that he too was aware of that.

Now it is an obvious idea to generalize Moritz' interpolation formula so as to find the potential directly by collocation (or generalized interpolation) from the gravity anomalies and thus drop the complicated integration procedure using Stokes' formula or a related formula. It is not less obvious that this method would not be confined to the use of gravity anomaly data, but that for example data related to satellites or vertical deflections could be included as well. In fact, I have derived such a formula in the first section and, more than that, I have also generalized this formula to make it possible, not by exact collocation but by the aid of a given variance-covariance matrix for the measurements, to find a potential which corresponds to improved measurements and which is more likely according to the auto-correlation hypothesis. This is what I call a smoothing procedure.

As Moritz has pointed out in [10] which is as amusing as it is interesting, there is a close relation between his form of interpolation and the classical least-squares adjustment. Therefore, I could re-deduce the results described in the first section from the point of view of adjustment described in the second section. The classical reciprocity between variance-covariance and weight corresponds here to a reciprocity between auto-correlation and norm metrics. This recalls the beautiful geometric method of deducing the formula for classical adjustment where the weight defines a Euclidean norm in the vector space of the measurements.

The technical difficulty here is that the vector space in which the adjustment has to be made is not a finite dimensional one, but, when the metric has been defined, a Hilbert space.

The Hilbert space should be within the natural sphere of interest of geodesists, since it is the most general substratum for the least-squares adjustment and also an indispensable tool of modern potential theory. But the Hilbert space is not enough. We must use a concept that exists in Hilbert spaces consisting of sufficiently well-behaved functions—and regular potentials are well-behaved in most respects. This is the concept of a reproducing kernel.

I have not found it possible in this paper to give an introduction to Hilbert spaces with a reproducing kernel, but good books exist on that subject. The book most relevant to the use I have made of Hilbert spaces with reproducing kernel is [9] which exists only in German, but the beautiful book [1] which contains little but enough about the subject and which contains many other interesting ideas not totally irrelevant to our problems, is also useful. On the other hand, I guess that most physical geodesists will be able to read this paper the first time with some profit without having ever seen the words "Hilbert spaces" and "reproducing kernel" together before.

The advantage we get by introducing the adjustment aspect is not only that we attain a closer connection to classical adjustment methods, which is only a formal gain, but also—and this is more essential—that we obtain a better insight in the set of potentials among which the solution is found. This set is exactly those potentials for which the norm (defining the metrics) is defined and finite, and—here it is again—all these potentials are regular outside a certain sphere in the interior of the Earth: The Bjerhammar sphere. Again: The coupling between interpolation and the Bjerhammar sphere.

And now we can go back to Moritz' formula. (I mean, here as before, Moritz' formula without height correction.) It presupposes the rotation invariance of the auto-correlation function, from which follows the rotation invariance of the auto-correlation function for the potential and again the rotation invariance for the domain of definition of the potential itself, independent of the way it is calculated: from Moritz' formula and Stokes' formula, or directly from the generalized Moritz' formula. Again: The Bjerhammar sphere.

There is a close connection between the problem of the Bjerhammar sphere and the problem of the convergence of series in spherical harmonics or better the question of approximation of potentials by series in spherical harmonics, and here I believe we have reached the very core of the foundation of physical geodesy.

The consequence must be that not only the first two sections of this paper but also very much of the work done by physical geodesists all over the world remain idle formalism until this approximation problem has been solved.

This is what I have tried to do in the third section. Here it appears that again the theory of Hilbert spaces with reproducing kernel is a valuable tool and that the central theorem—the Runge theorem—is a special case of a theorem, known as a theorem of the Runge type, which is well-known in the theory of linear partial differential equations and which concerns the approximation of solutions to a partial differential equation in some domain by solutions to the same equation in another domain containing the first one. The proof of Runge's theorem given in the Appendix is a reduction and specialization of Theorem 3.4.3 in [5].

As far as I can see, all the problems of approximations of potentials relevant to physical geodesy are solved or can be solved by the methods used in the third section and the Appendix. And it is curious that the investigations there seem to show that the adjustment method described in Section II (or, equivalently, the interpolation method from Section I) is the natural way to find the approximation, but the cause of this may be that I do not know about any other way to obtain an approximation to the potential of the Earth which can be proved relevant.

The last section is dedicated to the applications of the method of approximation treated in the first sections. I fear that most readers will feel disappointed by reading the last section as I myself felt disappointed when writing it. In the light of the many thoughts and feelings I have had as to the possibilities of application of the method it seems very poor to me. But in the less than 18 months elapsed from the first idea of the method originated until the paper was sent to the printers I have have had to concentrate so much on the foundation and the studies of the relevant mathematical literature, that the physical or geodetic aspects could only be vaguely glanced at. However, I feel that now the time is ripe for a discussion of this matter and that especially the question relating to the geodetic applications is best furthered by discussion and collaboration, and so I hope that the few hints in Section IV will be sufficient to start such a discussion.

Forse altri canterà con miglior plettro!

I. Covariance and Collocation

1. As starting point I shall take the least-squares prediction formula (7-63) from [4], page 268, and I assume that the reader is familiar with the statistical reasoning leading to this formula. To ensure a better understanding of what follows I shall make a few comments on the mathematical reasoning using Moritz' notation. The problem is to find the coefficients α_{P_i} in the linear prediction formula

$$\widetilde{\Delta g}_P = \sum_{i=1}^{n} \alpha_{P_i} \Delta g_i, \tag{4}$$

so that the mean square error m_P of the predicted anomaly at P attains a minimum. We have

$$m_P^2 = C_{PP} - 2\sum_{i=1}^{n} \alpha_{P_i} C_{P_i} + \sum_{i=1}^{n}\sum_{k=1}^{n} \alpha_{P_i}\alpha_{P_k} C_{ik}, \tag{5}$$

and the α_{P_i} are to be determined so that (5) attains a minimum.

With

$$\alpha_{P_i} = \beta_{P_i} + \gamma_{P_i} \tag{6}$$

(5) becomes

$$\begin{aligned} m_P^2 = C_{PP} &- 2\sum_{i=1}^{n} \beta_{P_i} C_{P_i} + \sum_{i=1}^{n}\sum_{k=1}^{n} \beta_{P_i}\beta_{P_k} C_{ik} \\ &+ 2\sum_{i=1}^{n} \gamma_{P_i}\left(\sum_{k=1}^{n} \beta_{P_k} C_{ik} - C_{P_i}\right) + \sum_{i=1}^{n}\sum_{k=1}^{n} \gamma_{P_i}\gamma_{P_k} C_{ik}. \end{aligned} \tag{7}$$

We may here choose the β_{P_i} as we like, and we shall choose them so that

$$\sum_{k=1}^{n} \beta_{P_k} C_{ik} = C_{P_i} \tag{8}$$

which is possible if $\{C_{ik}\}$ is non-singular. Then the second line in formula (7) is zero, and the first line has a fixed value independent of γ_{P_i}.

If now $\{C_{ik}\}$ is a positive definite matrix, then the third line of (7) is ≥ 0 and is $= 0$ only if $\gamma_{P_i} = 0$, i.e. (5) attains its minimum for α_{P_i} satisfying the normal equations

$$\sum_{k=1}^{n} \alpha_{P_k} C_{ik} = C_{P_i}. \tag{9}$$

If, on the other hand, $\{C_{ik}\}$ is not definite, then the third line may attain any positive or negative value, and m_P^2 cannot have any minimum.

Thus we see that for the least-squares problem to have a meaning the covariance function $C(P,Q)$[1] must be of positive type, i.e.

$$\sum_{i=0}^{n} \sum_{k=0}^{n} C(P_i, P_k) x_i x_k > 0, \tag{10}$$

for all sets of n points P_i and n corresponding numbers $x_i \neq 0$, for $i = 1, 2, \ldots, n$, for all natural numbers n. It is evident, that only in this case has $C(P,Q)$ a meaning as a covariance function.

The α_{P_i} having been found by (9), the predicted value can be found by (4) or by

$$\widetilde{\Delta g}_P = \sum_{i=1}^{n} \sum_{k=1}^{n} C_{ik}^{-1} C_{P_i} \Delta g_k. \tag{11}$$

If predicted values for several points are to be calculated, it is most economical first to solve the following set of "normal equations"

$$\sum_{i=1}^{n} \xi_i C_{ik} = \Delta g_k, \tag{12}$$

where ξ_i are independent of P. Then the set of solutions to (12) can be used to find the predicted values for all points by

$$\widetilde{\Delta g}_P = \sum_{i=1}^{n} \xi_i C_{P_i}. \tag{13}$$

It is evident that (11) follows again from (12) and (13), but (12) and (13) formulate a problem of interpolation and its solution:

---

[1] I regard the covariance function $C(P,Q)$ as a symmetric function of the two points P and Q and not merely as a function of the distance PQ. This will be of importance in the following part of the paper. It should perhaps be noted that C_{ik} means $C(P_i, P_k)$ and C_{P_i} means $C(P, P_i)$.

(13): Find $\widetilde{\Delta g}_P$ in the whole region as a linear combination of the functions $C(P, P_i)$ of P for $i = 1, 2, \ldots, n$, so that

(12): $\widetilde{\Delta g}_P = \Delta g_k$ for $k = 1, 2, \ldots, n$.

On the assumption that $C(P, Q)$ is of positive type this problem has always one and only one solution.

Having regarded the method from this slightly changed point of view one gets the courage to try to generalize.

Let us jump into the new problem:

By first using Moritz' prediction formula with a given covariance function $C(P, Q)$ we can find $\widetilde{\Delta g}_P$ for all points of some surface. Solution of the corresponding boundary value problem will then give us the disturbing potential T. But can we, by using another covariance function $K(P, Q)$ defined in a domain including the outer space of the Earth and describing the covariance between values of T at the points P and Q, find the potential T directly from the measured values of Δg_i?

Or to put the question in another way: Can we find the disturbing potential T by collocation (i.e. so that the corresponding gravity anomalies attain a given set of values at given points) so that T corresponds to the interpolated anomalies that could be found by Moritz' formula?

I shall prove that this is not only possible but also relatively simple.

Evidently we cannot find the covariance function directly, for to do so we should have a population of earths similar to our own and have the opportunity to measure any T for all these earths at all points. But let us for a moment assume that by a miracle we were given such a function; how then could we use it and what can we say a priori about its properties?

First of all we see that the whole theory about Moritz' prediction formula may be used unchanged. Given the values of the potential T at n points P_i, T can be predicted for all points P in the domain Ω where K is valid. The form of (13) will here be

$$\widetilde{T}_P = \sum_{i=1}^{n} \xi_i K(P, P_i) \tag{14}$$

where ξ_i is found from the normal equations

$$\sum_{k=1}^{n} K(P_k, P_i)\xi_k = T_P. \tag{15}$$

Using the Laplace operator Δ on both sides of (14) for $n = 1$ we find that

(a) for all P_i $K(P, P_i)$ satisfies the Laplace equation

$$\Delta_P K(P, P_i) = 0$$

at all points $P \in \Omega$ and is a regular potential at infinity as a function of P.

The last remark follows from T having the corresponding property. (By regularity at infinity for a potential ϕ I mean in this paper that 1) $\lim \phi = 0$ and 2) $\lim r \cdot \phi$ exists. It is evident that T has at least these two properties.)
From the definition of $K(P, Q)$ as a covariance function follows

(b) $K(P, Q) = K(Q, P)$. Then it is trivial that, for all P_i, $K(P_i, P)$ as a function of P is a regular potential in Ω.

(c) $K(P, Q)$ is a function of positive type.
This follows as in the theory explaining Moritz' formula.
As a harmonic function $K(P, Q)$ must be arbitrarily often differentiable with respect to the coordinates of the points P and Q in the domain Ω.
Now we shall investigate what can be deduced from a given $K(P, Q)$ about the covariance between derivatives of T.
From the definition of $K(P, Q)$ as a covariance function we have

$$M\{T_P, T_Q\} = K(P, Q). \tag{16}$$

For P, Q, R being three points in Ω follows

$$M\{T_P, T_R - T_Q\} = K(P, R) - K(P, Q). \tag{17}$$

Let the distance from Q to R be l, so that for some unity vector $\mathbf{e}$

$$R = Q + l\mathbf{e}; \tag{18}$$

then (17) gives

$$M\left\{T_P, \frac{T_{Q+l\mathbf{e}} - T_Q}{l}\right\} = \frac{K(P, Q + l\mathbf{e}) - K(P, Q)}{l}, \tag{19}$$

and for $l \to 0$

$$M\left\{T_P, \left(\frac{\partial T}{\partial l}\right)_Q\right\} = \left(\frac{\partial}{\partial l} K(P, Q)\right)_Q, \tag{20}$$

where $(\partial f / \partial l)_Q$ means the derivative of the function f in the direction $\mathbf{e}$ taken at the point Q.
Evidently (20) can be generalized to arbitrarily high derivatives in arbitrary directions and to all linear differential operators. Specially, we get

$$0 = M\{T_P, \Delta T_Q\} = \Delta_Q K(P, Q), \tag{21}$$

i.e. again the result that $K(P, Q)$ is harmonic as a function of Q.
The normal potential U being known, Δg can be found from T by a differential operator, let it be called $\mathcal{L}$, so that we get

$$M\{\Delta g_P, \Delta g_Q\} = \mathcal{L}_P \mathcal{L}_Q K(P, Q). \tag{22}$$

That is to say that if on a surface ω bounding Ω we know the covariance function $C(P, Q)$ for the gravity anomaly then we must have:

(d) $\mathcal{L}_P\mathcal{L}_Q K(P,Q) = C(P,Q)$ for $P, Q \in \omega$, and now it follows from (a), (b), and (d) that $K(P,Q)$ can be found from $C(P,Q)$ under certain conditions.

2. I hope that the reader now looks upon the possibility of finding $K(P,Q)$ with so much optimism that he is interested in learning how to use the covariance function.

I shall try to show it, using the same argumentation as that used in [4], p. 266ff.

We assume that we have found the anomalies Δg_i at n points

$$P_i = Q_i \qquad i = 1, 2, \ldots, n,$$

and that we use a linear prediction for the potential T at the point P:

$$\widetilde{T}_p = \sum_{i=1}^{n} \alpha_{P_i} \Delta g_i. \tag{23}$$

If the correct value for T at P is T_P, the error of prediction ϵ_P is

$$\epsilon_P = T_P - \widetilde{T}_P = T_P - \sum_{i=1}^{n} \alpha_{P_i} \Delta g_i. \tag{24}$$

Then we can find the error covariance

$$\sigma_{PQ} = M\{\epsilon_P \epsilon_Q\}. \tag{25}$$

$$\begin{aligned}
\sigma_{PQ} &= M\Big\{\Big(T_P - \sum_i \alpha_{P_i} \Delta g_i\Big)\Big(T_Q - \sum_k \alpha_{Q_k} \Delta g_k\Big)\Big\} \\
&= M\Big\{T_P T_Q - \sum_i \alpha_{P_i} T_Q \Delta g_i - \sum_k \alpha_{Q_k} T_P \Delta g_k + \sum_i \sum_k \alpha_{P_i} \alpha_{Q_k} \Delta g_i \Delta g_k\Big\} \\
&= K(P,Q) - \sum_{i=1}^{n} \alpha_{P_i} \mathcal{L}_{P_i} K(Q,P) - \sum_{k=1}^{n} \alpha_{Q_k} \mathcal{L}_{Q_i} K(P,Q) \\
&\quad + \sum_{i=1}^{n} \sum_{k=1}^{n} \alpha_{P_i} \alpha_{Q_k} \mathcal{L}_{P_i} \mathcal{L}_{Q_k} K(P,Q).
\end{aligned} \tag{26}$$

For $P = Q$ this becomes

$$\epsilon_P^2 = K(P,P) - 2\sum_{i=1}^{n} \alpha_{P_i} \mathcal{L}_{Q_i} K(P,Q) + \sum_{i=1}^{n} \sum_{k=1}^{n} \alpha_{P_i} \alpha_{P_k} \mathcal{L}_{P_i} \mathcal{L}_{Q_k} K(P,Q). \tag{27}$$

If ϵ_P^2 is to attain a minimum, α_{P_i} must satisfy the normal equations

$$\sum_{k=1}^{n} \alpha_{P_k} \mathcal{L}_{P_k} \mathcal{L}_{Q_i} K(P,Q) = \mathcal{L}_{Q_i} K(P,Q). \tag{28}$$

If the points $P_i \in \omega$, then the coefficients of the normal equations are $C(P_i, P_k)$, and the matrix is positive definite so that we can find the inverse matrix N_{ik} defined by

$$\sum_{j=1}^{n} N_{ij} \mathcal{L}_{P_j} \mathcal{L}_{Q_k} K(P,Q) = \delta_{ik}. \tag{29}$$

By the aid of N_{ik} the potential can be predicted for any point P in Q by the formula

$$\widetilde{T}_P = \sum_{i=1}^{n} \sum_{k=1}^{n} N_{ik} \mathcal{L}_{Q_i} K(P,Q) \Delta g_k. \tag{30}$$

The error covariance for the points P and Q is found by

$$\sigma_{PQ} = K(P,Q) - \sum_{i=1}^{n} \sum_{k=1}^{n} N_{ik} \mathcal{L}_P K(P,Q_i) \mathcal{L}_Q K(Q,P_k). \tag{31}$$

Also here we can turn the question in another way as in the case of formula (11):

We want to represent $\widetilde{T}_P$ as follows

$$\widetilde{T}_P = \sum_{i=1}^{n} \xi_i \mathcal{L}_{Q_i} K(P,Q_i) \tag{32}$$

for all points in Ω so that for $i = 1, 2, \ldots, n$ (ξ_i are constants)

$$\mathcal{L}_{P_k} \widetilde{T}_P = \Delta g_k. \tag{33}$$

This is precisely what is meant by collocation.

Now we get the normal equations

$$\sum_{i=1}^{n} \xi_i \mathcal{L}_{P_i} \mathcal{L}_{Q_i} K(P,Q) = \Delta g_k, \tag{34}$$

and once they are solved, T_P can be found by (32). Substitution of ξ_i from (34) in (32) gives again (30).

It follows from (32) and (a) that T_P is a potential, and from (33) and (34) that it gives the wanted values for the gravity anomalies at the points P_i. So far, we have in fact eliminated the statistical reasoning. The formula (31), on the other hand, follows only from the statistical hypothesis.

3. My treatment of the statistical aspect has so far been very loose. I shall now try to give it a better foundation.

Let us here, as in most of this paper, regard Ω as the domain outside a sphere with its centre O in the centre of gravity of the Earth and the potential T, regular in Ω, as a stochastic process on Ω and invariant with respect to rotations around the centre O. All functions of two points $P, Q \in \Omega$ that are invariant with respect to such rotations are functions of the distances r_P and r_Q of the points P and Q from O and of the angle w between OP and OQ. From this follows that $K(P,Q)$ is a function of not six but only three independent parameters, and it can be found from T as the mean of

$$T_{P'}T_{Q'} \tag{35}$$

over all pairs of points P' and Q' satisfying

$$\begin{aligned} OP' &= OP \\ OQ' &= OQ \\ \angle POQ' &= \angle POQ. \end{aligned} \tag{36}$$

In this way the means can be interpreted for simple interpolations (i.e. when no linear operators $\mathcal{L}$ are involved), and this is analogous to the interpretation of the means by Moritz. But in the case of operators not invariant with respect to rotation it does not work. Here a set of potentials Σ must be given so that the mean can be taken over point by point, and the set must have the ergodic property so that at each point the mean of $\phi \in \Sigma$ equals the mean of an individual ϕ over the sphere (with centre at O) passing through the point in question.

Let $\phi(P)$ be a potential regular in Ω. If R is a rotation (we regard only rotations about O) then $\phi(RP)$ is also a potential regular in Ω. We shall then define Σ_ϕ for a given potential ϕ as the set of potentials $\phi(RP)$ for all rotations R and interpret the mean at a point as the mean over Σ_ϕ using as a measure μ the invariant measure for the group of rotations G normalized so that $\mu(G) = 1$. When we use this interpretation of $M\{\cdot\}$, the reasoning leading to formula (30) runs without difficulty.

Now one could ask whether it is reasonable to represent the disturbing potential for a non-spherical planet as a rotation invariant stochastic process. My answer is that it may give a reasonable result if it is done in a reasonable way: The interpolation, i.e. the mathematical procedure that gives a potential T satisfying a finite set of conditions (from measurements), has an infinity of solutions. It must be theoretically exact, i.e. it must give one of these solutions, but for economical reasons we want to use a statistical method that can give us a solution that seems to us to use the observations as well as possible. Therefore, a rather rough approximation in the statistical hypothesis is not disastrous, but merely an economic question. The situation is exactly parallel to that in adjustment of geodetic networks, where it is essential that we use a high precision of the physical (or geometrical) constants, but much less important to use "exact" weight (or covariance) coefficients.

But the acceptance of a disturbing potential with some rotation invariant properties confronts us with a much severer problem: The T that we calculate must be defined and harmonic not only in the space outside to the surface of the Earth but also down to a Bjerhammar sphere, and we know that the physically existing T of the Earth cannot be extended to a potential with this property.

Now back to the formulae.

A rotation invariant $K(P,Q)$ restricted to a spherical surface in Ω with centre at O and radius R is a function of w only and can be expanded into a series in Legendre polynomials:

$$K(P,Q) = \sum_{n=0}^{\infty} A_n(2n+1)P_n(\cos w) \qquad \text{for} \quad r_P = r_Q = R. \tag{37}$$

Since $K(P,Q)$ is a potential as a function of P and as a function of Q, it can be expanded as follows:

$$K(P,Q) = \sum_{n=0}^{\infty}(2n+1)\frac{R^{2n+2}A_n}{r_P^{n+1}r_Q^{n+1}}P_n(\cos w) \qquad \text{for all } P,Q \in \Omega. \tag{38}$$

We see that $K(P,Q)$ is a function of r_P, r_Q and w only, and that it is symmetric and harmonic in Ω and regular at infinity. We shall just find the conditions for its being of positive type, and therefore we must investigate the expression

$$\begin{aligned}
&\sum_{i=1}^{N}\sum_{k=1}^{N} K(P_i,Q_k)x_i x_k \\
&\quad = \sum_i\sum_k K(i,k)x_i x_k \\
&\quad = \sum_{n=0}^{\infty}\left((2n+1)A_n R^{2n+2}\sum_i\sum_k \frac{P_n(\cos w_{ik})}{r_i^{n+1}r_k^{n+1}}x_i x_k\right) \\
&\quad = \sum_{n=0}^{\infty} A_n R^{2n+2}\Bigg(\sum_{m=0}^{n}\sum_i\sum_k \frac{\overline{R}_{nm}(\theta_i,\lambda_i)x_i}{r_i^{n+1}}\frac{\overline{R}_{nm}(\theta_k,\lambda_k)x_k}{r_k^{n+1}} \\
&\qquad\qquad + \sum_{m=1}^{n}\sum_i\sum_k \frac{\overline{S}_{nm}(\theta_i,\lambda_i)x_i}{r_i^{n+1}}\frac{\overline{S}_{nm}(\theta_k,\lambda_k)x_k}{r_k^{n+1}}\Bigg) \\
&\quad = \sum_{n=0}^{\infty} A_n R^{2n+2}\left(\sum_{m=0}^{n}\Big(\sum_i \frac{\overline{R}_{nm}(\theta_i,\lambda_i)x_i}{r_i^{n+1}}\Big)^2 + \sum_{m=1}^{n}\Big(\sum_i \frac{\overline{S}_{nm}(\theta_i,\lambda_i)x_i}{r_i^{n+1}}\Big)^2\right).
\end{aligned} \tag{39}$$

If $A_n \geq 0$ for all n, (39) shows that

$$\sum_i\sum_k K(i,k)x_i x_k \geq 0, \tag{40}$$

but if $A_n < 0$ for some n, then for some set of points P_i and weights x_i

$$\sum_i\sum_k K(i,k)x_i x_k < 0.$$

We may conclude that the necessary and sufficient condition of $K(P,Q)$ being of non-negative type is that for all the coefficients A_n in (38)

$$A_n \geq 0.$$

4. I think that the readers will agree that the notation used till now is rather clumsy. Therefore I have chosen already here to introduce another notation that only later will be theoretically justified, but which is much more handy.

As long as we are interested only in simple interpolation the new notation is nothing but the classical matrix notation and requires no justification.

As we can suppose that we are interested only in the interpolated values at a finite set of points we just need to know $K(P,Q)$ at a finite number of points, say m points, i.e. we may represent $K(P,Q)$ by a square matrix:

$K(P,Q)$:

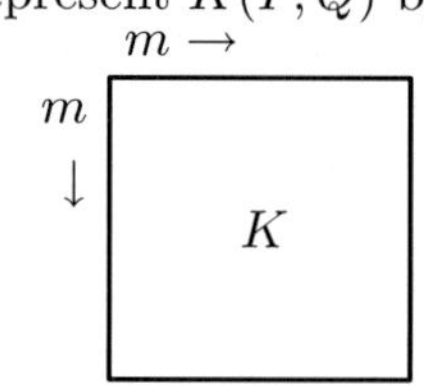

The square $n \times n$ matrix $K(P_i, Q_k)$ is still represented as an $n \times n$ matrix:

K_{ik}:

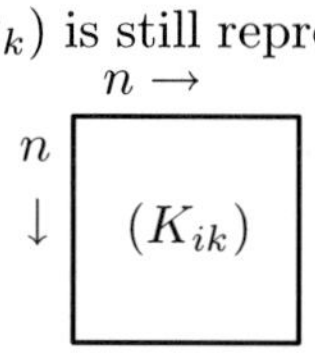

K_{ik} is obtained from $K(P,Q)$ using a projection matrix L. It is of dimensions $n \times m$ and consists of only 0's and 1's with precisely one 1 in each column.

L:

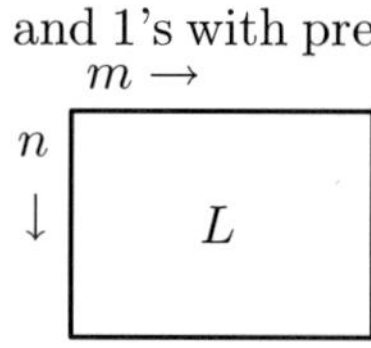

Then we may write:

$$(K_{ik}) = LKL^{\mathrm{T}}$$

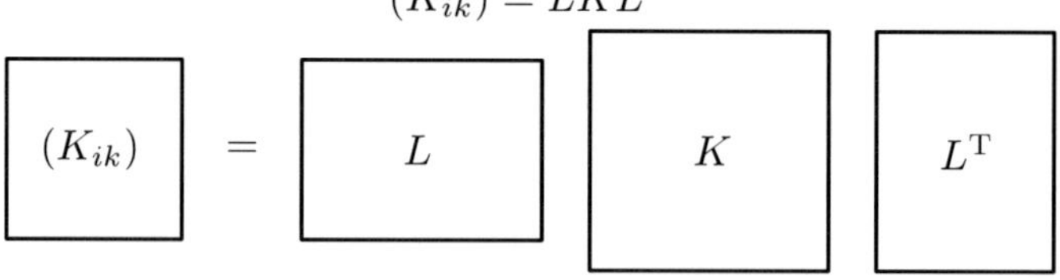

(α_{P_i}) is of the same form as L^{T}:

α_{P_i}:

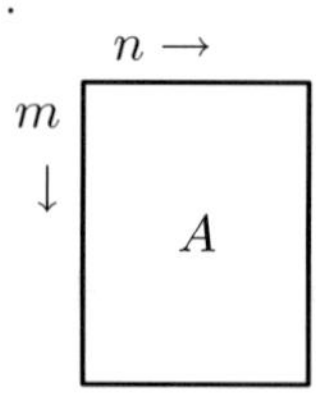

(T_P) is represented by a "long" vector:

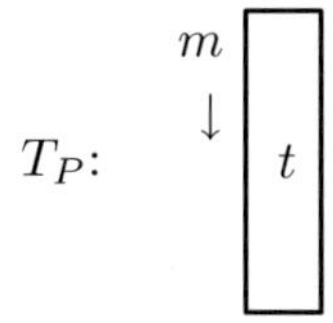

and Δg_i by a "short" one:

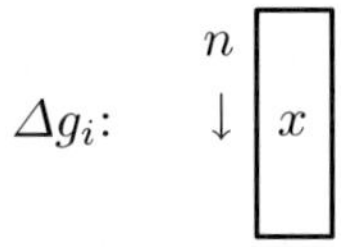

I call the vector representing Δg_i for x, because it may represent not only measurements of gravity anomalies, but also components of vertical deflections, perturbations of satellite orbits, and so on.

When we use the method not on a problem of simple interpolation, but on a problem of collocation, the dimensionality m is infinite, and L must be interpreted as a linear operator or, better, as a set of n linear operators operating at different points. However, I do not think that the interpretation will cause any difficulties; but it will be practical for the reader continuously to compare the following example with the formulae (23)–(34).

The problem to be solved using the new notation is this:

If an interpolation or collocation is used on many measurements, the requirement that the result must be exactly compatible with all the results of the measurements may have the effect that the result becomes very oscillating; therefore it is often better to use smoothing than exact interpolation.

Let us suppose that in addition to what was given in the problem treated in Subsection 2 of this section we are given an $n \times n$ variance-covariance matrix R for the n measurements, i.e.

$$x = y - v \tag{41}$$

where x is the vector of the measurements, y is the true value and v is the vector of the corrections.

We put as before, cf. (23),

$$\tilde{t} = Ax. \tag{42}$$

The error of prediction will now be

$$(\epsilon_P) = t - \tilde{t} = t - Ay + Av \tag{43}$$

and the error covariance

$$\begin{aligned}(\sigma_{PQ}) &= M\{(\epsilon_P)(\epsilon_Q)^{\mathrm{T}}\}\\ &= M\{(t - Ay + Av)(t^{\mathrm{T}} - y^{\mathrm{T}}A^{\mathrm{T}} + v^{\mathrm{T}}A^{\mathrm{T}})\}\\ &= M\{tt^{\mathrm{T}}\} - M\{ty^{\mathrm{T}}A^{\mathrm{T}}\} - M\{Ayt^{\mathrm{T}}\} + M\{tv^{\mathrm{T}}A^{\mathrm{T}}\} + M\{Avt^{\mathrm{T}}\}\end{aligned}$$

$$
\begin{aligned}
&+ M\{Ayy^{\mathrm{T}}A^{\mathrm{T}}\} - M\{Ayv^{\mathrm{T}}A^{\mathrm{T}}\} - M\{Avy^{\mathrm{T}}A^{\mathrm{T}}\} + M\{Avv^{\mathrm{T}}A^{\mathrm{T}}\} \\
&= K - KL^{\mathrm{T}}A^{\mathrm{T}} - ALK + 0 + 0 + ALKL^{\mathrm{T}}A^{\mathrm{T}} - 0 - 0 + ARA^{\mathrm{T}} \quad (44)
\end{aligned}
$$

or

$$(\sigma_{PQ}) = K - KL^{\mathrm{T}}A^{\mathrm{T}} - ALK + A(R + LKL^{\mathrm{T}})A^{\mathrm{T}}. \tag{45}$$

The condition that ϵ_P must attain a minimum gives the normal equations

$$A(R + LKL^{\mathrm{T}}) = KL^{\mathrm{T}} \tag{46}$$

or

$$(R + LKL^{\mathrm{T}})A^{\mathrm{T}} = KL^{\mathrm{T}}. \tag{47}$$

Here the matrix R is positive definite and LKL^{T} is always nonnegative definite; therefore the matrix of the normal equations

$$R + LKL^{\mathrm{T}}$$

is always positive definite. If we define N by

$$N(R + LKL^{\mathrm{T}}) = I, \tag{48}$$

where I is the $n \times n$ unit matrix, we get

$$A = KL^{\mathrm{T}}N, \qquad A^{\mathrm{T}} = NLK \tag{49}$$

and

$$\widetilde{t} = KL^{\mathrm{T}}Nx. \tag{50}$$

For the error covariance σ_{PQ} (45) and (49) give

$$(\sigma_{PQ}) = K - KL^{\mathrm{T}}NLK. \tag{51}$$

Also here it is most simple from a computational point of view first to solve the normal equations

$$(R + LKL^{\mathrm{T}})\xi = x, \tag{52}$$

and then to find $\widetilde{t}$ by

$$\widetilde{t} = KL^{\mathrm{T}}\xi. \tag{53}$$

The a posteriori variance-covariance matrix for the measurements may also be calculated:

$$
\begin{aligned}
L(\sigma_{PQ})L^{\mathrm{T}} &= LKL^{\mathrm{T}} - LKL^{\mathrm{T}}NLKL^{\mathrm{T}} = LKL^{\mathrm{T}}(I - N(R + LKL^{\mathrm{T}} - R)) \\
&= LKL^{\mathrm{T}}NR = (R + LKL^{\mathrm{T}} - R)NR = R - RNR. \quad (54)
\end{aligned}
$$

5. We return to the collocation problem (not the smoothing problem). Then $R = 0$ and (54) gives

$$L(\sigma_{PQ})L^{\mathrm{T}} = 0 \tag{55}$$

which is not surprising.

I have called σ_{PQ} the error covariance, because it has been called so by others up to now, although I should have preferred to call it the a posteriori covariance, and I shall try to explain why.

Imagine the situation that we have made a collocation as in Subsection 2 of this section and that afterwards we have made measurements at another set of points and now want to make the interpolation using both sets of measurements. I claim that it can be done in the following way:

Let the $\widetilde{t}$ found from the first collocation be called $\widetilde{t}_0$ and that found from the second one $\widetilde{t}$, and put

$$\widetilde{t} = \widetilde{t}_0 + \widetilde{s}. \tag{56}$$

Let the second set of measurements be called y, and let M play the role of L in the second set. Then $\widetilde{s}$ may be found using the ordinary method, if for K we use

$$K' = K - KL^{\mathrm{T}}NLK, \tag{57}$$

for L we use N, and for x we use

$$y - \widetilde{t}_0,$$

i.e.

$$\widetilde{s} = (K - KL^{\mathrm{T}}NLK)M^{\mathrm{T}}N'(y - \widetilde{t}_0), \tag{58}$$

where

$$N' = \left(M(K - KL^{\mathrm{T}}NLK)M^{\mathrm{T}}\right)^{-1}. \tag{59}$$

This may be verified directly by the well-known technique of manipulation with partitioned matrices, but it will not be done here as I do not think the result will be of any computational importance; nevertheless, I find it rather interesting, especially when it is formulated as follows:

If one has used the prediction method on the result of some measurements and after that made some new measurements (at other points), then the following two procedures will give the same results:

1. Use the prediction method on all measurements with the original covariance function.
2. Use the prediction method on the improvements (the new measured values minus the predicted values at the same points) with the a posteriori covariance function.

II. The Least-Squares Method in Hilbert Spaces

1. In the first section I have carried the generalization of Moritz' prediction formula to a point—or perhaps a little beyond a point—where some difficulties, e.g. with respect to a simple and consistent notation, seem to indicate the need for a more powerful mathematical apparatus.

I have several times stressed that the prediction method may be looked upon in two different aspects: the original prediction aspect, where one formally asks for the predicted value at a single point and, on the other hand, the collocation aspect where one asks for an interpolating function as an entity. Perhaps we can say that the first aspect is a discrete or finite one, while the second is a continuous or infinite one; therefore, it is not very surprising that we have to use methods from the functional analysis in order to make more extensive use of the second aspect.

The form of the normal equations for the second aspect (12) seems to indicate that the interpolating function is a result of an adjustment problem. When we introduced the new notation, we saw that in the case of ordinary interpolation the notation was the ordinary matrix notation, provided that we asked for the value of the interpolating function at a finite set of points only. Let us first investigate this special case to find out which quantity is minimized.

Let us try to find an m-dimensional vector t so that n, say the first n of its coordinates, equal the n coordinates of the n-dimensional vector x and so that

$$t^{\mathrm{T}} G t = \min \tag{60}$$

where G is an $m \times m$-dimensional positive definite matrix. The first condition can be written

$$Lt = x, \tag{61}$$

where the $m \times n$ matrix L can be partitioned according to

$$\begin{bmatrix} I_n & 0 \end{bmatrix} \tag{62}$$

where I_n is the $n \times n$-dimensional unity matrix, and 0 is the $n \times (m-n)$-dimensional zero matrix. The problem is now a classical least-square adjustment problem with the equations of condition (61) and the weight matrix G. Its solution is

$$t = G^{-1} L^{\mathrm{T}} (L G^{-1} L^{\mathrm{T}})^{-1} x. \tag{63}$$

If

$$G = K^{-1}, \tag{64}$$

then we have (50), i.e. x is the solution to the prediction problem for $K = G^{-1}$. We can explain what we have seen here by saying that the correlation function defines the metric of the vector space of which t is an element, and that the matrix defining this metric is the weight matrix corresponding to the given correlation.

When we have found the t that minimizes

$$t^{\mathrm{T}} K^{-1} t, \tag{65}$$

we can also find the value of the minimum:

$$\begin{aligned}\min_t t^{\mathrm{T}} K^{-1} t &= x^{\mathrm{T}}(LKL^{\mathrm{T}})^{-1} LKK^{-1}KL^{\mathrm{T}}(LKL^{\mathrm{T}})^{-1}x \\ &= x^{\mathrm{T}}(LKL^{\mathrm{T}})^{-1}LKL^{\mathrm{T}}(LKL^{\mathrm{T}})^{-1}x = x^{\mathrm{T}}(LKL^{\mathrm{T}})^{-1}x. \end{aligned} \tag{66}$$

This result is very interesting because it is independent of the set of $m-n$ points at which we wanted to find the prediction, and because the result is not limited to simple interpolation but may be generalized to the case where the unit matrix I_n in (62) is replaced with another nonsingular $n \times m$ matrix.

If we interpret K^{-1} as the matrix that defines the metric in the m-dimensional vector space of the t's, the norm of t is defined by

$$\|t\| = \left(t^{\mathrm{T}} K^{-1} t\right)^{1/2}, \tag{67}$$

and using (66) we can define the norm of the interpolating function t for the problem given by L, K and x:

$$\|t\| = \left(x^{\mathrm{T}}(LKL^{\mathrm{T}})^{-1}x\right)^{1/2}. \tag{68}$$

Now we shall find a more explicit expression of the norm. Suppose that $K(P,Q)$ can be expressed as follows

$$K(P,Q) = \sum_n k_n \phi_n(P)\phi_n(Q) \qquad \text{for} \quad P,Q \in \Omega, \tag{69}$$

where k_n are positive constants, $\phi_n(P)$ is a set of functions harmonic in Ω and the range of n is finite or infinite. As we are here interested in simple interpolation, we want to find such a function $t(P)$ that

$$t(P_i) = x_i \qquad i = 1,\, 2,\, \ldots,\, N. \tag{70}$$

As in (12) we put

$$t(P) = \sum_{i=1}^{N} \xi_i K(P,Q_i) = \sum_n k_n \phi_n(P) \sum_{i=1}^{N} \xi_i \phi_n(Q_i) = \sum_n t_n \phi_n(P), \tag{71}$$

where

$$t_n = k_n \sum_{i=1}^{N} \xi_i \phi_n(Q_i) \tag{72}$$

are the coefficients in the expansion of $t(P)$ into a series of $\phi_n(P)$.

The conditions (70) give the normal equations ((52) for $R=0$)

$$LKL^{\mathrm{T}}\xi = x, \tag{73}$$

i.e.

$$\begin{aligned}\|t\|^2 &= \xi^{\mathrm{T}} LKL^{\mathrm{T}} (LKL^{\mathrm{T}})^{-1} LKL^{\mathrm{T}} \xi = \xi^{\mathrm{T}} LKL^{\mathrm{T}} \xi \\ &= \sum_{i=1}^{N} \sum_{k=1}^{N} \xi_i \xi_k K(P_i, P_k) = \sum_i \sum_k \xi_i \xi_k \sum_n k_n \phi_n(P_i) \phi_n(P_k) \\ &= \sum_n k_n \sum_i \sum_k \xi_i \xi_k \phi_n(P_i) \phi_n(P_n) = \sum_n k_n \Big(\sum_i \xi_i \phi_i(P_i) \Big)^2 = \sum_n \frac{t_n^2}{k_n}.\end{aligned} \tag{74}$$

If for all f for which

$$f = \sum_n f_n \phi_n(P) \tag{75}$$

converges at all points of Ω and for which

$$\sum_n \frac{f_n^2}{k_n} \tag{76}$$

also converges, we define

$$\|f\| = \left(\sum_n \frac{f_n^2}{k_n} \right)^{1/2}, \tag{77}$$

then our definition is consistent with (33). We shall adopt this definition, and call the set of all such f H_K.

H_K is a linear space, i.e. for $f, g \in H_K$ and for a being any number $f + g \in H_K$ and $af \in H_K$.

Let us for $f, g \in H_K$ define the scalar product $\langle f, g \rangle$ by

$$2\langle f, g \rangle = \|f + g\|^2 - \|f\|^2 - \|g\|^2. \tag{78}$$

It is evident that $\langle f, g \rangle = \langle g, f \rangle$, that the scalar product is bilinear, and that

$$\|f\|^2 = \langle f, f \rangle. \tag{79}$$

If $\langle f, g \rangle = 0$, we say that f and g are orthogonal.

For

$$f = \phi_n + \phi_m$$

we find by (78)

$$\|f\|^2 = \|\phi_n + \phi_m\|^2 = \|\phi_n\|^2 + \|\phi_m\|^2 + 2\langle \phi_n, \phi_m \rangle \tag{80}$$

and using (77)

$$\frac{1}{k_n} + \frac{1}{k_m} = \frac{1}{k_n} + \frac{1}{k_m} + 2\langle \phi_n, \phi_m \rangle \qquad \text{for } m \neq n \tag{81}$$

and

$$\frac{4}{k_n} = \frac{2}{k_n} + 2\langle \phi_n, \phi_n \rangle \qquad \text{for } m = n. \tag{82}$$

Equation (81) gives

$$\langle \phi_n, \phi_m \rangle = 0 \qquad \text{for } m \neq n. \tag{83}$$

and (82) gives

$$\|\phi_n\|^2 = \langle \phi_n, \phi_n \rangle = \frac{1}{k_n} \tag{84}$$

which shows that $\phi_1, \phi_2, \ldots$ is a system of orthogonal functions.

As a function of ϕ the covariance function $K(P,Q)$ is an element of the space H_K and for f being any element of H_K we can calculate

$$\begin{aligned} \langle K(P,Q), f(Q) \rangle &= \Big\langle \sum_n k_n \phi_n(P) \phi_n(Q), \sum_m f_m \phi_m(Q) \Big\rangle \\ &= \sum_n \sum_m k_m f_m \phi_n(P) \langle \phi_n(Q), \phi_m(Q) \rangle \\ &= \sum_n k_n f_n \phi_n(P) \frac{1}{k_n} = \sum_n f_n \phi_n(P) = f(P). \end{aligned} \tag{85}$$

Specially for $f(Q) = K(Q,R)$ we get

$$\langle K(P,Q), K(Q,R) \rangle = K(P,R). \tag{86}$$

What we have found so far may be summarized as follows: The covariance function given by (69) defines a Hilbert space H_K consisting of the functions f satisfying (75) and (77), and $K(P,Q)$ is the reproducing kernel for H_K. From the theory of Hilbert spaces with kernel function it follows that it is not necessary to demand the convergence of (75)—it follows from that of (76).

As to the rotation invariant case we have found the form of the covariance function (38):

$$\begin{aligned} K(P,Q) &= \sum_{n=0}^{\infty} (2n+1) A_n \Big(\frac{R^2}{r_P r_Q} \Big)^{n+1} P_n(\cos w) \\ &= \sum_{n=0}^{\infty} A_n \Big(\frac{R^2}{r_P r_Q} \Big)^{n+1} \sum_{m=-n}^{n} E_{nm}(\theta_P, \lambda_P) E_{nm}(\theta_Q, \lambda_Q) \end{aligned} \tag{87}$$

where $A_n \geq 0$, and I have defined

$$\begin{aligned} E_{nm}(\theta, \lambda) - \overline{R}_{nm}(\theta, \lambda) \qquad & \text{for } m \geq 0, \\ E_{nm}(\theta, \lambda) = \overline{S}_{nm}(\theta, \lambda) \qquad & \text{for } m \leq 0. \end{aligned} \tag{88}$$

H_K consists here of all functions of the form

$$f(P) - \sum_n{}' \sum_{m=-n}^{n} f_{nm} \Big(\frac{R}{r_P} \Big)^{n+1} E_{nm}(\theta, \lambda), \tag{89}$$

for which

$$\|f\|^2 = \sum_n{}' \sum_{m=-n}^{n} \frac{f_{nm}^2}{A_n} \tag{90}$$

converges. $\sum_n'$ denotes here and below that the sum is to be taken over the set of n for which $A_n > 0$.

Equations (83) and (84) give here

$$\left\langle \left(\frac{R}{r}\right)^{i+1} E_{ij}(\theta,\lambda), \left(\frac{R}{r}\right)^{k+1} E_{kl}(\theta,\lambda) \right\rangle = \begin{cases} \frac{1}{A_i} & \text{for } i = k \text{ and } j = l, \\ 0 & \text{in all other cases.} \end{cases} \tag{91}$$

2. In this section the two classical least-squares adjustment problems will be generalized to Hilbert spaces.

In both cases we have two Hilbert spaces H_1 and H_2, and each of them may independent of the other be finite- or infinite-dimensional. The scalar products and the norms in the different spaces will be distinguished by lower indices, so that for instance $\|\cdot\|_2$ is the norm in H_2. The two spaces may be identical.

First I shall draw the attention of the reader to some definitions concerning linear operators in Hilbert spaces.

A linear operator $A\colon H_1 \to H_2$ is a function on H_1 to H_2 so that if A is defined for two elements x_1 and $y_1 \in H_1$ in such a way that

$$\begin{aligned} Ax_1 &= x_2 \\ Ay_1 &= y_2 \end{aligned} \tag{92}$$

where x_2 and $y_2 \in H_2$, then A is also defined for $x_1 + y_1$, and

$$A(x_1 + y_1) = x_2 + y_2, \tag{93}$$

and for ax_1, where a is any real number, and

$$A(ax_1) = aAx_1. \tag{94}$$

A linear operator $A\colon H_1 \to H_2$ is said to be bounded if it is defined for every $x \in H_1$ and there exists such a positive number b that for all $x \in H_1$

$$\|Ax\|_2 \le b\|x\|_1 \tag{95}$$

(in fact the fulfilment of the second condition follows from that of the first).

If $H_1 = H_2$, the bounded linear operators $A\colon H_1 \to H_2$ are analogous to square matrices; if $H_1 \ne H_2$, they are analogous to rectangular matrices. There is also the following analogy to transposition:

If $A\colon H_1 \to H_2$ is a bounded operator, then there exists one and only one bounded linear operator $A^{\mathrm{T}}\colon H_2 \to H_1$ so that for all $x_1 \in H_1$ and $x_2 \in H_2$

$$\langle Ax_1, x_2\rangle_2 = \langle x_1, A^{\mathrm{T}} x_2\rangle_1. \tag{96}$$

When I use the word operator, I generally mean bounded linear operator.

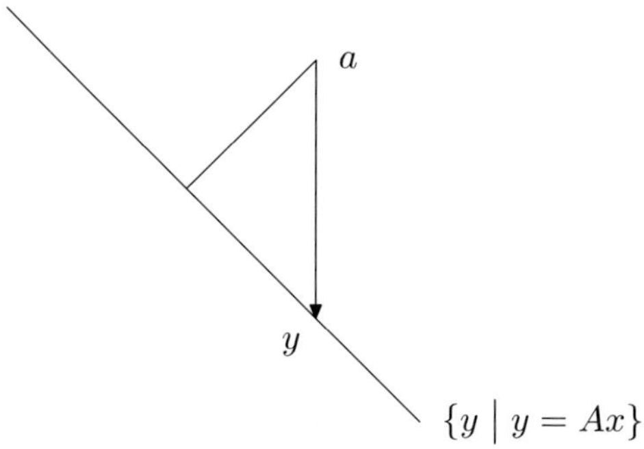

The first problem. Given a bounded operator $A\colon H_1 \to H_2$ and an element $a \in H_2$. Find such an element $x \in H_1$ that $\|z\|_2$, where

$$z = Ax - a, \tag{97}$$

is as small as possible.

$\|z\|_2$ attains a minimum if, and only if, z is orthogonal to Ay for all $y \in H_1$:

$$\begin{aligned} \langle z, Ay\rangle_2 &= 0 \\ \langle A^{\mathrm{T}} z, y\rangle_1 &= 0 \end{aligned} \qquad \text{for all } y \in H_1 \tag{98}$$

that is

$$\begin{aligned} 0 = A^{\mathrm{T}} z &= A^{\mathrm{T}} Ax - A^{\mathrm{T}} a, \\ A^{\mathrm{T}} Ax &= A^{\mathrm{T}} a. \end{aligned} \tag{99}$$

If the operator $A^{\mathrm{T}} A$ is invertible, the unique solution is

$$x = (A^{\mathrm{T}} A)^{-1} A^{\mathrm{T}} a \tag{100}$$

and

$$\begin{aligned} \|z\|_2^2 &= \langle A(A^{\mathrm{T}} A)^{-1} A^{\mathrm{T}} a - a, A(A^{\mathrm{T}} A)^{-1} A^{\mathrm{T}} a - a\rangle_2 \\ &= \langle A(A^{\mathrm{T}} A)^{-1} A^{\mathrm{T}} a, A(A^{\mathrm{T}} A)^{-1} A^{\mathrm{T}} a\rangle_2 - 2\langle A(A^{\mathrm{T}} A)^{-1} A^{\mathrm{T}} a, a\rangle_2 + \langle a, a\rangle_2 \\ &= \langle A^{\mathrm{T}} A(A^{\mathrm{T}} A)^{-1} A^{\mathrm{T}} a, (A^{\mathrm{T}} A)^{-1} A^{\mathrm{T}} a\rangle_1 - 2\langle A(A^{\mathrm{T}} A)^{-1} A^{\mathrm{T}} a, a\rangle_2 + \langle a, a\rangle_2 \\ &= \langle A^{\mathrm{T}} a, (A^{\mathrm{T}} A)^{-1} A^{\mathrm{T}} a\rangle_1 - 2\langle (A^{\mathrm{T}} A)^{-1} A^{\mathrm{T}} a, A^{\mathrm{T}} a\rangle_1 + \langle a, a\rangle_2 \\ &= \langle a, a\rangle_2 - \langle a, A(A^{\mathrm{T}} A)^{-1} A^{\mathrm{T}} a\rangle_2 \\ &= \langle a, a\rangle_2 - \langle A(A^{\mathrm{T}} A)^{-1} A^{\mathrm{T}} a, A(A^{\mathrm{T}} A)^{-1} A^{\mathrm{T}} a\rangle_2 \\ &= \|a\|_2^2 - \|A(A^{\mathrm{T}} A)^{-1} A^{\mathrm{T}} a\|_2^2 \end{aligned} \tag{101}$$

because

$$\left(A(A^{\mathrm{T}} A)^{-1} A^{\mathrm{T}}\right)^2 = A(A^{\mathrm{T}} A)^{-1} A^{\mathrm{T}} A(A^{\mathrm{T}} A)^{-1} A^{\mathrm{T}} = A(A^{\mathrm{T}} A)^{-1} A^{\mathrm{T}}. \tag{102}$$

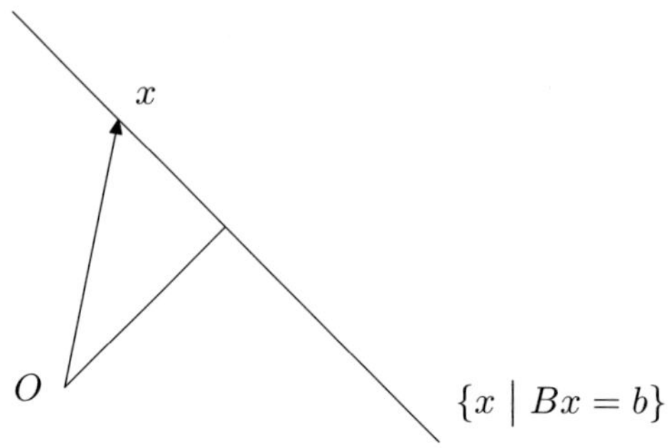

The operator of the normal equations (99) is non-negative definite:

$$\langle x, A^{\mathrm{T}}Ax\rangle_2 \geq 0 \qquad \text{for all } x \in H_2 \tag{103}$$

because

$$\langle x, A^{\mathrm{T}}Ax\rangle_2 = \langle Ax, Ax\rangle_1 = \|Ax\|_1^2 \geq 0. \tag{104}$$

The operator $A^{\mathrm{T}}A$ is symmetric:

$$(A^{\mathrm{T}}A)^{\mathrm{T}} = A^{\mathrm{T}}A. \tag{105}$$

The second problem. Given a bounded operator $B\colon H_2 \to H_1$ and an element $b \in H_1$. Find such an element $x \in H_2$ that

$$Bx = b \tag{106}$$

and that

$$\|x\|_2 \tag{107}$$

is as small as possible.

Any $x \in H_2$ can be written as

$$x = B^{\mathrm{T}}\xi + y \tag{108}$$

with a suitable $\xi \in H_1$ and $y \in H_2$ so that y is orthogonal to $B^{\mathrm{T}}s$ for all $s \in H_1$, that is

$$0 = \langle B^{\mathrm{T}}s, y\rangle_2 = \langle s, By\rangle_1 \qquad \text{for all } s \in H_1 \tag{109}$$

or

$$By = 0. \tag{110}$$

Now (108) and (110) give

$$Bx = BB^{\mathrm{T}}\xi + By = BB^{\mathrm{T}}\xi. \tag{111}$$

The condition of x, written in the form (108), satisfying (106) is that ξ is determined by

$$BB^{\mathrm{T}}\xi = b. \tag{112}$$

If BB^{T}, which definitely is symmetric and non-negative definite, is invertible, we have

$$\xi = (BB^{\mathrm{T}})^{-1}b. \tag{113}$$

In (108) x is expressed as a sum of two orthogonal elements of which the first is given; the norm of x therefore attains its minimum for $y = 0$, and we have the solution:

$$x = B^{\mathrm{T}}(BB^{\mathrm{T}})^{-1}b. \tag{114}$$

For the minimum value $\|x\|_2$ we have

$$\begin{aligned}\|x\|_2^2 = \langle B^{\mathrm{T}}(BB^{\mathrm{T}})^{-1}b, B^{\mathrm{T}}(BB^{\mathrm{T}})^{-1}b\rangle_2 &= \langle BB^{\mathrm{T}}(BB^{\mathrm{T}})^{-1}b, (BB^{\mathrm{T}})^{-1}b\rangle_1 \\ &= \langle b, (BB^{\mathrm{T}})^{-1}b\rangle_1. \end{aligned} \tag{115}$$

As in the finite-dimensional case we can also here define and solve the more sophisticated mean squares problems.

There is a method of including in the two problems treated here problems that look more general by using direct sums of Hilbert spaces.

For two (finite- or infinite-dimensional) Hilbert spaces H_1 and H_2 we can define a third Hilbert space H_+:

$$H_+ = H_1 \oplus H_2 \tag{116}$$

called the direct sum of the spaces H_1 and H_2. It consists of all ordered pairs

$$(x_1, x_2) \tag{117}$$

of elements $x_1 \in H_1$ and $x_2 \in H_2$, and the scalar product in H_+ is defined as

$$\langle (x_1, x_2), (y_1, y_2)\rangle_+ = \langle x_1, y_1\rangle_1 + \langle x_2, y_2\rangle_2. \tag{118}$$

If we have a fourth Hilbert space H_3 and an operator $A\colon H_+ \to H_3$, it will often be practical to partition A:

$$A = \begin{bmatrix} A_1 & A_2 \end{bmatrix}, \tag{119}$$

where $A_1 : H_1 \to H_3$ and $A_2 : H_2 \to H_3$.

If, in the same way, $B\colon H_3 \to H_+$, then

$$B = \begin{bmatrix} B_1 \\ B_2 \end{bmatrix} \tag{120}$$

where $B_1 : H_3 \to H_1$ and $B_2 : H_3 \to H_2$. The analogy to matrix notation is striking.

3. To be able to use the adjustment formulae we must know how to obtain explicit forms of the operators A and B and here the covariance function or kernel $K(P, Q)$ may help us very much.

Formula (85) shows that $K(P, Q)$ can operate on a function $f(P)$. We shall now write (85) in this way:

$$Kf = f. \tag{121}$$

We see that in reality this operator is the unity operator in the Hilbert space H_K.

If $P_1, P_2, \ldots, P_N$ is any ordered set of points in Ω, then the operator L defined by

$$L_i = \langle K(P_i, Q), f(Q) \rangle = f(P_i) \tag{122}$$

is a bounded linear operator $L\colon H_K \to E_N$, where E_N is the N-dimensional Euclidean space, which to an element of H_K assigns the vector consisting of its values at the points P_i. This can be generalized, defining L by

$$L_i = \langle \mathcal{L}_{P_i} K(P, Q), f(Q) \rangle = \mathcal{L}_{P_i} f(P), \tag{123}$$

where $\mathcal{L}_i$ are N-linear functionals, e.g. differential operators operating at discrete points.

If (V_i) is a vector in E_N, then L^{T} is defined:

$$\langle \mathcal{L}_{Q_i} K(P, Q), V \rangle_N = \sum_i \mathcal{L}_{Q_i} K(P, Q) V_i, \tag{124}$$

and it follows that

$$LL^{\mathrm{T}} V = \sum_k \mathcal{L}_{P_i} \mathcal{L}_{Q_k} K(P, Q) V_k \tag{125}$$

i.e. LL^{T} is the same operator (matrix) as LKL^{T} in for example (46). This is not surprising, as we have seen that the operator K is equivalent to the unity operator in H_K.

Now we can solve the problem which was previously solved by reasoning on covariance as an example of the second problem of adjustment.

H_2 is now the infinite-dimensional Hilbert space H_K, and H_1 is the N-dimensional space of measurements E_N, where N is the number of scalar measurements involved.

$B\colon H_2 \to H_1$ becomes here $L\colon H_K \to E_N$, and b is the N-vector of the results of the measurements. The normal equations are

$$LL^{\mathrm{T}} \xi = b, \tag{126}$$

and now they really are normal equations. The solution is given by (114) or by

$$x = L^{\mathrm{T}} (LL^{\mathrm{T}})^{-1} b. \tag{127}$$

We notice that we can compute the solution to the problem without explicitly using the norm in the potential space, but only the covariance function K. As we see from (115), the norm $\|x\|$ of the result can also be calculated without our knowing the explicit expression of the norm in the potential space.

An instructing example of the adjustment technique is the smoothing problem from the first section.

Here we have again the measurement equations

$$(Lf)_i \equiv \mathcal{L}_i f(P) = b_i, \qquad i = 1, 2, \ldots, n, \tag{128}$$

but now we do not want them to be satisfied exactly. We ask for such a potential f that

$$Lf - v = b, \tag{129}$$

and

$$\|f\|^2 + v^{\mathrm{T}} P v = \min, \tag{130}$$

where P is a positive definite $n \times n$ matrix.

This problem may be treated as an adjustment problem of the second type.

Our unknown quantities consist of the potential f and the vector v, and we may look upon them as a single quantity

$$x = f \oplus v \tag{131}$$

or

$$x = \begin{bmatrix} f \\ v \end{bmatrix}. \tag{132}$$

If we have another element y of the same type

$$y = \begin{bmatrix} g \\ u \end{bmatrix} \tag{133}$$

we can define the scalar product

$$\langle x, y\rangle_2 = \langle f, g\rangle + v^{\mathrm{T}} P u \tag{134}$$

and the norm

$$\|x\|_2 = \langle x, x\rangle_2^{1/2}. \tag{135}$$

Now we have defined the space H_2.

H_1 is the space of the measurements, i.e. the n-dimensional Euclidean space, and the operator $B\colon H_2 \to H_1$ is defined by (129) or by

$$Bx = \begin{bmatrix} L & -I_n \end{bmatrix} \begin{bmatrix} f \\ v \end{bmatrix} = Lf - v, \tag{136}$$

where I_n is the unity matrix in H_1.

The problem is now: Find such an element $x \in H_2$ that

$$Bx = b \tag{137}$$

and that

$$\|x\|_2 \tag{138}$$

is as small as possible.

We first have to find the normal equations

$$BB^{\mathrm{T}}\xi = b \tag{139}$$

but what is B^{T}?

From the definition of transposed operators it follows that $B^{\mathrm{T}}: H_1 \to H_2$ is given by the equation

$$\langle Bx_2, x_1\rangle_1 = \langle x_2, B^{\mathrm{T}}x_1\rangle_2 \tag{140}$$

for all $x_1 \in H_1$ and $x_2 \in H_2$, or if we write

$$x_2 = \begin{bmatrix} f_2 \\ v_2 \end{bmatrix}. \tag{141}$$

and

$$B^{\mathrm{T}}x_1 = \begin{bmatrix} y \\ z \end{bmatrix}, \qquad y \in H, \quad z \in H, \tag{142}$$

and use (134) and (136)

$$\langle Lf_2 - v_2, x_1\rangle_1 = \langle f_2, y\rangle + v_2^{\mathrm{T}}Pz \tag{143}$$

or

$$\langle f_2, L^{\mathrm{T}}x_1\rangle - \langle v_2, x_1\rangle_1 = \langle f_2, y\rangle + \langle v_2, Pz\rangle_1. \tag{144}$$

Equation (144) will only hold for all x_2, that is for all f_2 and v_2, if

$$y = L^{\mathrm{T}}x_1 \tag{145}$$

and

$$z = -P^{-1}x_1 \tag{146}$$

i.e.

$$B^{\mathrm{T}}x_1 = \begin{bmatrix} L^{\mathrm{T}}x_1 \\ -P^{-1}x_1 \end{bmatrix} = \begin{bmatrix} L^{\mathrm{T}} \\ -P^{-1} \end{bmatrix} x_1. \tag{147}$$

Then

$$BB^{\mathrm{T}} = \begin{bmatrix} L & -I_n \end{bmatrix} \begin{bmatrix} L^{\mathrm{T}} \\ -P^{-1} \end{bmatrix} = LL^{\mathrm{T}} + P^{-1}, \tag{148}$$

where P^{-1} is the variance-covariance matrix R of the measurements. If this matrix is a diagonal matrix, as it is commonly assumed to be, it will have an ameliorating influence on the condition number of the normal equations

$$(R + LL^{\mathrm{T}})\xi = b. \tag{149}$$

The solution to the problem is now given by

$$x = B^{\mathrm{T}}\xi \tag{150}$$

or more explicitly

$$f = L^{\mathrm{T}}\xi, \qquad v = -R\xi. \tag{151}$$

Equation (115) gives here

$$\|x\|_2^2 = \|f\|^2 + v^{\mathrm{T}}Pv = b^{\mathrm{T}}\xi = b^{\mathrm{T}}(R + LL^{\mathrm{T}})^{-1}b. \tag{152}$$

The reader should carefully compare this solution of the smoothing problem with that given in the first section (41)–(54).

The same problem could also be solved as a problem of the first type.

III. Hilbert Spaces with Kernel Function and Spherical Harmonics

1. As in the first approximation the Earth is spherical, it is tempting to look for its gravitation potential among the potentials which are regular outside some sphere. There has been some discussion among geodesists as to the permissibility thereof, but before answering this important question we shall first study such sets of potentials and their connection with the spherical harmonics.

Let Σ be the part of the space outside a sphere with radius R and surface σ, which surface is not included in Σ. We shall be interested in several sets of potentials ϕ all of which are regular in Σ including infinity so that

$$\lim_{P\to\infty} \phi(P) = 0. \tag{153}$$

The first set S of these potentials consists of those which are continuous in $\Sigma + \sigma$, i.e. those which have continuous boundary values on the surface σ of the sphere. For such potentials we can define a scalar product

$$\langle \phi, \psi \rangle = \frac{1}{4\pi} \int_\sigma \phi(P)\psi(P)\, d\sigma, \qquad \text{for} \quad \phi, \psi \in S. \tag{154}$$

This is the mean value of the product of the boundary values for the two potentials on the surface of the sphere. The corresponding norm is

$$\|\phi\| = \langle \phi, \psi \rangle^{1/2} = \left(\frac{1}{4\pi} \int_\sigma \phi(P)^2\, d\sigma \right)^{1/2}. \tag{155}$$

If for a potential $\phi \in S$, $\|\phi\| = 0$, then ϕ must be zero on σ, because ϕ is continuous and

$$\int_\sigma \phi(P)^2\, d\sigma = 0. \tag{156}$$

But a potential which is regular in Σ and continuous in $\Sigma + \sigma$ is zero at all points of Σ if it is zero on σ. So $\|\phi\| = 0$ if, and only if $\phi = 0$: (155) defines really a norm, and the set S is a pre-Hilbert space.

It is well-known, see [4], p. 34–35, that, if the boundary values of a potential $\phi \in S$ on σ are known, then ϕ can be found in Σ by Poisson's integral:

$$\phi(P) = \frac{R(r_P^2 - R^2)}{4\pi} \int_\sigma \frac{\phi(Q)}{l^3} \, d\sigma, \qquad P \in \Sigma,\ Q \in \sigma, \tag{157}$$

where

$$l = \sqrt{r_P^2 + R^2 - 2Rr_P \cos\psi}, \tag{158}$$

and that "Poisson's kernel" can be expressed by spherical harmonics in the following way

$$\frac{R(r_P^2 - R^2)}{l^3} = \sum_{n=0}^{\infty} (2n+1) \left(\frac{R}{r_P}\right)^{n+1} P_n(\cos\psi). \tag{159}$$

Poisson's kernel is a function of the two points P and Q, of which one is on σ and the other in Σ. We can define a symmetrical kernel $K(P,Q)$ by putting:

$$K(P,Q) = \frac{r_P^2 r_Q^2 - R^4}{RL^3} \tag{160}$$

where

$$L = \sqrt{\frac{r_P^2 r_Q^2}{R^2} - 2r_P r_Q \cos\psi + R^2}. \tag{161}$$

As $K(P,Q)$ is the result of substituting

$$R^2 \quad \text{for} \quad R$$

and

$$r_P r_Q \quad \text{for} \quad r_P,$$

in the formula for Poisson's kernel, the same substitution in (159) will give the expansion of $K(P,Q)$ into spherical harmonics:

$$K(P,Q) = \sum_{n=0}^{\infty} (2n+1) \left(\frac{R^2}{r_P r_Q}\right)^{n+1} P_n(\cos\psi). \tag{162}$$

$K(P,Q)$ is defined for P and $Q \in \Sigma$ and also for $P \in \sigma$ and $Q \in \Sigma$ (or $P \in \Sigma$ and $Q \in \sigma$), but for both P and $Q \in \sigma$ $K(P,Q)$ is zero for $P \neq Q$ and not defined for $P = Q$.

From (162) it follows, at least formally, that for either P or Q being fixed $K(P,Q)$ is a regular potential as a function of the other variable. A straightforward differentiation will verify that. Therefore, we have that for a fixed $P \in \Sigma$, $K(P,Q)$ as a function of Q is a member of the set S.

Now we may calculate the scalar product of $K(P,Q)$ and a potential $\phi(Q) \in S$. Here we shall only use the values of $K(P,Q)$ for $Q \in \sigma$, in which case $r_Q = R$ and $K(P,Q)$ has the same values as Poisson's kernel. Therefore, the scalar product is exactly the right member of (157), and we have

$$\big\langle K(P,Q), \phi(Q)\big\rangle = \phi(P), \qquad \text{for } P \in \Sigma. \tag{163}$$

i.e. $K(P,Q)$ is the reproducing kernel for the set S of potentials.

With the metric defined in (155) S is not a Hilbert space, but only a pre-Hilbert space, i.e. not every sequence

$$\{\psi_n\}, \qquad \psi_n \in S \quad \text{for } n = 0,\, 2,\, \ldots$$

for which

$$\lim_{n,m\to\infty} \|\psi_n - \psi_m\| = 0,$$

has a limit $\psi \in S$. Therefore we shall complete S to a Hilbert space H, which can be proved to consist of potentials regular in Σ and having square integrable boundary values on σ. For this Hilbert space $K(P,Q)$ is the reproducing kernel. Now the values of the integrals in (154), (155), etc. must be understood as the limits for $r > R$, $r \to R$ of the corresponding integrals over spheres with radius r.

If we define the functions with two indices

$$\{\phi_n^m(P)\} = \begin{cases} m &= -n,\, -n+1,\, \ldots,\, n-1,\, n \\ n &= 0,\, 1,\, 2,\, \ldots \end{cases}$$

by

$$\phi_n^m(P) = \begin{cases} \left(\frac{R}{r_P}\right)^{n+1} \overline{R}_{nm}(Q_P, \lambda_P) & \text{for } m \geq 0 \\ \left(\frac{R}{r_P}\right)^{n+1} \overline{S}_{nm}(Q_P, \lambda_P) & \text{for } m < 0 \end{cases} \tag{164}$$

where $\overline{R}_{nm}$ and $\overline{S}_{nm}$ are the fully normalized harmonics, see [4], p. 31, then we can write (162) as follows:

$$K(P,Q) = \sum_{n=0}^{\infty} \sum_{m=-n}^{n} \phi_n^m(P)\phi_n^m(Q), \tag{165}$$

which shows that the spherical harmonics $\{\phi_n^m\}$ form a complete orthonormal system for the Hilbert space H. This means that every $\phi \in H$ may be represented by a series expansion:

$$\phi(P) = \sum_{n=0}^{\infty} \sum_{m=-n}^{n} a_n^m \phi_n^m(P), \qquad \text{for } P \in \Sigma \tag{166}$$

where

$$a_n^m = \langle \phi, \phi_n^m \rangle. \tag{167}$$

This, however, is to be understood in the sense of convergence in the Hilbert space metric

$$\lim_{N\to\infty}\left\|\phi(P)-\sum_{n=0}^{N}\sum_{m=-n}^{n}a_n^m\phi_n^m(P)\right\|=0, \tag{168}$$

and we want a theorem which secures uniform convergence of the series. Fortunately the theory of reproducing kernels may help us here. For every $\psi \in H$ we can write

$$\begin{aligned}|\phi(P)| &= |\langle\psi(Q),K(P,Q)\rangle| \le \|\phi\|\cdot\|K(P,Q)\|_Q\\ &= \|\phi\|\langle K(P,Q),K(P,Q)\rangle_Q^{1/2} = \|\phi\|K(P,P)^{1/2}.\end{aligned} \tag{169}$$

Here we have used the fact that $K(P,Q)$ is a reproducing kernel, Schwartz' inequality, the expression of the norm by the scalar product (the index Q signifies that the norm and the scalar product are to be understood with respect to Q) and once more the fact that $K(P,Q)$, is a reproducing kernel.

Let us use (169) on

$$\begin{aligned}\left\|\phi(P)-\sum_{n=0}^{N}\sum_{m=-n}^{n}a_n^m\phi_n^m(P)\right\| &= \left\|\sum_{n=N+1}^{\infty}\sum_{m=-n}^{n}a_n^m\phi_n^m(P)\right\|\\ &\le\left(\sum_{n=N+1}^{\infty}\sum_{m=-n}^{n}\left(a_n^m\right)^2\right)^{1/2}\end{aligned} \tag{170}$$

to get

$$\left|\phi(P)-\sum_{n=0}^{\infty}\sum_{m=-n}^{n}a_n^m\phi_n^m(P)\right|\le\left(\sum_{n=N+1}^{\infty}\sum_{m=-n}^{n}\left(a_n^m\right)^2\right)^{1/2}K(P,P)^{1/2}. \tag{171}$$

A simple calculation gives that

$$0<K(P,P)^{1/2}\le\frac{Rr_P}{r_P^2-R^2}\sqrt{2} \tag{172}$$

and then (171) shows that the series (166) converges uniformly for all P so that

$$r_P\ge r_0>R. \tag{173}$$

We now have to go back to the formulae (166) and (167).

From the general theory for Hilbert spaces we know that from (166) and (167) follows

$$\|\phi\|=\left(\sum_{n=0}^{\infty}\sum_{m=-n}^{n}\left(a_n^m\right)^2\right)^{1/2} \tag{174}$$

and that for every sequence $\{a_n^m\}$ for which

$$\sum\sum\left(a_n^m\right)^2$$

converges such an element $\phi \in H$ exists that (166), (167), and (174) hold.

If we have a ϕ given by (166), it might be of interest to know if the series converges for points P so that $r_P < R$.

Let us put

$$A_n = \left(\sum_{m=-n}^{n}\left(a_n^m\right)^2\right)^{1/2} \tag{175}$$

then we have from (174)

$$\sum_{n=0}^{\infty} A_n^2 = \|\phi\|^2, \tag{176}$$

and therefore we must have

$$\lim_{n\to\infty} A_n\rho^n = 0 \tag{177}$$

for $0 \leq \rho \leq 1$. Let the least upper bound of ρ for which (177) is valid be called ρ_0. (ρ_0 may be ∞.)

Then (177) is valid for every ρ so that $0 < \rho < \rho_0$.

If $\rho_0 > 1$, then take two members ρ_1 and ρ_2 so that

$$0 < \rho_1 < \rho_2 < \rho_0. \tag{178}$$

Then we have

$$\sum_{n=0}^{\infty} A_n\rho_1^n < \infty \tag{179}$$

for since

$$\lim_{n\to\infty} A_n\rho_2^n = 0 \tag{180}$$

there will be some N so that

$$|A_n\rho_2^2| < 1 \qquad \text{for } n > N. \tag{181}$$

We may then write

$$\sum_{n=0}^{\infty}\left(A_n\rho_1^n\right)^2 \equiv \sum_{n=0}^{N}\left(A_n\rho_1^2\right)^2 + \sum_{n=N+1}^{\infty}\left(\frac{\rho_1}{\rho_2}\right)^{2n} = \sum_{n=0}^{N}\left(A_n\rho_1^n\right)^2 + \frac{(\rho_1^2/\rho_2^2)^{N+1}}{1-\rho_1^2/\rho_2^2} \tag{182}$$

and may also state that

$$\sum_{n=0}^{\infty}\sum_{m=-n}^{n}(a_n^m\rho_1^n)^2$$

will converge for every ρ_1 so that (172) or

$$0 < \rho_1 < \rho_0 \tag{183}$$

holds.

Let us now use functions $\{\psi_n^m\}$ defined as $\{\phi_n^m\}$ by (164) only with R/ρ_1 substituted for R; then we have the result:

$$\psi(P) = \sum_{n=0}^{\infty} \sum_{m=-n}^{n} (\rho_1^{n+1} a_n^m) \phi_n^m(P) \qquad \text{for} \quad r_P > \frac{R}{\rho_1}. \tag{184}$$

If we call the sphere with centre at the centre of σ and radius R/ρ_0 the limit sphere, we can express our result as follows:

The series expansion of a potential into spherical harmonics with O as origin will converge uniformly on the surface of and in the space outside any sphere with O as centre and so that all the singularities of the potential are in the interior of the sphere. There exists such a radius R' that for $R > R'$ the series will converge uniformly on the surface of and outside any sphere with radius R and centre O. This, however, is not true for $R < R'$.

The last (the negative) part of the theorem will be proved a few pages further on.

2. Geodesists are interested in exterior potential fields; therefore I have only treated such fields here. But by means of inversion with respect to a sphere we get corresponding results for interior spherical regions, which will recall the well-known theorems on the convergence of power series in the complex plane. There we have a limiting circle and uniform convergence in circles inside and divergence at all points outside the circle. I have the impression that many geodesists believe that it should, correspondingly, be so that the spherical harmonics series diverge at all points inside the limit sphere, but they cannot give any proof of it.

I shall not give the proof—on the contrary, I shall give an example that shows that the conjecture is false.

Consider the following potential

$$\phi = \frac{1-x}{(1-x)^2 + y^2} \tag{185}$$

in a three-dimensional space. It corresponds to a uniform mass distribution on the line

$$x = 1 \qquad \text{and} \qquad y = 0 \tag{186}$$

and is only a function of x and y. It is elementary to show that it can be expanded into a series

$$\phi = \sum_{n=0}^{\infty} a_n r^n \cos n\theta \cos n\lambda = \sum_{n=0}^{\infty} a_n (x^2 + y^2)^{n/2} \cos n\lambda \tag{187}$$

or

$$\phi = \sum_{n=0}^{\infty} b_n r^n P_n^n(\sin\theta) \cos n\lambda \tag{188}$$

$$P_n^n(\sin\theta) = (-1)^n 1 \cdot 3 \cdot 5 \cdots (2n+1) \cos n\theta. \tag{189}$$

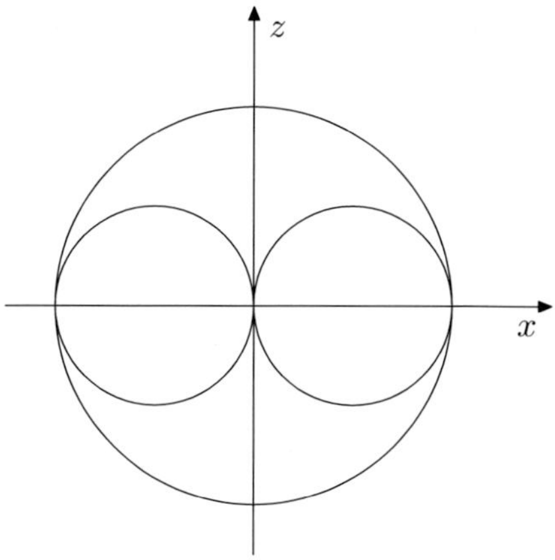

Neither the potential ϕ nor the coefficients in (187) or (188) depend on the coordinate z; therefore the series (188) will be convergent in the cylinder with $x = 0$, $y = 0$ as axis and with radius 1. If we make an inversion with respect to the sphere

$$x^2 + y^2 + z^2 = 1, \tag{190}$$

then

$$\Phi(x, y, z) = \frac{1}{r}\phi\Big(\frac{x}{r^2}, \frac{y}{r^2}, \frac{z}{r^2}\Big) = \frac{r(r^2 - x)}{(r^2 - x)^2 + y^2} \tag{191}$$

is also a potential, and it is regular at infinity. The series expansion of Φ corresponding to (188) becomes

$$\Phi = \sum_{n=0}^{\infty} b_n \frac{1}{r^{n+1}} P_n^n(\sin\theta)\cos n\lambda, \tag{192}$$

which therefore will converge at all points of the space outside the torus into which the cylinder is transformed. The torus is that described by the circle:

$$(x - \tfrac{1}{2})^2 + z^2 = \tfrac{1}{4}$$
$$y = 0,$$

when its plane is rotated around the z-axis. By using the idea of this example it is not difficult to construct other expansions into spherical harmonics that converge in regions that are not spherical. I have nevertheless still the feeling that normally the limit sphere bounds the region of convergence, but this question requires a closer study.

The criterion given here for the limit sphere is merely a theoretical one, as the limit radius can only be found when all the coefficients of the series are known. Therefore, we must have a criterion that can give more practical information.

From the above it follows that outside the limit sphere the series represents a regular potential, provided that the radius of the limit sphere is smaller than the radius of the smallest sphere containing the points at which the originally given potential was undefined or irregular. In other words, the series gives an analytic continuation of the potential. It is a well-known fact that the potential of a homogeneous sphere can be analytically continued to the whole space except the centre and that the normal potential of an oblate ellipsoid of revolution can also be so continued to the whole space except the "focal disc." Naturally the continuation of the outer potential into the gravitating body has nothing to do with the (irregular) potential that exists physically in the body. Two potentials which are regular in some domain and are identical in some open set of points in that domain are known to be identical in the whole domain of definition. Therefore, we may define a maximal potential as a regular potential defined in a connected region so that there does not exist any larger connected region in which the potential could be defined and be regular, and we may claim that to each potential there corresponds one and only one maximal potential continuation of the given. In the vicinity of every point of the boundary of the region of definition of a maximal potential there are points at which the potential is singular. Consequently, we may claim:

To every potential regular in the vicinity of infinity and to every sphere Σ containing all the singular points of the corresponding maximal potential there exists a series expansion of the form (166) which converges uniformly on and outside every sphere concentric with Σ so that no point of its surface is in the vicinity of any of the singular points. If we call the minimum value of the radii of such spheres R_0, then the series must be divergent at some points of every sphere with radius less than R_0 and concentric with Σ.

The remaining part of the claim runs like this:

1. For fixed θ and λ a series as (73) is a power series in $1/r$; therefore, if it diverges for some value of $1/r$, say α, it will diverge for every value of r so that $1/r > \alpha$ and the same θ and λ, or in other words, if it diverges at one point, it will diverge at all points of the line segment connecting that point and the origin.
2. From the definition of the limit sphere it follows that there are singularities in every vicinity of some of its points: therefore, given any $\epsilon > 0$ we can find a sphere concentric with Σ and with radius $r > R_0 - \epsilon$, so that there are singular points on its surface, and then the series cannot be uniformly convergent on such a sphere (for if it were, it would represent a continuation of the potential which was supposed to be maximal). The surface of a sphere is a closed set of points; therefore, convergence at all points would simply be uniform convergence.
3. From 2) it follows that there are spheres with radii arbitrarily near R_0 where the series diverges (at least at some points), and from 1) it follows that all the spheres with radii less than R_0 will also have divergence points on their surfaces.

The potential which is of most interest to geodesists is the disturbing potential T; therefore, it would be interesting to know whether it is reasonable to hope that the limit sphere for the corresponding maximal potential would be located below the surface of the Earth. Using an idea from [11] I shall show that the basis for such a hope is slender. More precisely, I shall show that if to a gravitating body for which the corresponding maximal potential is defined below the surface we locally add some mass distribution, e.g. a grain of sand, above the surface at a place where the original maximal potential is defined below the surface, then the resulting maximal potential will have a singularity in the interior of the added mass distribution.

The proof is *almost* trivial. The resulting potential is the sum of the original maximal potential Φ and the potential ϕ of the added mass distribution and is regular at least where both are defined. Let us enclose the added mass distribution with a single closed surface σ leaving the singularities of Φ outside (this is possible according to what we have supposed). ϕ is regular outside σ and must therefore have a singularity inside σ (since every regular potential defined in the whole space vanishes). On the other hand Φ is regular inside σ, and therefore $\Phi + \phi$ must have a singularity inside σ, and as we can select σ approximately, we have the theorem.

Here I must warn the reader that if he has not noticed the importance of the word "locally," then he has not really understood the proof. (The added mass distribution must not cover the whole surface of the original body, for then we should not be able to find σ.)

If as an example of an added mass distribution I mentioned a grain of sand and not, as Moritz did, a mountain, it was in order to push the discussion on the series of harmonic functions near the gravitating masses ad absurdum. A popular way of expressing the theorem would be: Even if the series were convergent at the surface of the Earth, a displacement of a single grain of sand should spoil the convergence.

The consequence of this must be that the convergence of series of spherical harmonics near the surface of the Earth is such an unstable property that it has no physical meaning at all.

I know that here many geodesists will argue that we may use the series without knowing anything about their convergence. I must confess that I do not understand what they mean by the little word "use."

Let us take an example. Let us regard the "Poisson kernel" (159). It has a singularity at the "north pole"—$r_P = R$, $\psi = 90°$—and is regular at all other points in the space. For the singular point the series gives

$$\sum_{n=0}^{\infty}(2n+1). \tag{193}$$

The sum of the first N members is N^2, and this is all right; for all other points of the sphere, $r_P = R$, the kernel is zero. But for the "south pole"—$\psi = -90°$—the series gives

$$\sum_{n=0}^{\infty}(-1)^n(2n+1), \tag{194}$$

and the sum of the first N members is $(-1)^{N+1}N$.

For a point having $r_P = R/\rho$ and $\psi = -90°$ $(\rho > 1)$ the series gives

$$\sum_{n=0}^{\infty}(-1)^n(2n+1)\rho^n, \tag{195}$$

and the sum of the first N members is

$$\frac{1-\rho+(-1)^N\rho^{2N}\left(2N+1+(2N-1)\rho^2\right)}{(1+\rho)^2} \tag{196}$$

How can you "use" such a result?

I see very well that if we multiply the kernel by a small constant and add a dominating "normal potential," then it is a simple task in the resulting series to 'filter' the disturbing part with the increasing coefficients from the well-behaved part with decreasing coefficients. But if we happen to be interested mainly in the disturbing potential, how are we then to "use" the series?

I do like this example very much, so I ask the reader to have patience enough to follow another experiment with it.

The series (159) may be written like this:

$$F_0 = \sum_{n=0}^{\infty}(2n+1)\left(\frac{R}{r_P}\right)^n P_n(\cos\psi). \tag{197}$$

We define another series

$$F_\lambda = \sum_{n=0}^{\infty}\frac{2n+1}{1+2^{n+1}\lambda}\left(\frac{R}{r_P}\right)^{n+1} P_n(\cos\psi), \quad \lambda \geq 0. \tag{198}$$

It is evident that for $\lambda = 0$, $F_\lambda = F_0$, and it seems likely that for small λ, F_λ should approximate F_0; in fact we have:

$$\begin{aligned}|F_0 - F_\lambda| &\leq \sum_{n=0}^{\infty}(2n+1)\left(1-\frac{1}{1+2^{n+1}\lambda}\right)\left(\frac{R}{r_P}\right)^{n+1}|P_n(\cos\theta)\\ &\leq \sum_{n=0}^{\infty}(2n+1)\frac{2^{n+1}\lambda}{1+2^{n+1}\lambda}\left(\frac{R}{r_P}\right)^{n+1}.\end{aligned} \tag{199}$$

For every fixed $r_P > R$ the series in the second line is uniformly convergent for $\lambda \geq 0$, and, consequently, it represents a continuous function. As this function is zero for $\lambda = 0$, we have

$$\lim_{\lambda \to 0} F_\lambda = F_0 \qquad \text{for} \quad r_P > R. \tag{200}$$

The interesting thing is that for $\lambda > 0$ the series for F_λ (198) is convergent outside a sphere with radius $R/2$ and not R as in the case of the series for F_0.

It is not difficult to see that the method used in this example can be used generally to solve the following problem:

Given a potential ϕ defined in the space outside a sphere σ. Find a sequence of potentials $\{\phi_n\}$ regular outside a sphere concentric with σ and with half the radius so that for all points outside σ

$$\lim_{n \to \infty} \phi_n = \phi. \tag{201}$$

If from the series expansions for the potentials $\{\phi_n\}$ we take only the members up to the n'th degree ϕ'_n, then we also have

$$\lim_{n \to \infty} \phi'_n = \phi, \tag{202}$$

and here we have an approximation of ϕ by "polynomials" of spherical harmonics (i.e. series with a finite number of members).

We have seen a very important new aspect of the instability of convergence for series of spherical harmonics. We saw before (Moritz' theorem) that in the vicinity of every potential which can be expressed by a series convergent at the surface of the Earth there is another potential regular outside the Earth but for which the series diverges at the surface. Now we see that *perhaps* there exists also another theorem (Runge's theorem) expressing that in the vicinity of every potential ϕ regular in the space outside the Earth there is a potential for which the series of spherical harmonics converges down to the surface of some sphere in the interior of the Earth; that is to say: ϕ can be approximated arbitrarily well by polynomials of spherical harmonics.

In fact there exists a Runge's theorem for physical geodesy (as to the name I have given it, see [1], p. 275–278), but as the proof of it is rather technical, I have given it in the Appendix and shall only state the result here:

> *Runge's theorem:* Given any potential regular outside the surface of the Earth and any sphere in the interior of the Earth. For every closed surface surrounding the Earth (which surface may be arbitrarily near the surface of the Earth) there exists a sequence of potentials regular in the whole space outside the given sphere and uniformly converging to the given potential on and outside the given surface.

This theorem is extremely important. In fact it permits a mathematical treatment of physical geodesy by giving a good compensation for the possibility of using series of harmonic functions converging down to the surface of the Earth.

3. Runge's theorem merely establishes the existence of a sequence with the wanted properties. I shall now, at least theoretically, show how such a sequence can be found by means of the adjustment method from Section II.

Let the given potential be ϕ and the domain on which it is given be Ω. The part of the space outside the sphere σ is called Σ, and we have $\Omega \in \Sigma$. Let us then define a metric $\langle\ ,\ \rangle_\Sigma$ for potentials regular in Σ, and let us call the corresponding Hilbert space H_Σ. We suppose that H_Σ has a reproducing kernel K. We know that it has one if the metric in H_Σ is that defined in the first part of this section (155). Let ω be a closed smooth surface in Ω surrounding and arbitrarily near to the boundary of Ω. Then we can define the norm

$$\|\phi\|_\Omega = \left(\frac{1}{4\pi}\int\limits_\omega \phi^2\, d\omega\right)^{1/2} \tag{203}$$

for potentials regular in Ω. By this norm and the corresponding scalar product we have defined a Hilbert space H_Ω consisting of potentials regular in Ω.

Then the problem is:

Find $\psi \in H_\Sigma$ so that

$$\|\psi\|_\Sigma^2 + \lambda\|\psi - \phi\|_\Omega^2 = \min, \tag{204}$$

where λ is a not yet specified constant. Let us rewrite (204) as

$$\left\|\tfrac{1}{\sqrt{\lambda}}\phi - 0\right\|_\Sigma^2 + \left\|\psi - \phi\right\|_\Omega^2 = \min. \tag{205}$$

Then the problem is an adjustment problem of the first type treated in Section II (formulae (97)–(105)), provided that we put

$$\begin{aligned} H_1 &= H_\Sigma \\ H_2 &= H_\Sigma \oplus H_\Omega . \end{aligned} \tag{206}$$

The operator $A\colon H_1 \to H_2$ is defined as follows:

$$\begin{aligned} A &= \begin{bmatrix} A_1 \\ A_2 \end{bmatrix} \\ A_1 &: H_\Sigma \to H_\Sigma \\ A_1 &= \frac{1}{\sqrt{\lambda}} \\ A_2 &: H_\Sigma \to H_\Omega \\ A_2\psi &= \psi' \qquad \text{where } \psi' \text{ is the restriction of } \psi \text{ to } \Omega. \end{aligned} \tag{207}$$

$$a = \begin{bmatrix} 0 \\ \phi \end{bmatrix} \tag{208}$$

The normal equations (99):

$$A^{\mathrm{T}}A\psi = A^{\mathrm{T}}a \tag{209}$$

become here

$$\begin{bmatrix} A_1^{\mathrm{T}} & A_2^{\mathrm{T}} \end{bmatrix} \begin{bmatrix} A_1 \\ A_2 \end{bmatrix} \psi = \begin{bmatrix} A_1^{\mathrm{T}} & A_2^{\mathrm{T}} \end{bmatrix} \begin{bmatrix} 0 \\ \phi \end{bmatrix} \tag{210}$$

or

$$\left(A_1^{\mathrm{T}} A_1 + A_2^{\mathrm{T}} A_2\right)\psi = A_2^{\mathrm{T}} \phi. \tag{211}$$

Here A_1 is a scalar operator, i.e. an operator indicating multiplication by a scalar, and thus identical with its transpose so that

$$A_1^{\mathrm{T}} A_1 \psi = \tfrac{1}{\lambda}.\psi \tag{212}$$

For the restriction ψ' of ψ to Ω we have

$$\psi'(Q) = \langle K(Q,P), \psi(P)\rangle_\Sigma \qquad \text{for } Q \in \Omega,\ P \in \Sigma, \tag{213}$$

which follows from the trivial fact that

$$\psi'(Q) = \psi(Q), \qquad \text{for } Q \in \Omega, \tag{214}$$

and from the defining property of the reproducing kernel $K(Q,P)$.

To find the transpose A_2^{T} we remember the definition of the transposed operator (96) and write for $\xi \in H_\Omega$:

$$\begin{aligned} \langle A^{\mathrm{T}}\xi(Q), \psi(Q)\rangle = \langle \xi(Q), A(\psi Q)\rangle &= \int_\omega \xi(Q)\langle K(Q,P), \psi(P)\rangle_\Sigma \, d\omega_Q \\ &= \left\langle \int_\omega \xi(Q) K(Q,P)\, d\omega_Q, \psi(P) \right\rangle_\Sigma \end{aligned} \tag{215}$$

so that

$$A^{\mathrm{T}}\xi(Q) = \int_\omega \xi(Q) K(Q,P)\, d\omega_Q \qquad \text{for } Q \in \Omega,\ P \in \Sigma \tag{216}$$

follows.

Now we can write (211):

$$\frac{1}{\lambda}\psi(P) + \int_\omega \psi(Q) K(Q,P)\, d\omega_Q = \int_\omega \phi(Q) K(Q,P)\, d\omega_Q, \qquad Q \in \Omega,\ P \in \Sigma. \tag{217}$$

Let us define $\xi \in H_\Omega$ by

$$\xi(Q) = \phi(Q) - \psi(Q); \tag{218}$$

then for $P, Q \in \Omega$, (206) becomes

$$\xi(P) + \lambda \int_\omega K(Q,P)\xi(Q)\, d\omega_Q = \phi(P). \tag{219}$$

Equation (219), which is analogous to the normal equations, is an integral equation. As we shall see, $\xi(P)$ can be found from (219), and then $\psi(Q)$ can be found by

$$\psi(Q) = \phi(Q) - \xi(Q) \tag{220}$$

for $Q \in \Omega$; for $P \in \Sigma$ $\psi(P)$ can be found from (217):

$$\psi(P) = \lambda \int_\omega K(Q,P)\xi(Q)\,d\omega_Q \qquad \text{for } P \in \Sigma,\ Q \in \Omega. \tag{221}$$

The integral equation (219) is a Fredholm integral equation of the second kind with bounded continuous positive definite symmetric kernel. Normally such an integral equation is written with a minus and not a plus before the λ, and then one of the many elegant theorems on equations of this kind shows that all the eigenvalues are positive or zero; therefore (219) can have no eigensolutions for positive λ's. So for a given ϕ it will have a unique solution ξ for every positive λ, and then by (221) also ψ will be uniquely determined, and the least-squares problem (204) has been solved.

But what does that mean?

Let us start with a "small" $\lambda > 0$, and let then λ increase. Then we may expect to find potentials $\psi \in H_\Sigma$ which, in H_Ω, approximate ϕ better and better so that in the part of Σ outside Ω ψ increase and, if we have luck,

$$\|\psi - \phi\|_\Omega \to 0 \qquad \text{for} \quad \lambda \to \infty. \tag{222}$$

I shall prove that this is in fact so, and for the proof I shall use the theory of integral equations and Runge's theorem.

Let the homogeneous integral equation

$$\Phi(P) - \lambda \int_\omega K(Q,P)\Phi(Q)\,d\omega_Q = 0; \qquad P, Q \in \Omega \tag{223}$$

have the eigenvalues $\{\lambda_n\}$ and the corresponding eigenfunctions $\{\phi_n\}$; we may suppose that

$$0 < \lambda_1 \leq \lambda_2 \leq \lambda_3 \leq \cdots \tag{224}$$

and that the eigenfunctions are normalized and orthogonal so that

$$\langle \phi_n, \phi_m \rangle_\Omega = \begin{cases} 1 & \text{for } m = n \\ 0 & \text{for } m \neq n. \end{cases} \tag{225}$$

The eigenfunctions $\{\phi_n\}$ are only defined in Ω, but they may also be defined in Σ and on σ by

$$\phi_n(P) = \lambda_n \int_\omega K(Q,P)\phi_n(Q)\,d\omega_P, \qquad P \in \Sigma + \sigma, \quad Q \in \Omega \tag{226}$$

because $K(Q,P)$ is regular for $Q, P \in \sigma + \Sigma$ provided that not both Q and $P \in \sigma$. From Mercer's theorem it follows that

$$K(P,Q) = \sum_n \frac{\phi_n(P)\phi_n(Q)}{\lambda_n} \tag{227}$$

for $P, Q \in \Omega$ and by the method of proof used for Mercer's theorem (227) can be proved for $P, Q \in \Sigma$.

Now $K(P,Q)$ is the reproducing kernel for H_Σ and, therefore,

$$\phi_n(P) = \left\langle \sum_m \frac{\phi_m(P)\phi_n(Q)}{\lambda_m}, \phi_n(Q) \right\rangle_\Sigma = \sum_m \frac{\phi_m(P)}{\lambda_m} \langle \phi_m(Q), \phi_n(Q) \rangle_\Sigma . \tag{228}$$

As $\{\phi_n\}$ are orthogonal in H_Ω, they must be linearly independent, and so (228) implies

$$\frac{1}{\lambda_m} \langle \phi_m, \phi_n \rangle_\Sigma = \begin{cases} 1 & \text{for } m = n \\ 0 & \text{for } m \neq n, \end{cases} \tag{229}$$

or

$$\langle \phi_m, \phi_n \rangle = \begin{cases} \lambda_n & \text{for } m = n \\ 0 & \text{for } m \neq n, \end{cases} \tag{230}$$

i.e. the functions $\{\phi_n/\sqrt{\lambda_n}\}$ form a system of orthonormal functions in H_Σ. If we write

$$\phi'_n = \frac{\phi_n}{\sqrt{\lambda_n}}, \tag{231}$$

then (227) becomes

$$K(P,Q) = \sum_n \phi'_n(P)\phi'_n(Q). \tag{232}$$

A well-known theorem from the theory of Hilbert spaces with reproducing kernel establishes that, if a reproducing kernel can be expressed by (232), i.e. by a set of orthonormal functions, then this set will be complete. Therefore the set of functions $\{\phi_n\}$ is complete in H_Σ.

The system $\{\phi_n\}$ will also be complete in H_Ω. For if it were not, there would exist such an element $\eta \in H_\Omega$ that

$$\|\eta\|_\Omega = 1 \qquad \text{and} \quad \langle \eta, \phi_n \rangle_\Omega = 0 \quad \text{for all } \phi_n. \tag{233}$$

From Runge's theorem it follows that for any given $\epsilon > 0$ there exists such an element $\mu \in H_\Sigma$ that

$$|\eta(Q) - \mu(Q)| < \epsilon \qquad \text{for } Q \in \omega. \tag{234}$$

Being an element of H_Σ $\mu(Q)$ may be expressed as

$$\mu(Q) = \sum_n a_n \phi_n, \tag{235}$$

where the sum converges uniformly on ω. Therefore,

$$\left\|\eta(Q) - \sum_n a_n\phi_n(Q)\right\|_\Omega^2 = 1 + \sum a_n^2 > 1, \tag{236}$$

(from (233)). On the other hand,

$$\left\|\eta(Q) - \sum_n a_n\phi_n(Q)\right\|_\Omega^2 = \int_\omega \left(\eta(Q) - \sum_n a_n\phi_n(Q)\right)^2 d\omega \leq \epsilon^2 A, \tag{237}$$

where A is the area of the surface ω. If we choose $\epsilon < A^{-1/2}$, we have a contradiction, and, consequently, $\{\phi_n\}$ will be complete also in H_Ω.

Now we can go back to the integral equation (219). Here we can express ϕ and ξ by the complete set $\{\phi_n\}$:

$$\xi = \sum_n x_n\phi_n, \qquad \phi = \sum_n f_n\phi_n \tag{238}$$

so that, using (227), we have

$$\sum_n x_n\phi_n + \lambda\sum_n \frac{x_n}{\lambda_n}\phi_n = \sum_n f_n\phi_n \tag{239}$$

or

$$x_n = \frac{\lambda_n}{\lambda_n + \lambda} f_n, \tag{240}$$

$$\xi = \sum_n \frac{\lambda_n}{\lambda_n + \lambda} f_n\phi_n. \tag{241}$$

ψ is defined by (218) so that

$$\|\phi(Q) - \psi(Q)\|_\Omega = \left(\sum_n \left(\frac{\lambda_n}{\lambda_n + \lambda}\right)^2 f_n^2\right)^{1/2}. \tag{242}$$

(We should remember that $\psi(Q)$ is a function of λ.)

The series on the right side of (242) is uniformly convergent in λ for $\lambda > 0$ since

$$\sum f_n^2 = \|\phi\|_\Omega^2, \tag{243}$$

and

$$\frac{\lambda_n}{\lambda_n + \lambda} < 1 \qquad \text{for } \lambda > 0. \tag{244}$$

However, every member of the series converges to zero when $\lambda \to \infty$, and, therefore,

$$\lim_{\lambda\to\infty} \|\phi(Q) - \psi(Q)\|_\Omega = 0, \tag{245}$$

from which uniform convergence follows in the usual way (169).

If we put

$$\psi = \sum_n P_n \phi_n, \tag{246}$$

we find from (221)

$$P_n = \frac{\lambda}{\lambda_n + \lambda} f_n \tag{247}$$

or

$$\psi(P) = \sum_n \frac{\lambda}{\lambda_n + \lambda} f_n \phi_n(P) \tag{248}$$

and

$$\|\psi\|_\Sigma^2 = \sum_n \left(\frac{\lambda}{\lambda_n + \lambda}\right)^2 f_n^2. \tag{249}$$

Only if the series $\sum f_n^2$ is convergent, i.e. if the definition of ϕ can be extended to Σ, does the series in (249) converge for $\lambda \to \infty$. Only in this case does

$$\lim_{\lambda \to \infty} \psi = \phi \tag{250}$$

hold in H_Σ. $\psi(P)$ does not in general converge for $\lambda \to \infty$ (for $P \in \Sigma$ but not $\in \Omega$), as can be seen from the example (198).

Now we have—at least theoretically—solved the problem of approximation of potentials down to the surface of the Earth by potentials regular down to a Bjerhammar sphere and thus given a sound mathematical foundation of the method described in the previous two sections. Moreover—as far as I can see—only this method presents a way in which one can find such approximations from physical measurements of the effects of the potential.

The proof given here of the convergence of the function ψ for $\lambda \to \infty$ might be used as a model for proofs of the convergence of the results of the application of the adjustment method on concrete problems in physical geodesy, provided that the number and the quality of the measurements increase until we have enough exact measurements. However, I shall not give such a proof for any practical case here, because I do not really see the value thereof. I believe that sufficient information about the reliability of the results can be found by means of the statistical method mentioned in the first section. The important information that Runge's theorem gives us in this connection is the method of approximation, which does not introduce any form of systematic error in the result.

4. Before leaving the question about the convergence of series of spherical harmonics I should like to advance a few naive considerations.

It has often been said that generally the convergence of series of spherical harmonics is slow. (I often wonder if those who say so have ever tried to calculate e^{-100} using directly the very well-known power series for e^x which converges for all x.)

The reason why the series of spherical harmonics used here are so slowly converging is rather that the functions we want to represent are very complicated (i.e. contain a large amount of information) than that the spherical harmonics are not well suited for the set of functions in which we are interested.

If we try to describe some function defined on a sphere by a series of spherical harmonics of up to the 36th degree, then we must have $36^2 = 1296$ parameters, but we cannot expect that details of a magnitude less than $180°/36 = 5°$ can be sharply mapped. If we want a more detailed mapping, the price to be paid is more parameters and this is relatively independent of the type of function used for the mapping. By local interpolation it is of course possible to map much smaller details by a suitable choice of the 1296 first coefficients in the series, but then we are to expect a very "wild" behaviour of the series outside the local domain in which it has been forced to follow the details.

After this warning I shall say something about criteria for the choice of kernels and the corresponding metrics.

There exists an infinite number of metrics symmetric with respect to rotation for sets of potentials regular outside a given Bjerhammar sphere and having corresponding reproducing kernels. One of them was treated in the first few pages of this section. There is another metric which has been mentioned sometimes in literature. It is defined by the scalar product:

$$\langle \phi, \psi \rangle = \frac{1}{4\pi} \int_{\Sigma} \operatorname{grad} \phi \operatorname{grad} \psi \, d\Sigma, \tag{251}$$

where the domain of integration is the whole exterior space. Using Green's theorem and $\Delta\psi = 0$ we find, however:

$$\langle \phi, \psi \rangle = -\frac{1}{4\pi} \int_{\sigma} \phi \frac{\partial \psi}{\partial r} \, d\sigma. \tag{252}$$

This property of the special metric has made it suitable for the study of the classical boundary value problems related to Laplace's equation. But for our purpose it is not of very great interest, since it has the drawback that the expression of the reproducing kernel is a rather complicated one containing a logarithm.

A metric that has a slightly more complicated scalar product but a much simpler reproducing kernel is defined by the following scalar product:

$$\langle \phi, \psi \rangle_L = \frac{1}{4\pi} \int_{\Sigma} \frac{1}{r} \operatorname{grad} \phi \operatorname{grad} \psi \, d\Sigma, \tag{253}$$

and has the reproducing kernel

$$K(P, Q) = \frac{2R}{L},$$

where L is given by (161)

$$L = \sqrt{\frac{r_P^2 r_Q^2}{R^2} - 2 r_P r_Q \cos\psi + R^2}.$$

If P is a point in Σ, then

$$P' = \left(\frac{x_P}{R^2}, \frac{y_P}{R^2}, \frac{z_P}{R^2}\right) \tag{254}$$

is the point in the interior of the sphere at which P is mapped by inversion with respect to the sphere. We have

$$L = \frac{r_P}{R}\sqrt{r_Q^2 - 2 r_Q r_{P'} \cos\psi + r_{P'}^2}, \tag{255}$$

i.e. for a fixed P, $1/L$ is proportional to $1/QP'$, and therefore $1/L$ is a regular potential in Σ as a function of one of the points P, Q, the other being fixed, and

$$\frac{2R}{L} = \frac{2R^2}{r_P}\frac{1}{QP'} = \frac{R^2}{2r_P}\frac{1}{r'_P}\sum_{n=0}^{\infty}\left(\frac{r_P}{r_Q}\right)^{n+1} P_n(\cos\psi) = 2\sum_{n=0}^{\infty}\left(\frac{R^2}{r_P r_Q}\right)^{n+1} P_n(\cos\psi). \tag{256}$$

I shall prove that $2R/L$ is the reproducing kernel corresponding to $\langle\ ,\ \rangle_L$. Since

$$\phi_i^k = \left(\frac{R}{r}\right)^{i+1} E_i^k \qquad \text{for} \quad k = -i, -i+1, \ldots, i \quad \text{and} \quad i = 0, 1, \ldots, \tag{257}$$

where E_i^k are $2i+1$ fully normalized spherical harmonics for $i = 0, 1, 2, \ldots,$ is a complete orthogonal system, a scalar product is known once its effect on $\{\phi_i^k\}$ is known. We find

$$\begin{aligned}\langle \phi_i^k, \phi_j^l \rangle_L &= \frac{1}{4\pi}\int_\Sigma \frac{1}{r}\operatorname{grad}\phi_i^k \operatorname{grad}\phi_j^l \, d\Sigma \\ &= \frac{1}{4\pi}\int_\Sigma \frac{1}{r}\left(\frac{\partial}{\partial r}\phi_i^k \frac{\partial}{\partial r}\phi_j^l + \operatorname{grad}_2\phi_i^k \operatorname{grad}_2\phi_j^l\right) d\Sigma\end{aligned}$$

(where grad_2 is the two-dimensional gradient on the sphere)

$$\begin{aligned}&= \frac{1}{4\pi}\int_\Sigma \frac{1}{r}\left(\frac{d}{dr}\left(\frac{R}{r}\right)^{i+1}\frac{d}{dr}\left(\frac{R}{r}\right)^{i+1} E_i^k E_j^l \right.\\ &\qquad\left. + \left(\frac{R}{r}\right)^{i+1}\left(\frac{R}{r}\right)^{j+1}\operatorname{grad}_2 E_i^k \operatorname{grad}_2 E_j^l\right) d\Sigma\end{aligned}$$

(now we shall make use of Green's theorem for the surface of the sphere)

$$= \frac{1}{4\pi} \int_{\Sigma} \frac{1}{r} \left(\frac{(i+1)(j+1)}{r^2} \left(\frac{R}{r}\right)^{i+j+2} E_i^k E_j^l - \left(\frac{R}{r}\right)^{i+j+2} E_j^l \Delta_2 E_i^k \right) r \sin\theta \, d\theta \, d\lambda$$

(where Δ_2 is the Laplace-Beltrami operator on the sphere)

$$= \frac{1}{4\pi} \int_{\Sigma} \frac{1}{r} \left(\frac{R}{r}\right)^{i+j+2} \big((i+1)(j+1) + i(i+1)\big) E_i^k E_j^l \sin\theta \, d\theta \, d\lambda$$

$$= \frac{(i+1)(i+j+1)}{4\pi} \int E_i^k E_j^l \sin\theta \, d\theta \, d\lambda \int_R^\infty \frac{R^{i+j+2}}{r^{i+j+3}} \, dr$$

$$= (i+1)(i+j+1)\delta_{ij}\delta_{kl} \frac{1}{2i+2} = \begin{cases} \frac{2i+1}{2} & \text{for } i = j,\ k = l \\ 0 & \text{in all other cases.} \end{cases}$$

We can now write:

$$\left\langle \frac{2R}{L}, \phi_j^k \right\rangle_L = \left\langle 2 \sum_{i=0}^{\infty} \left(\frac{R^2}{r_P r_Q} \right)^{i+1} P_i(\cos\psi), \phi_j^l \right\rangle_L$$

$$= \left\langle 2 \sum_{i=0}^{\infty} \frac{1}{2i+1} \phi_i^k(P) \phi_i^k(Q), \phi_j^l(Q) \right\rangle_L = \phi_j^l(P), \tag{258}$$

and so we have proved that $2R/L$ is the reproducing kernel in the Hilbert space H_L with the scalar product $\langle\ ,\rangle_L$.

This section will be concluded by a short discussion of the important problem: How to choose the metric (or the kernel) for practical computations.

The most obvious idea is perhaps to use as kernel the finite series of spherical harmonics which corresponds to Kaula's expansion of the correlation function $C(P,Q)$, see [6]. (In Kaula's publications one of the coefficients in this expansion is negative, but it cannot be so; is it an iterated printing error?)

But if one uses a kernel with only a finite number of members, one has limited the solution to a finite-dimensional space consisting of potentials expressible by spherical harmonics of the same degrees as those occurring in the kernel. To get good local results one, consequently, has to use a very large number of members.

Another possibility is to use a kernel with a simple closed expression, e.g. $2R/L$ (or $2R/L$ multiplied by a suitable constant). If the radius R of the Bjerhammar sphere is chosen a few per cent smaller than the mean radius of the Earth, a good local approximation to Kaula's correlation function is achieved.

As far as I can see the best thing to do is to use a combination of these two ideas, i.e. $2R/L$ multiplied by some constant plus a correction consisting of a finite sum of spherical harmonics so that the corresponding correlation function is sufficiently similar to Kaula's correlation function. The scalar product will then be a constant multiple of $\langle\ ,\ \rangle_L$ plus a correction which is simple to calculate, but as the scalar product is not explicitly used in the calculations I shall not give the result here. From a theoretical point of view the important thing is that since this correction is finite, the Hilbert space corresponding to the corrected kernel consists of the same elements as does H_L.

IV. Application of the Method

Already today Molodenskiy's problem is looked upon as the classical problem of physical geodesy; it is therefore reasonable to start the discussion on the application of the kernel-method with this problem.

As I do not find that the mathematical aspects of Molodenskiy's problem are clearly formulated in literature, I shall first propose another formulation of the problem.

Molodenskiy's problem is the problem of finding a better approximation of the potential of the Earth—the normal potential—from a given approximation by geodetic measurements.

Let us assume that we have measured the following data for points on the physical surface of the Earth:

1. the astronomical geographical coordinates, i.e. the direction of the plumb line,
2. the potential of gravity of the Earth W, and
3. the gravity g.

Let us also assume that we have given a normal potential U; we can then for each point P on the surface of the Earth find such a point Q that

(a) the normal potential at Q equals the potential at P:

$$U_Q = W_P, \tag{259}$$

and that

(b) the geodetic normal coordinates of Q are equal to the astronomical coordinates of P, which can be expressed by the vector equation

$$\frac{1}{\gamma_Q}(\operatorname{grad} U)_Q = \frac{1}{g_P}(\operatorname{grad} W)_P. \tag{260}$$

Both terms in this equation represent unity vectors defining the directions in question. The locus of the points Q is the telluroid.

We want to find the vector $\overrightarrow{QP}$ and $T = W - U$. Here T, the disturbance potential, is a regular potential, because the potential of the centrifugal force is the same in W and U.

Since $T = W - U$, (259) can now be written

$$T_P + U_P - U_Q = 0, \tag{261}$$

and (260) can be written

$$(\operatorname{grad} T)_P + (\operatorname{grad} U)_P - \frac{g_P}{\gamma_Q}(\operatorname{grad} U)_Q = 0, \tag{262}$$

or

$$(\operatorname{grad} T)_P + (\operatorname{grad} U)_P - (\operatorname{grad} U)_Q = \frac{g_P - \gamma_Q}{\gamma_Q}(\operatorname{grad} U)_Q. \tag{263}$$

Now

$$g_P - \gamma_Q = \Delta g \tag{264}$$

is the gravity anomaly at the point in question, so that (263) becomes:

$$(\operatorname{grad} T)_P + (\operatorname{grad} U)_P - (\operatorname{grad} U)_Q = \frac{\Delta g}{\gamma}(\operatorname{grad} U)_Q. \tag{265}$$

If we introduce a Cartesian system of coordinates (x_1, x_2, x_3) and call the components of the vector $\overrightarrow{QP}$ p_1, p_2, and p_3, we can write (261) and (265) in linear approximation

$$T + \sum_{i=1}^{3} p_i \frac{\partial U}{\partial x_i} = 0, \tag{266}$$

and

$$\frac{\partial T}{\partial x_j} + \sum_{i=1}^{3} p_i \frac{\partial^2 U}{\partial x_i \partial x_j} = \frac{\Delta g}{\gamma}\frac{\partial U}{\partial x_j}, \qquad j = 1,\, 2,\, 3. \tag{267}$$

These four equations must be satisfied for all points of the telluroid, where all variables except T, $\partial T/\partial x_j$, and p_j are known.

If T and $\partial T/\partial x_j$ were known at a point Q of the telluroid, then (266) and (267) would be four linear equations in the three variables p_j, which could be determined if, and only if, (266) and (267) were compatible; let $i = 1, 2, 3$ run over columns, let $j = 1, 2, 3$ run over rows, hence the determinants on both sides become 4 by 4:

$$\begin{vmatrix} T & \dfrac{\partial U}{\partial x_i} \\[2ex] \dfrac{\partial T}{\partial x_j} & \dfrac{\partial^2 U}{\partial x_i \partial x_j} \end{vmatrix} = \frac{g}{\gamma} \begin{vmatrix} 0 & \dfrac{\partial U}{\partial x_i} \\[2ex] \dfrac{\partial U}{\partial x_j} & \dfrac{\partial^2 U}{\partial x_i \partial x_j} \end{vmatrix} \tag{268}$$

in other words (268) has to be satisfied for all points of the telluroid.

Equation (268) is the correct form of the boundary conditions for T in Molodenskiy's problem.

The boundary value problem for T is not one of the classical boundary value problems for potentials, i.e. problems where the potential, the normal derivative of the potential, or a linear combination thereof is given at the boundary. Ours is the so-called oblique derivative problem where a linear combination of the potential and the derivative of it in some direction is given at the boundary. It can be proved that the direction in question is that of the normal line through the point Q, the normal line being the curve consisting of points having the same normal coordinates as Q; the normal lines are approximately vertical. If this direction does not at any point of the boundary coincide with the direction of a tangent to the boundary at the same point and if for the whole boundary the direction is to the same side of the boundary surface, then we have the regular oblique derivative problem, provided, however, that the boundary surface and the coefficients in equation (268) satisfy some very weak regularity conditions.

The oblique derivative problem is in general very complicated, but if it is regular it has been proved ([5], p. 265, and [2], p. 82) that the theorem called Fredholm's alternative applies. Since we have a certain liberty in choosing the mathematical model for the surface of the Earth, we can and shall assume that we have to do with the regular oblique derivative problem.

Fredholm's alternative runs as follows: Either

(a) there is no regular potential T different from zero which satisfies the homogeneous boundary value equation corresponding to (268) (i.e. (268) with the right term equal to zero); if so, (268) has for all right terms a unique solution T that is a regular potential (outside the boundary surface), or
(b) the homogeneous problem has a finite number n of linearly independent solutions; if so, the inhomogeneous problem is solvable only if the right term satisfies n linearly independent linear homogeneous conditions, and then it has n linearly independent solutions so that the difference between two arbitrary solutions is a solution to the corresponding homogeneous problem.

In pure mathematics it makes good sense to work with clear alternatives—in applied mathematics and in numerical mathematics the facts are more blurred. There we often have a situation where it is practically impossible to tell whether we are in case (a) or case (b); we have the same situation when we are to solve a system of linear algebraic equations so that the coefficient matrix has a "small" eigenvalue: the system is unstable—a "small" change in the input values may cause a change in the result that is not "small."

To find out which case applies to Molodenskiy's problem, we shall first consider the simplified situation where we have a non-rotating planet.

Here the normal potential U is regular also at infinity as are all its derivatives with respect to the Cartesian coordinates; thus,

$$\frac{\partial U}{\partial x_1}, \quad \frac{\partial U}{\partial x_2}, \quad \frac{\partial U}{\partial x_3}$$

are regular potentials.

If in the left determinant of (268) we substitute $\partial U/\partial x_h$ for T, then the first and the $h+1$'th column are identical and the determinant vanishes, i.e. $\partial U/\partial x_h$ is a solution to the homogeneous problem corresponding to Molodenskiy's problem for $h = 1, 2, 3$; that is to say we are in case (b) with n being at least three.

Let us first suppose that $n = 3$. Then, if the boundary value problem has a solution T_0—and so it has if the gravity anomalies satisfy three linear equations

$$T = T_0 + a\frac{\partial U}{\partial x_1} + b\frac{\partial U}{\partial x_2} + c\frac{\partial U}{\partial x_3} \tag{269}$$

is a solution for all values of a, b and c, and we may choose the constants so that the gravity centre for T coincides with the gravity centre for U, i.e. so that the three first-order terms of the expansion of T into spherical functions in the vicinity of infinity vanish.

For $n > 3$ the anomalies must satisfy more than three conditions, but if there are solutions, there will always be such which have gravity centres coinciding with that of the Earth. However, the solutions will be $n - 3$ times indeterminate.

Now we shall discuss the interesting case where the Earth is rotating. Let us fix the coordinate system so that the origin is at the gravity centre of the Earth and so that the x_3-axis coincides with the axis of rotation. Then

$$\frac{\partial U}{\partial x_3}$$

is still a solution to the homogeneous problem, whereas

$$\frac{\partial U}{\partial x_1} \quad \text{and} \quad \frac{\partial U}{\partial x_2}$$

are only formal solutions; they are not zero at infinity; they are not even bounded. Therefore, we can only say that n is *at least* one and, if the problem has a solution T_0, we cannot generally obtain coincidence between the gravity centres of T and the Earth. So we have again the situation that in order to obtain a usable solution, we must have a set of Δg satisfying (at least) three conditions. And even more: It is known that also two of the first-order terms in the expansion must be zero, see [4], p. 62, so in reality we have (at least) five conditions that should be satisfied.

The result of this investigation must be that Molodenskiy's problem is ill-posed, in the terminology of J. Hadamard, see [8]. For a problem to be well-posed it should according to Hadamard have one and only one solution for arbitrarily given data, and small variations of the given data should cause reasonably small variations in the solution. The correct formulation of Molodenskiy's problem would be to ask for a potential T that is regular outside the telluroid, that satisfies five conditions at infinity (the vanishing of the first-order terms and two of the second order terms in the expansion into spherical harmonics) and that satisfies equation (268) where in the right term $\Delta g + V$ is substituted for Δg, where

$$\int_{\omega} p(Q)V^2(Q)\,d\omega = \min,$$

the integral to be taken over the telluroid and p being a given positive weight function. This form of boundary value problem might be called a least-squares boundary value problem.

I shall not follow up this idea here, since we are not in possession of a continuous field of boundary data, but I have tried to point out that adjustment methods, also from an abstract theoretical point of view, are more realistic than the classical approach.

In the computation for the determination of the potential it may be practical to include results concerning the deflection of the vertical on one hand and to make it possible to find deflections of the vertical on the other hand. Therefore, we must find the differential operators that give these quantities from the representation of the disturbing potential, and I shall here derive those corresponding to the same model that led to the formula (268) for the differential operator giving Δg.

At the point P the direction of the physical vertical is $-\operatorname{grad} W$ and the direction of the normal vertical is $-\operatorname{grad} U$. We are interested in the projection on a horizontal plane through P of the difference between the unity vectors in the two verticals. This difference is

$$\begin{aligned} -\frac{1}{g}\operatorname{grad} W + \frac{1}{\gamma}\operatorname{grad} U &= -\frac{1}{g}\operatorname{grad} W + \frac{1}{\gamma}(\operatorname{grad} W - \operatorname{grad} T) \\ &= \left(\frac{1}{\gamma} - \frac{1}{g}\right)\operatorname{grad} W - \frac{1}{\gamma}\operatorname{grad} T. \end{aligned} \tag{270}$$

Let us use a coordinate system with the z-axis in the direction of the physical vertical, the x-axis in the west-east direction and the y-axis in the south-north direction. The horizontal plane through P has the equation $z =$ constant, i.e. the projection on this plane of $\operatorname{grad} W$ is zero, and we have for the west-east and the south-north components of the vertical deflection:

$$\xi = \frac{1}{\gamma}\frac{\partial T}{\partial x} \qquad \text{and} \qquad \eta = -\frac{1}{\gamma}\frac{\partial T}{\partial y}. \tag{271}$$

If the interpolation method is used for the interpolation of vertical deflections using gravity anomalies, we have at least two advantages over the classical method: 1) the theoretical advantage that all the measurements enter into the calculation in the same way, 2) the practical advantage that there is no integration in the process; the anomalies at the measured stations enter directly into the calculations, which, therefore, can be automatized.

I find it a very attractive thought that the problem of local interpolation of vertical deflections can be solved by the smoothing method so that the measurements of vertical deflections and of gravity anomalies enter formally into the calculations in the same way. This problem is in fact the first one on which we have planned to use the present theory at the Danish Geodetic Institute.

But still more attractive is it to use this method on an integrated adjustment of dynamical satellite measurements and measurements referring to the potential at the surface of the Earth.

I cannot write down yet practical formulae to be used in such calculations; I think that special research work is needed on this problem and I can only offer some theoretical comments on the question.

It is important to remember that the disturbing potentials T and R used for measurements relative to the surface of the Earth and to the satellites respectively are not the same, but as the difference between them is known, this fact does not cause severe difficulties. Also the slightly more complicated case where the difference is not completely known but is dependent on one or more unknown parameters may be dealt with by letting these parameters enter into the adjustment as unknowns. This problem is perhaps not unrealistic.

As a starting point for the discussion of the explicit form of the normal equations I take the formulae from [7], p. 29:

$$
\begin{aligned}
\frac{da}{dt} &= \frac{2}{na}\frac{\partial R}{\partial M} \\
\frac{de}{dt} &= \frac{1-e^2}{na^2 e}\frac{\partial R}{\partial M} - \frac{(1-e^2)^{1/2}}{na^2 e}\frac{\partial R}{\partial \omega} \\
\frac{d\omega}{dt} &= -\frac{\cos i}{na^2(1-e^2)^{1/2}\sin i}\frac{\partial R}{\partial i} + \frac{(1-e^2)^{1/2}}{na^2 e}\frac{\partial R}{\partial e} \\
\frac{di}{dt} &= \frac{\cos i}{na^2(1-e^2)^{1/2}\sin i}\frac{\partial R}{\partial \omega} - \frac{1}{na^2(1-e^2)^{1/2}\sin i}\frac{\partial R}{\partial \Omega} \\
\frac{d\Omega}{dt} &= \frac{1}{na^2 e(1-e^2)^{1/2}\sin i}\frac{\partial R}{\partial i} \\
\frac{dM}{dt} &= n - \frac{1-e^2}{na^2 e}\frac{\partial R}{\partial e} - \frac{2}{na}\frac{\partial R}{\partial a}.
\end{aligned}
\tag{272}
$$

Here, as in the previous problems, the operations on the right sides of the equations (272) are not to be performed directly on the disturbing potential R but on the reproducing kernel with respect to the first point P or the second point Q. If the kernel is given explicitly, e.g. as $2\alpha R/L$ it is not difficult to express it by the elements corresponding to the two points P and Q and perform the differentiations. If the kernel contains correction members in the form of spherical harmonics, the differential coefficients with respect to the elements can be derived in the traditional way or by using the generalized spherical harmonics, see [12] and [3].

Now, if we could measure directly the rates of variation of the elements in short intervals of time, then the problem would be solved, but we can only find the resulting perturbations during long intervals of time, and, consequently, we have to use integrations or mean values. This situation, however, is not peculiar to our problem, so it should be possible to overcome also that difficulty.

The most severe draw-back of the method is that it results in very large systems of normal equations—one equation for each measurement—and that these equations are not sparse, as are for instance the normal equations used for the adjustment of geodetic networks. It is a consolation that the matrix of the normal equations is positive definite, so that the equations may be solved without using pivoting, and that the adjustment procedure—like adjustment procedures in general—is relatively simple to automatize.

I believe it is necessary to find some trick that may reduce the number of normal equations or at least the number of coefficients different from zero; I have some ideas in this respect, but I think it is too early to go into computational details.

The adjustment technique introduced here is a typical data-processing method giving a formally correct result which is absolutely independent of the meaning given (or not given) to the input data. It is in my opinion a dangerous draw-back of this method, as of all adjustment methods, that it gives an answer to even the most foolish question if it is only asked in a formally correct way.

Therefore, I hope that the time gained by this and other forms of mechanization of tedious calculations will not be used exclusively for the production of more figures but for a better formulation of the problem, so that the questions we ask may be more realistic, from the physical as well as from the numerical point of view. I think that some of the thoughts expressed in this paper may be helpful in that respect.

Appendix. Proof of Runge's Theorem

I want to prove that any potential regular in an open bounded region Ω can be approximated by potentials regular in an open sphere Σ containing Ω in its interior. The region Ω is supposed to be bounded by a surface ω which is sufficiently regular, e.g. by having a finite curvature all over. This condition could be weakened very much, but I do not think it would be of much interest in this connection. It is, however, important that $\Sigma - \Omega$ is connected.

The theorem will be proved through a series of lemmas.

Lemma 4.1 *For every function f continuous in $\Omega + \omega$ and 0 on ω we have that from*

$$\int_{\Omega} f\psi \, d\Omega = 0 \tag{273}$$

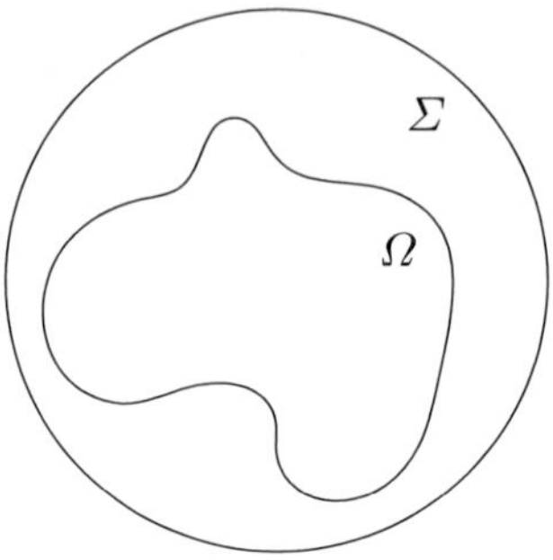

Fig. 4.1. Runge's theorem

for all potentials ψ regular in Σ follows

$$\int_{\Omega} f\phi \, d\Omega = 0. \tag{274}$$

for all potentials ϕ regular in Ω.

It is obvious that, if we can prove the lemma for certain potentials ψ regular in Σ, then the lemma will be true as it stands. The set of ψ I shall use is that represented by continuous single layer distributions on the surface σ of Σ, i.e. for potentials representable by

$$\psi_P = \int_{\sigma} \frac{1}{r_{PQ}} \kappa(Q) \, d\sigma_Q, \tag{275}$$

where

$$P \in \Sigma \quad \text{and} \quad Q \in \sigma$$

and r_{PQ} is the distance between the points P and Q. κ is the density of the single layer distribution on σ.

For (273) to be satisfied for all ψ represented by (275) it is necessary and sufficient that

$$F(Q) \equiv \int_{\Omega} f(P) \frac{1}{r_{PQ}} \, d\Omega_P = 0 \qquad \text{for all } Q \in \sigma. \tag{276}$$

F is here a regular potential in the space outside ω. In Ω F is a solution to the Poisson equation $\Delta F = f$.

Now (276) says that F is zero on the surface of the sphere and that it has the finite mass

$$M = \int_{\Omega} f \, d\Omega \tag{277}$$

i.e. F is zero at infinity; therefore it will vanish in the space outside the sphere, but then it must be zero in the whole space outside ω and on ω.

For every potential ϕ regular in Ω we may, consequently, write:

$$\int_{\Omega} f\phi\, d\Omega = \int_{\Omega} \Delta F\phi\, d\Omega = \int_{\Omega} F\Delta\phi\, d\Omega = 0, \tag{278}$$

(here we have used Green's formula) and the lemma is proved.

Let us now consider the Hilbert space H consisting of functions, not necessarily potentials, f defined in Ω so that the integral

$$\|f\|^2 = \int_{\Omega} f^2\, d\Omega \tag{279}$$

is finite. The scalar product in H is

$$\langle f_1, f_2\rangle = \int_{\Omega} f_1 f_2\, d\Omega. \tag{280}$$

Since functions of the type of f in Lemma 4.1 are dense in H, we have

Lemma 4.2 *In the Hilbert space H any element orthogonal to every potential ψ regular in Σ and restricted to Ω is orthogonal to every element ϕ of H that is a regular potential in Ω.*

Now, according to the elementary theory of Hilbert spaces Lemma 4.2 implies:

Theorem 4.1 *Any element ϕ of H which is a regular potential in Ω can be approximated* in the strong topology *in H by restriction to Ω of potentials ψ regular in Σ.*

But what we wanted was not a theorem on approximation in the strong topology in H, but in the uniform topology on all closed subsets of Ω.

If for a moment we assume that the elements ϕ of H which are regular potentials in Ω form a Hilbert space, say H_0, with the reproducing kernel $K(P,Q)$, then we can deduce the following theorem using a technique used already in Section III.

Theorem 4.2 *Any element ϕ of H_0 (i.e. any ϕ of H which is a regular potential in Ω) can be approximated uniformly on all closed subsets of Ω by restriction to Ω of potentials ψ regular in Σ, so that also any derivative of ϕ with respect to the coordinates in Ω is uniformly approximated by the corresponding derivatives of the ψ's on the same closed subsets of Ω.*

Corollary 4.1 *By using inversion with respect to the sphere Σ this theorem gives a strengthened form of Runge's theorem, at least for $\phi \in H_0$*

Proof of Theorem 4.2: For a function $f(P)$ that is an element of a Hilbert space with reproducing kernel $K(P,Q)$, we have, cf. (169):

$$\begin{aligned} |f(P)| &= |\langle f(Q), K(Q,P)\rangle| \leq \|f\| \|K(Q,P)\| \\ &= \|f\| \langle K(P,Q), K(Q,P)\rangle_Q^{1/2} = \|f\| K(P,P)^{1/2} \end{aligned} \tag{281}$$

and

$$\begin{aligned} \left|\left(\frac{\partial f}{\partial x_P}\right)_P\right| &= \left|\left\langle f(Q), \frac{\partial}{\partial x_P} K(Q,P)\right\rangle\right| \leq \|f\| \left\|\frac{\partial}{\partial x_P} K(Q,P)\right\|_Q \\ &= \|f\| \left\langle \frac{\partial K(P,Q)}{\partial x_P}, \frac{\partial K(Q,P)}{\partial x_P}\right\rangle_Q^{1/2} \end{aligned} \tag{282}$$

as well as similar formulae for the higher derivatives. From the properties of the reproducing kernel and the boundedness of Ω it follows that

$$\langle K(P,Q), K(Q,P)\rangle_Q^{1/2}$$

is finite. From the same premises and from the fact that $K(P,Q)$, a function of P or Q, is a potential regular in Ω it follows that the same applies to

$$\left\langle \frac{\partial K(P,Q)}{\partial x_P}, \frac{\partial K(Q,P)}{\partial x_P}\right\rangle_Q^{1/2}$$

and also to the higher derivatives of the kernel.

Now, from Theorem 4.1 it follows that, given a $\phi \in H_0$ we can find a sequence $\{\phi_n\}$ of potentials ψ_n regular in Σ so that for every $\epsilon > 0$ there is such an N that

$$\|\phi - \psi_n\| < \epsilon \qquad \text{for } n > N. \tag{283}$$

By putting $f = \phi - \psi_n$ Theorem 4.2 follows from (281), (282), etc.

In order to get rid of the restriction that ϕ must have a finite H-norm, we may use the spaces H_p instead of H. For any twice continuously differentiable positive function defined on Ω H_p is given by the scalar product

$$\langle f, g\rangle_p = \int_\Omega p(P) f(P) g(P)\, d\Omega \tag{284}$$

and the corresponding norm

$$\|f\|_p^2 = \left(\int_\Omega p f^2\, d\Omega\right)^{1/2}. \tag{285}$$

Given any potential ϕ regular in Ω, we can find such a p that

$$\phi \in H_p.$$

(Take $p = (1+\phi^2)^{-1/2}$).

The reader is invited to prove Lemma 4.1, Lemma 4.2 and Theorem 4.1 for H_p instead of H, which is quite simple. Then the restriction is removed and Runge's theorem is proved as soon as we have proved:

Lemma 4.3 *The subset of H_p consisting of potentials regular in Ω is a Hilbert space with reproducing kernel.*

But we shall first prove

Lemma 4.4 *Any Hilbert space consisting exclusively of potentials regular in Ω has a reproducing kernel.*

If ϕ is an element of the Hilbert space in question and P is a fixed point in Ω, then $\phi(P)$ is finite. The linear operator A_P from the given Hilbert space to the real numbers which to ϕ assigns the value $\phi(P)$ of ϕ at P is therefore defined for all ϕ of the Hilbert space. From a well-known theorem from functional analysis (Hellinger and Toeplitz' Theorem or the Closed Graph Theorem) it follows that A_p is a bounded operator or, in our case where the range is the real numbers, a bounded linear functional, and from this it follows again from one of the fundamental theorems of Hilbert spaces with reproducing kernel that the Hilbert space in question has a reproducing kernel.

As to the proof of Lemma 4.3 it remains only to be shown that the subset of H_p consisting of potentials regular in Ω forms a Hilbert space, i.e. a *closed* linear subspace of H_p. That it is a linear subspace is evident; the only difficulty is to prove that the subspace is closed.

Let us use M to denote the set of twice continuously differentiable functions defined on Ω and zero outside some closed subset of Ω. It is evident that $M \in H_p$ and that any function of H_p which is orthogonal to every $f \in M$ is equivalent to zero, so that a necessary and sufficient condition of $\phi \in H_p$ being a regular potential in Ω is that

$$\int_{\Omega} pf \Delta\phi \, d\Omega = 0 \qquad \text{for all } f \in M. \tag{286}$$

From (286) follows

$$\int_{\Omega} \Delta(pf)\phi \, d\Omega = 0 \qquad \text{for all } f \in M, \tag{287}$$

Green's formula having been used.

If ϕ is twice differentiable, (286) follows from (287), and as the $\phi \in H_p$ for which (287) holds form a closed linear subset of H_p, Lemma 4.3 follows from the famous Weyl's lemma, which shows that the ϕ for which (287) holds are not only twice but arbitrarily often differentiable.

Finally, I shall outline a short proof of Weyl's lemma.

Let S be the unit sphere, and let us suppose that $\Phi(P)$ is an n times continuously differentiable function which is zero outside S and which depends only on the distance of P from the origin. Let us also suppose that

$$\int_S \Phi \, d\Omega = 1. \tag{288}$$

For every function $u(P) \in H_p$ and every $\epsilon > 0$ we can now define:

$$u_\epsilon(P) = \int_S \Phi(Q) u(P - \epsilon Q) \, d\Omega_Q = \epsilon^{-3} \int \Phi\Big(\frac{P-Q}{\epsilon}\Big) u(Q) \, d\Omega_Q, \tag{289}$$

where the latter integral is taken over the domain where the integrand is different from zero.

It is easy to prove that u_ϵ is n times continuously differentiable, and that

$$\lim_{\epsilon \to 0} u_\epsilon(P) = u(P) \tag{290}$$

almost everywhere.

We shall now prove that, given two positive numbers ϵ_1 and ϵ_2, it follows from (287) that

$$\phi_{\epsilon_1}(P) = \phi_{\epsilon_2}(P) \tag{291}$$

for all $P \in \Omega$ so that the distance from P to the boundary of Ω is larger than both ϵ_1 and ϵ_2.

Let us define the function F by

$$F(P) = \int \bigg(\epsilon_1^{-3} \Phi\Big(\frac{P-Q}{\epsilon_1}\Big) - \epsilon_2^{-3} \Phi\Big(\frac{P-Q}{\epsilon_2}\Big) \bigg) \frac{1}{r_Q} \, d\Omega_Q. \tag{292}$$

For F we have immediately

$$\Delta F(P) = \epsilon_1^{-3} \Phi\Big(\frac{P}{\epsilon_1}\Big) - \epsilon_2^{-3} \Phi\Big(\frac{P}{\epsilon_2}\Big) \tag{293}$$

and

$$F(P) = 0 \tag{294}$$

for P outside the spheres with centres at the origin and radii ϵ_1 and ϵ_2.

Therefore, for a given $Q \in \Omega$ and for ϵ_1 and ϵ_2 sufficiently small $\Delta F(P-Q)$ is a function of P in the set M, and so (287) implies:

$$\int F(P-Q) \phi(Q) \, d\omega_Q = 0, \tag{295}$$

which in its turn is equivalent to (291).

Equations (290) and (291) now give

$$\phi_\epsilon(P) = \phi(P) \tag{296}$$

almost everywhere for ϵ sufficiently small (dependent on P).

But a consequence of (296) is that ϕ is equivalent to a function which is n times continuously differentiable. For $n = 2$ this means that we may deduce (286) from (287), and we have Weyl's lemma.

Acknowledgements

A night in 1965 in the lobby of a small hotel at Uppsala Dr. Erik Tengström opened my eyes to some difficulties in the mathematical foundation of physical geodesy and encouraged me to attempt to settle some of these questions. (At that time I must have underestimated the lot of mathematical literature I should have to study before I could really attack the problems—else I should not have accepted the task.) During the past year in his circular letters to the members of the special study group on mathematical methods in geodesy Professor Helmut Moritz has given a sharp analysis of related problems. Without the inspirations from these two exponents of our science and the personal discussions I have had with Professor Moritz in Berlin last spring this publication would never have seen the light of day.

I am also glad to have the opportunity to express my gratitude to my colleagues at the Danish Geodetic Institute for their kind patience in listening to me when I could not help speaking about my ideas.

Last but not least my thanks will be addressed to the director of the Institute, Professor Einar Andersen, D.Sc., for the favourable conditions under which I could study and work and for the confidence he has shown me in publishing my paper.

January, 1969.

References

[1] Philip J. Davis. *Interpolation and Approximation.* Blaisdell Publishing Company, Waltham, Massachusetts, 1963.

[2] Gaetano Fichera. *Linear Elliptic Differential Systems and Eigenvalue Problems*, volume 8 of *Lecture Notes in Mathematics.* Springer, Berlin, 1965.

[3] I. M. Gelfand and Z. Šapiro. Representations of the group of rotations of 3-dimensional space and their applications. *American Mathematical Society Translations*, 2:207–316, 1956.

[4] Weikko A. Heiskanen and Helmut Moritz. *Physical Geodesy.* W. H. Freeman and Company, San Francisco, 1967.

[5] L. Hörmander. *Linear Partial Differential Operators.* Springer, Berlin, 1964.

[6] W. M. Kaula. Statistical and harmonic analysis of gravity. *Journal of Geophysical Research*, 64, 1959.

[7] William M. Kaula. *Theory of Satellite Geodesy. Applications of Satellites to Geodesy.* Blaisdell Publishing Company, Waltham, Massachusetts, 1966.

[8] M. M. Lavrentiev. *Some Improperly Posed Problems of Mathematical Physics*, volume 11 of *Springer Tracts in Natural Philosophy*. Springer, Berlin, 1967.

[9] Herbert Meschkowski. *Hilbertsche Räume mit Kernfunktion*, volume 113 of *Die Grundlehren der mathematischen Wissenschaften in Einzeldarstellungen mit besonderer Berücksichtigung der Anwendungsgebiete*. Springer, Berlin, 1962.

[10] H. Moritz. Schwerevorhersage und Ausgleichungsrechnung. *Zeitschrift für Vermessungswesen*, 90:21–40, 1965.

[11] Helmut Moritz. Über die Konvergenz der Kugelfunktionsentwicklung für das Aussenraumpotential an der Erdoberfläche. *Österreichische Zeitschrift für Vermessungswesen*, 49:11–26, 1961.

[12] E. Stiefel. Expansion of spherical harmonics on a satellite orbit based on spinor algebra. *Mathematische Methoden der Himmelsmechanik und Astronautik*, pages 341–350, 1965.

5

A Remark on Approximation of T by Series in Spherical Harmonics

I believe that the problem about the convergence or not convergence of series in spherical harmonic down to the surface of the earth is posed in a wrong way, and I shall therefore try to stress a more realistic aspect of it.

First I shall mention two theorems. We have already heard about the first of them to day:

Moritz' theorem: in every vicinity of a potential for which the series expansions in spherical harmonics converges down to the surface of the earth there is a potential regular outside the earth for which the expansion in spherical harmonics diverges at some point at the surface of the earth.

The other theorem is exactly as positive as the first one is negative. It has been known by mathematicians for years but the communication between mathematics and geodesy is not so fast to day as it was a hundred years ago. For our use it can be stated as follows:

Runge's theorem: in every vicinity of a potential regular outside the earth and for every given sphere with centre in the centre of the earth there is a potential which is regular down to the surface of the given sphere, or more precisely:

Given any potential regular outside the surface of the earth and any sphere in the interior of the earth, for every closed surface surrounding the earth (which surface may be arbitrarily near the surface of the earth) there exists a sequence of potentials regular in the whole space outside the given sphere and uniformly converging to the given potential on and outside the given surface, and even more, the derivatives of all orders converge also uniformly (each of them) in the same region.

This theorem is at least interesting but it will be extremely useful if the problem could be solved to find an approximation, the existence of which is secured by the theorem, of a potential (e.g. T) which is (partly) known from measurements. This problem is not a trivial one but in the forthcoming "Meddelelse No. 44" from the Danish Geodetic Institute: "A Contribution to the Mathematical Foundation of Physical Geodesy" is given a method which could be called the method of least-squares collocation which should solve this problem. It can be looked at as a combination of Moritz interpolation formula and some of Bjerhammar's ideas and should at the same time give the latter that mathematical coherence which some of us may have missed in them before.

The publication, which will be available at the Danish Geodetic Institute from the 5th of September 1969, will also contain a proof of Runge's theorem.

Another promising approach to the problem is Kolmogorov's theory on ϵ-entropy an approximation of functions.

In the discussion following the remark I had to stress that Runge's theorem does not claim that a potential regular outside the earth can be continued down to a sphere inside the earth but only that it can be approximated arbitrarily well by a potential regular outside the sphere. On the other hand this distinction has only a meaning for a mathematically given potential but not for a physically given one, exactly as it has no meaning to say that a physical quantity is expressed by a rational or an irrational number.

6

On the Geometry of Adjustment

Zusammenfassung

Der Artikel behandelt die wichtige Frage der Abhängigkeit der Lösung eines Ausgleichungsproblems von der Gewichtsmatrix P. Das Problem wird zunächst qualitativ behandelt, wobei unter Variation von P die Menge der möglichen Lösungen eines gegebenen Ausgleichungsproblems sowohl unter der Voraussetzung unkorrelierter wie korrelierter Messungen beschrieben wird. Bei der qualitativen Darstellung wird eine Formel für die maximale Änderung der Lösung als Resultat einer Änderung von P angegeben.

Die angewandte Methode ist vollständig geometrisch; es ist kurios, daß keine der klassischen Formeln der Ausgleichungsrechnung benötigt wird.

Der Verfasser wünscht mit diesem Artikel darauf hinzuweisen, daß die geometrische Methode in der Theorie der Ausgleichung nicht nur wie bekannt die Ableitung der fundamentalen Formeln der Ausgleichungsrechnung gestattet, sondern auch ein sehr wirksames Mittel für die Lösung tieferliegender Probleme in dieser Disziplin ist. Die gefundenen Resultate könnten schwerlich mit traditionellen Methoden hergeleitet werden.

Geometrically the process of classical adjustment can be described as follows: The result of the n observations may be represented as a point in an n-dimensional vector space, the *observation space*. The locus of observation points consistent with the physical model for the observations constitutes an m-dimensional linear subspace, the *permissible subspace*, of the observation space. If the permissible subspace is given on parameter form—i.e. its coordinates in the observation space are given as n linear expressions in m parameters then we have the classical case of adjustment in elements; if it is given by $n - m$ linear equations in the coordinates then we have the adjustment with correlates, but from the geometrical point of view there is no difference between these cases—or between these cases and the hybrid cases also mentioned in the literature.

The problem of adjustment is now, given the observation point x to find that point y in the permissible subspace for which the "distance" to the observation point is minimum. The square of this distance is expressed by a quadratic form in the coordinate differences of these two points

$$d^2 = (x - y)^{\mathrm{T}} P(x - y). \tag{1}$$

Here P is the weight matrix, a symmetric positive definite $n \times n$ matrix which is the reciprocal to the matrix of variance-covariance. Now it is a well-known fact that such a positive definite matrix by the equation (1) determines a Euclidean metric on the n-dimensional vector space, the observation space. Partisans of tensor analysis would say that P is the contravariant metric tensor and that the variance-covariance matrix is the covariant metric tensor. It is now evident that y must be the orthogonal projection of x into the permissible subspace—here orthogonal means orthogonal with respect to the Euclidean metric determined by P through (1)—and so we have seen that from the geometrical point of view least-squares adjustment is nothing else than orthogonal projection.

It is clear that we could use this geometrical insight in deducing all the well-known formulas in adjustment theory in a very simple way but here I shall concentrate on the problem on the dependence of y on P.

The first question in which I am interested is: Given x which values of y are possible? It is evident that there are two limitations on y:

1. y must be a point of the permissible subspace, and
2. if the observations are formally correct i.e. if x is already in the permissible subspace then y is identically equal to x.

I shall prove that there are no other limitations: Given a point x of the observation space but not in the permissible subspace and a point y in the permissible subspace there exists a positive definite matrix P such that y is the result of the least-squares adjustment of the observations represented by x with P as weight matrix.

Suppose it is possible to find P and an orthogonal coordinate system in the observation space such that the first m coordinate axes are vectors in the permissible subspace and the next coordinate axis is parallel to $x - y$ (orthogonality is here meant with respect to P), then $x-y$ would be orthogonal to permissible subspace and y would be the orthogonal projection of x into this subspace and P would be a solution.

The possibility of choosing P and the coordinate system is secured by the following lemma: Given n linearly independent vectors in an n-dimensional vector space there exists a positive definite matrix P so that in the metric determined by P and (1) these vectors constitute an orthonormal set.

Let A be the $n \times n$ matrix consisting of the given n vectors as column vectors, then A is non singular, and the condition that the vectors shall be

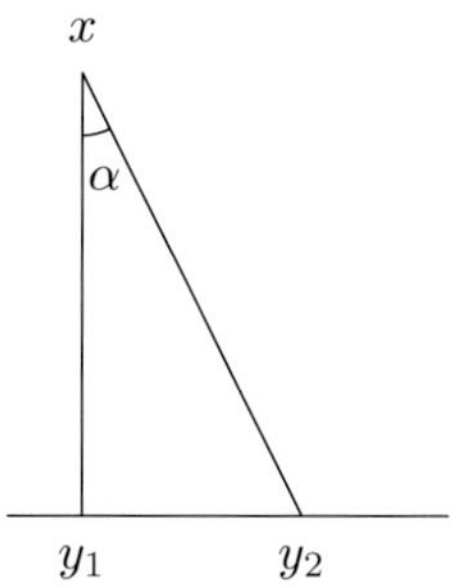

Fig. 6.1. Drawing in P_1 metric: Observation point x and its orthogonal projections y_1 and y_2 onto the permissible subspace with respect to the Euclidean metrics P_1 and P_2, respectively.

orthonormal with respect to P can be expressed as follows

$$A^{\mathrm{T}}PA = I \qquad \text{or} \qquad P = (AA^{\mathrm{T}})^{-1},$$

and this is a positive definite matrix.

This simple result concerning the possible results of an adjustment seems to be little known to geodesists—in fact from time to time geodesists have tried to prove almost the opposite result.

The simplest not trivial adjustment problem is that where there are given two correlated observations of the same physical quantity, in this case you can easily by calculation verify our result and you will see on the other hand that for a diagonal weight matrix—i.e. for not correlated observations—the result will always lie between the values of the two observations. This fact may also be generalized to the general case of adjustment with diagonal weight matrix. The proof is quite elementary but perhaps a little complicated therefore I shall only quote the result.

The set of points of the permissible subspace corresponding to a given value of one of the observations corresponds to a hyperplane of this subspace. So to the n observations there will correspond n hyperplanes in the permissible subspace and these hyperplanes will generally bound several polyhedra of different dimensionality in the permissible subspace but there will be one of these polyhedra which contains all the other polyhedra and the possible domain for y is exactly the interior of this maximum polyhedron. It is perhaps astonishing that this polyhedron is not necessarily convex.

As you see the situation is very much different in this case where the observations are supposed to be uncorrelated from that in the general case.

Now we have obtained a qualitative view over the possibilities for the solution of adjustment problems. Let us try after that to complement this with a more quantitative insight in the same problem: If the weight matrix is changed so and so much how much can the result then change?

Let us try to formulate the problem.

We have two adjustments. For both of them as well the permissible subspace as the observation space are identical. The values of the observations are in both cases the n dimensional vector x. The two weight matrices are P_1 and P_2, and the results y_1 and y_2 are regarded as two functions of x or as one function of P and x

$$y = f(P, x),$$

so that

$$y_1 = f(P_1, x) \qquad \text{and} \qquad y_2 = f(P_2, x).$$

We want to define a measure for the effect of changing the weight matrix from P_1 to P_2. The effect on a given set of observations (corresponding to the vector x) could be measured by the distance between y_1 and y_2, this distance being measured by P_1 or by P_2. This distance is not bounded as a function of x; it is proportional to the vector of the residuals i.e. the distance from x to the permissible subspace, therefore the distance y_1y_2 divided by the vector of the residuals—both distances measured in the same metric P_1 or P_2—would be a more natural measure for the discussed effect.

In Figure 6.1 I have tried to illustrate the meaning. The figure is drawn using the P_1 metric. The above mentioned ratio is

$$\frac{|y_2 - y_1|_{P_1}}{|x - y_1|_{P_1}} = \tan\alpha,$$

here $|\cdot|_{P_1}$ means the P_1-length and α is the angle under which y_1y_2 is seen from the point x.

The angle at y_2 (i.e. the angle y_1y_2x) is P_2-orthogonal but in the P_1-metric it is $\pi/2 - \alpha$, that is the shifting of the metric from P_2 to P_1 has changed this angle from $\pi/2$ to $\pi/2 - \alpha$. In the Appendix I have proved that α has to satisfy the inequalities

$$-\frac{\nu^{1/2} - \nu^{-1/2}}{2} \le \tan\alpha \le \frac{\nu^{1/2} - \nu^{-1/2}}{2},$$

where ν is the condition number of the matrix

$$Q = P_1^{-1/2} P_2 P_1^{-1/2}$$

i.e. the ratio of the largest to the smallest of the eigenvalues of Q. Here $P_1^{1/2}$ is that positive definite matrix which satisfies

$$\left(P_1^{1/2}\right)^2 = P_1.$$

This matrix exists certainly and is non-singular.

Recapitulating we have almost proved the following theorem: Given an adjustment problem by

1. an observation space,
2. a permissible subspace of it,

3. a weight matrix P_0 and
4. an observation point not in the permissible subspace.

Then for all weight matrices P for which the spectrum of $P_0^{-1/2} P P_0^{-1/2}$ is contained in the closed interval $[a, b]$, $b \geq a > 0$, the following inequality is valid for the angle α under which the displacement of the result of the adjustment effected by the use of the weight matrix P instead of P_0 is seen from the observation point in the P_0 metric:

$$|\tan \alpha| \leq \frac{1}{2}\left(\sqrt{\frac{b}{a}} - \sqrt{\frac{a}{b}}\right), \tag{2}$$

and there exists at least one P such that this inequality is valid with the equality sign.

The last assertion that the bound given here is the best possible does not follow immediately from the foregoing but it should not be difficult for the interested reader to prove it.

For the ratio between the lengths of the residual vectors according to the two weight matrices P_1 and P_2 we find the inequalities:

$$\lambda_{\max}^{-1/2} \leq \frac{|x - y_1|_{P_1}}{|x - y_2|_{P_2}} \leq \lambda_{\max}^{1/2} \tag{3}$$

where $\lambda_{\max}$ is the largest eigenvalue of the matrix

$$R = P_2^{-1/2} P_1 P_2^{-1/2}.$$

Proof: According to Figure 6.1 we have

$$|x - y_1|_{P_1} \leq |x - y_2|_{P_1} \leq \lambda_{\max}^{1/2} |x - y_2|_{P_2},$$

where the last inequality follows from the first part of the Appendix, consequently

$$\frac{|x - y_1|_{P_1}}{|x - y_2|_{P_2}} \leq \lambda_{\max}^{1/2},$$

which is the second inequality of (3).

In the same way we find that

$$\frac{|x - y_2|_{P_2}}{|x - y_1|_{P_1}} \leq \lambda_1^{1/2}$$

where λ_1 is the largest eigenvalue of

$$Q = P_1^{-1/2} P_2 P_1^{-1/2}.$$

Then the first part of (3) follows from the fact that the eigenvalues of R are the reciprocals of those of Q: The eigenvalues of R are the reciprocals to those of

$$R^{-1} = P_2^{1/2} P_1^{-1} P_2^{1/2}$$

and the matrices Q and R^{-1} have the same eigenvalues because these two matrices are similar:

$$\begin{aligned} Q &= \big(P_1^{-1/2}P_2^{1/2}\big)\big(P_2^{1/2}P_1^{-1/2}\big) \\ &= \big(P_1^{-1/2}P_2^{1/2}\big)\big(P_2^{1/2}P_1^{-1/2}\big)\big(P_1^{-1/2}P_2^{1/2}\big)\big(P_1^{-1/2}P_2^{1/2}\big)^{-1} \\ &= \big(P_1^{-1/2}P_2^{1/2}\big)R^{-1}\big(P_1^{-1/2}P_2^{1/2}\big)^{-1}. \end{aligned}$$

This similarity has also the consequence that what we have proved here holds good with very small modifications if we let P (P_1 and P_2) mean the variance-covariance matrix instead of the weight matrix.

A special case of the problem on the influence on the result of an adjustment of a change of the variance-covariance metric has been treated in several articles during the last years, it is that of the effect of ignoring the correlations between the observations. As the reader will see immediately the matrix Q will in that case be equal to the correlation matrix (having ones in the diagonal). If the condition number of the correlation matrix is large it may in fact be dangerous to ignore the correlations as our results show.

Appendix

Let us have a look at the geometric implications of a shift of metric from that determined by the positive definite matrix P_1 to that determined by P_2.

First let us see how much the length of a vector ϕ can be changed: Let us find the maximum and minimum lengths of a vector ϕ according to the metric P_2 if ϕ is a unit vector according to the metric P_1 i.e. find

$$\max_{\phi^{\mathrm{T}}P_1\phi=1} \big(\phi^{\mathrm{T}}P_2\phi\big)^{1/2} \tag{4}$$

and

$$\min_{\phi^{\mathrm{T}}P_1\phi=1} \big(\phi^{\mathrm{T}}P_2\phi\big)^{1/2}. \tag{5}$$

Using the transformation of coordinates defined by

$$\psi = P_1^{1/2}\phi$$

and putting

$$Q = P_1^{-1/2}P_2P_1^{-1/2}$$

the problem is reduced to the following: Find

$$\max_{|\psi|=1}\big(\psi^{\mathrm{T}}Q\psi\big)^{1/2} \qquad \text{and} \qquad \min_{|\psi|=1}\big(\psi^{\mathrm{T}}Q\psi\big)^{1/2} \tag{6}$$

where $|\psi| = (\psi^{\mathrm{T}}\psi)^{1/2}$. But as Q is symmetric and positive definite it is a well-known fact that the maximum and minimum in (4) and consequently also in (6) are respectively the square roots of the largest and the smallest eigenvalues of Q.

Next let us find out how much an angle which is right according to one metric can differ from a right angle according to the second metric i.e. the two vectors ϕ and ψ are supposed to be 2-orthogonal:

$$\phi^{\mathrm{T}} P_2 \psi = 0, \tag{7}$$

call the 1-angle between them $\pi/2 - \alpha$ such that

$$\sin\alpha = \cos\left(\tfrac{\pi}{2} - \alpha\right) = \frac{\phi^{\mathrm{T}} P_1 \psi}{(\phi^{\mathrm{T}} P_1 \phi)^{1/2} (\psi^{\mathrm{T}} P_1 \psi)^{1/2}} \tag{8}$$

what is the maximum for α determined by (8) over all pairs of vectors ϕ and ψ satisfying (7)?

Using the same coordinate transformation as above we may reformulate the problem as: Find the maximum value of

$$\sin\alpha = \phi^{\mathrm{T}} \psi$$

under the constraints

$$\phi^{\mathrm{T}}\phi = \psi^{\mathrm{T}}\psi = 1 \qquad \text{and} \qquad \phi^{\mathrm{T}} Q \psi = 0.$$

We shall solve it by the method of Lagrangean multipliers. Define

$$f(\phi, \psi) \equiv \phi^{\mathrm{T}}\psi - \mu\phi^{\mathrm{T}}\phi - \nu\psi^{\mathrm{T}}\psi - \lambda\phi^{\mathrm{T}} Q \psi,$$

then a necessary condition for maximum (or minimum) is

$$df \equiv d\phi^{\mathrm{T}}(\psi - 2\mu\phi - \lambda Q\psi) + d\psi^{\mathrm{T}}(\phi - 2\nu\psi - \lambda Q\phi) = 0 \qquad \text{for all } d\phi^{\mathrm{T}} \text{ and } d\psi^{\mathrm{T}},$$

or

$$\psi - 2\mu\phi - \lambda Q\psi = 0, \qquad \phi - 2\nu\psi - \lambda Q\phi = 0. \tag{9}$$

By multiplication at left of the first of these equations by ϕ^{T} and of the second one by ψ^{T} using the conditions we obtain

$$2\mu = 2\nu = \phi^{\mathrm{T}}\psi = \sin\alpha$$

so that (9) may be written as

$$\psi - \phi\sin\alpha = \lambda Q\psi, \qquad \phi - \psi\sin\alpha = \lambda Q\phi. \tag{10}$$

Adding and subtracting of these equations give:

$$\frac{1 - \sin\alpha}{\lambda}\left(\phi + \psi\right) = Q(\phi + \psi), \qquad \frac{1 + \sin\alpha}{\lambda}\left(\phi - \psi\right) = Q(\phi - \psi). \tag{11}$$

(11) expresses that $\phi + \psi$ respectively $\phi - \psi$ are eigenvectors for the matrix Q corresponding to the eigenvalues

$$\Lambda_+ = \frac{1 - \sin\alpha}{\lambda} \tag{12}$$

respectively

$$\Lambda_- = \frac{1 + \sin\alpha}{\lambda}. \tag{13}$$

Elimination of λ between (12) and (13) gives

$$\frac{1-\sin\alpha}{1+\sin\alpha} = \frac{\Lambda_+}{\Lambda_-} \qquad \text{or} \qquad \sin\alpha = \frac{1-\frac{\Lambda_+}{\Lambda_-}}{1+\frac{\Lambda_+}{\Lambda_-}}.$$

Different combinations of the eigenvalues of Q will generally give different values for α, and the largest and the smallest possible values for α are among them, the largest value for α will correspond to the combination $\Lambda_+ =$ the smallest eigenvalue of Q and $\Lambda_- =$ the largest eigenvalue of Q.

If we put

$$\nu = \frac{\Lambda_-}{\Lambda_+}$$

for this combination we have

$$\sin\alpha_{\max} = \frac{\nu-1}{\nu+1} \qquad \text{and} \qquad \tan\alpha_{\max} = \frac{\nu^{1/2}-\nu^{-1/2}}{2}$$

in the same way we find for the smallest possible value of α:

$$\tan\alpha_{\min} = \frac{\nu^{-1/2}-\nu^{1/2}}{2}$$

so that we have

$$-\frac{\nu^{1/2}-\nu^{-1/2}}{2} \leq \tan\alpha \leq \frac{\nu^{1/2}-\nu^{-1/2}}{2}.$$

7

Remarks to the Discussion Yesterday

There is one point in applied mathematics about which I cannot stop wondering. It is that we have to work with notions such as continuity, differentiability etc. inherited from pure mathematics, although these concepts have no meaning in a numerical model, where we work with finite representations of numbers and functions.

It is perhaps an important aspect of human conditions that we have to balance on two levels: one ideal and abstract and one practical and approximative. This is a philosophical question but it is evident that it is so, when we shall apply mathematics on physical and geodetic problems. When we advance our numerical model for a geodetic problem we must apply our knowledge of the abstract mathematical background, but the practical methods we use need not be direct maps of the mathematical model.

As far as I can see Molodenskiy's problem and method belong to the abstract level of physical geodesy and it should never be used for calculations of the potential of the earth. The mathematical beauty of Molodenskiy's theory is connected with the fact that it (under certain conditions) gives a method for determining of the potential from exactly the necessary and sufficient data. But in physical geodesy we should exactly as in geometric geodesy use a least-squares method permitting us to let all relevant geodetic observations enter into the determination of coordinates and potential in a natural way. I have engaged myself in the work to create such a least-squares model which I call "integrated geodesy" towards which the least-squares collocation is the first step.

But even if we accept a method for calculation, it is important at the abstract level to find out what information we may extract from special forms of data and here theories such as Molodenskiy's will always be of interest.

I believe that Moritz' paper on the convergence of Molodenskiy's series is a step toward a better understanding of Molodenskiy's problem first of all because he faces the fact that the integral operator in Molodenskiy's integral equation is of Calderon-Zygmund type. Too many authors have ignored this fact with disastrous consequences for their results.

The idea behind the Molodenskiy's series is to regard the boundary value problem as a perturbation of Stoke's problem. The question Moritz asks (and to some extent answers) can therefore be expressed as follows: how long can the solution of the perturbed problem be developed in a Taylor series in the perturbation parameter?

I have asked myself the opposite question: what can make the convergence of such a series break down?

I shall follow Moritz' example and treat the problem using matrices in order to avoid functional analysis. It is a matter of routine work to make the considerations theoretically valid.

We have to solve the linear problem

$$A_t \phi = \alpha \tag{1}$$

where A_t is a matrix $A_0 + tB$, for $t = 1$, and we know the solution for $t = 0$.

It is a well-known fact that A_t^{-1} depends analytically on t at all points in the complex t-plane for which the matrix A_t is non-singular. (Here the word "singular" is used in quite another meaning then when we speak about "singular kernels.") Therefore the matrix A_t^{-1} (and so also the solution of (1)) may be expanded in a Taylor series (after powers of t) converging inside a circle with centre $t = 0$ and radius r, where r equals the absolute value of that t which is nearest to 0 among the values of t for which A_t is singular. That is, the problem about the radius of convergence of Molodenskiy's series is equivalent to an eigenvalue problem for a generalized complex Molodenskiy problem.

It should be well-known that Molodenskiy's problem as a boundary value problem for a harmonic function has three (linearly independent) eigensolutions connected with the three coordinates for the mass centre of the earth and that Molodenskiy's integral equation in a way eliminates these three eigensolutions and so it seems that obstruction of the convergence of Molodenskiy's series should be connected with the existence in certain cases of nontrivial eigensolutions—at least for the "complexified" Molodenskiy's problem.

I shall come back to the eigenvalue problem (for the real case) at a later occasion, probably in a letter to study group 4.31. I am very interested in this problem and it is a consolation for me that the convergence problem for Molodenskiy's series which is even more complicated after all has advanced.

It was asked yesterday how we shall react to the result that our statistical model is non-ergodic. To this let me say first that it is always rash to built up a new theory before one really understands the problem, and I do not think that we shall understand it before Lauritzen's book exists printed in its definitive form. Therefore the following has to be taken with due reservation.

1. As far as I understand Lauritzen does not claim that the covariance form does not exist but that we are principally unable to find it.

2. It is a widespread feeling—and it may be easily proved—that if the data are good enough and really relevant then the result will be only weakly dependent on the covariance form.
3. On the other hand the estimation of the a posteori variances depend strongly on the covariance form used.

My conclusion are: let us continue to use and improve the statistical methods but *with precaution* using different reasonable covariance forms, and let us see if the results are independent of these. Let us not believe in the estimates for the variances but apply a control based on prediction of data which have not entered in the calculations. At the theoretical level let us analyse more thoroughly the nonstatistical aspects of the least squares methods and study Molodenskiy's problem and other similar problems in order to find out what are the most relevant data for the solution of different geodetic problems.

Lauritzen's result was not as we hoped it should be. But it is better for it opens a way for us to a lot of new interesting problems. Moreover it seems to be true.

Remarks

Moritz: I agree with Prof. Tengström that it would be preferable to be able to consider the earth's surface as known, but even then the problem remains an oblique-derivative problem with difficulties basically similar to Molodensky's problem, although on a reduced scale.

Bursa: The applying of Molodensky's theory is connected with the smoothing of the actual Earth's surface in an extent. The smoothing should be done before. The problem of a uniform smoothing is up till now open, but it is not so difficult than the regularization in the classical theory is. Therefore, the main progress done by Molodensky's solution is that the actual Earth, not regularized (however, smoothed), and its external gravity field, be investigated.

8

Letters on Molodenskiy's Problem

I. The Simple Molodenskiy Problem

1. Mathematical Formulation

In the three-dimensional Euclidean space, the following are given

1. A closed surface ω with two times continuously differentiable coordinates which is *star-shaped* with respect to origin O, i.e., every half line from O meets ω in exactly one point
2. A continuous function f on ω.

Let Ω be the open domain outside ω. What we want is then a function T on $\Omega + \omega$ which is 1) differentiable on $\Omega + \omega$, 2) harmonic on Ω, 3) regular at infinity, i.e., for $|x|$ a sufficiently large T is representable as

$$T = \frac{h_0(x)}{|x|} + \frac{h_1(x)}{|x|^3} + \cdots + \frac{h_n(x)}{|x|^{2n+1}} + \cdots, \tag{1}$$

here and in the following $h_n(x)$ are (solid) homogeneous harmonic polynomials of degree n, and 4)

$$\sum_{i=1}^{3} x^i \frac{\partial T}{\partial x^i} + 2T = f \qquad \text{on } \omega. \tag{2}$$

As we shall see later, we have wanted too much; therefore we have to reduce our demands slightly.

If for such a T $h_1(x) = 0$ then T is said to be an *admissible solution* of the simple Molodenskiy problem. (Contrary to some authors, we do not demand that also $h_0(x) = 0$. More about this question will be addressed in Section II[1]).

[1] There are four Molodenskiy letters; each of them is identical to a section in this chapter.

2. The Necessary and Sufficient Conditions for Solution

It is well-known (and it will be proved below) that if T is harmonic in Ω then the left side of (2) is also harmonic in Ω; that is, at all points of $\Omega + \omega$, T satisfies the equation

$$\sum_{i=1}^{3} x^i \frac{\partial T}{\partial x^i} + 2T = F, \qquad x \in \Omega \tag{3}$$

where F is the solution of the Dirichlet problem: Find F, continuous in $\Omega+\omega$, harmonic in Ω, regular at infinity and $F = f$ on ω. In that way we have almost reduced the problem to one of solving a partial differential equation of order one (3): Under the given assumptions Dirichlet's problem has exactly one solution F, and our problem therefore corresponds to F to find a *harmonic* function T satisfying (3).

In (3) put $T_n = h_n(x)/|x|^{2n+1}$. The ratio $h_n(x)/|x|^{2n+1}$ is homogeneous of degree $-(n+1)$; therefore $\sum_i x^i \partial T_n/\partial x^i = -(n+1)T_n$ and

$$\sum_i x^i \frac{\partial T_n}{\partial x^i} + 2T_n = -(n-1)T_n;$$

that is, the differential operator $T \to \sum x^i(\partial T/\partial x^i) + 2T$ has eigenvalues $\lambda_n = -(n-1)$, $n = 0, 1, \ldots$ and for every eigenvalue λ_n it has $2n+1$ linearly independent eigensolutions $h_n(x)/|x|^{2n+1}$ corresponding to the $2n+1$ linearly independent homogeneous harmonic polynomials of degree n. As these functions for $n = 0, 1, \ldots$ span the space of harmonic functions on Ω (Runge's theorem) 1) there do not exist other eigenvalues (and eigensolutions) and 2) we have proved that $\sum x^i(\partial T/\partial x^i) + 2T$ is harmonic if T is harmonic.

From 1) follows for $n = 1$ that the only solutions of

$$\sum_i x^i \frac{\partial T}{\partial x^i} + 2T = 0$$

are linear combinations of the three functions $x^i/|x|^3$, $i = 1, 2, 3$, and we see that if the simple Molodenskiy problem has a solution then the sum of this solution and such a linear combination is also a solution, and also that if it has a solution then it has also an admissible solution.

Now let σ be the surface of a fixed sphere with its center in O so that ω is in its interior and let Σ be the exterior of the same sphere. Then the expansion (1) for T is valid for every $x \in \Sigma$. Application of our differential operator on T gives again for $x \in \Sigma$

$$\sum x^i \frac{\partial T}{\partial x^i} + 2T = \frac{h_0(x)}{|x|} - \frac{h_2(x)}{|x|^5} - \frac{2h_3(x)}{|x|^7} - \cdots - \frac{(n-1)h_n(x)}{|x|^{2n+1}} - \cdots. \tag{4}$$

We observe that (3) has a solution which is regular in Σ if and only if the coefficients to the spherical harmonics of the first degree in the expansion of F vanish, i.e., if the mass center of the potential F is the point O, but the methods we have used so far do not permit us to decide whether the solution we have found in Σ can be extended so as to be a solution in Ω.

(3) may be written as

$$\frac{1}{|x|}\frac{\partial}{\partial|x|}\left(|x|^2T\right) = F,$$

so that a formal solution of (3) should be

$$T = \frac{1}{|x|^2}\int |x|F(x)\,d|x|,$$

more precisely as $T(\infty)$ should be 0:

$$T(x) = -\left[\frac{1}{|y|^2}\int |y|F(y)\,d|y|\right]_{y=x}^{y=\infty} \tag{5}$$

where the integration is to take place along the half line from O through the point $x \in (\Omega + \omega)$. The complete solution of (3) contains a 'constant' of integration C. That C is 'constant' must mean that it is constant on every such half line; that is it depends only on $x/|x|$. Therefore we write $C = C(x/|x|)$. The total contribution of C to the solution is

$$T_0(x) = \frac{C\left(\frac{x}{|x|}\right)}{|x|^2}$$

which is positively homogeneous of degree -2. Therefore if $T_0(x)$ shall be harmonic, C must be a linear function of x divided by $|x|$, or

$$T_0(x) = \frac{h_1(x)}{|x|^3}$$

which are the zero solutions we have found before.

In order to help one understand this, one may control (5) by applying it to the functions

$$F_n(x) = \frac{h_n(x)}{|x|^{2n+1}}.$$

Simple calculations show that for $n \neq 1$ the admissible results are

$$T = \frac{1}{1-n}\frac{h_n(x)}{|x|^{2n+1}}$$

as it should be. For $n = 1$ we obtain

$$T = \frac{h_1(x)}{|x|^3}\ln|x|$$

which is not harmonic, in fact

$$\Delta\left(\frac{h_1(x)}{|x|^3}\ln|x|\right) = -\frac{5h_1(x)}{|x|^5},$$

If $|x|^3 F(x)$ is bounded (for $x \in \overline{\Omega}$) it is clear that the right side of (5) exists (for $x \in \overline{\Omega}$). Under the same assumptions we have

$$-\left[\frac{1}{|y|^2}\int |y|F(y)\,d|y|\right]_{y=x}^{y=\infty} = -\frac{1}{|x|^2}\int_x^\infty |y|F(y)\,d|y|$$

and the expression on the right side is harmonic when $F(y)$ is harmonic. This is most easily proved by using spherical coordinates:

$$\Delta F(x) = \frac{1}{r^2}\frac{\partial}{\partial r}\left(r^2\frac{\partial F}{\partial x}\right) + \frac{1}{r^2}\left(\frac{1}{\sin\theta}\frac{\partial}{\partial\theta}\left(\sin\theta\frac{\partial F}{\partial\theta}\right) + \frac{1}{\sin^2\theta}\frac{\partial^2 F}{\partial\lambda^2}\right) = 0$$

or

$$\frac{1}{\sin\theta}\frac{\partial}{\partial\theta}\left(\sin\theta\frac{\partial F}{\partial\theta}\right) + \frac{1}{\sin^2\theta}\frac{\partial^2 F}{\partial\lambda^2} = -\frac{\partial}{\partial r}\left(r^2\frac{\partial F}{\partial r}\right)$$

so that

$$\begin{aligned}
&\Delta\left(-\frac{1}{|x|^2}\int_x^\infty |y|F(y)\,d|y|\right)\\
&= -\frac{1}{r^2}\frac{\partial}{\partial r}\left[r^2\frac{\partial}{\partial r}\left(\frac{1}{r^2}\int_{|y|=r}^\infty |y|F(y)\,d|y|\right)\right]\\
&\qquad -\frac{1}{r^4}\int_r^\infty |y|\left(\frac{1}{\sin\theta}\frac{\partial}{\partial\theta}\left(\sin\theta\frac{\partial F}{\partial\theta}\right) + \frac{1}{\sin^2\theta}\frac{\partial^2 F}{\partial\lambda^2}\right)d|y|\\
&= -\frac{1}{r^2}\frac{\partial}{\partial r}\left(-rF - \frac{2}{r}\int_r^\infty |y|F(y)\,d|y|\right) + \frac{1}{r^4}\int_r^\infty |y|\frac{\partial}{\partial|y|}\left(|y|^2\frac{\partial F}{\partial|y|}\right)d|y|\\
&= -\frac{1}{r^2}\left(-F - r\frac{\partial F}{\partial r} + \frac{2}{r^2}\int_r^\infty |y|F(y)\,d|y| + 2F\right)\\
&\qquad + \frac{1}{r^4}\left(\left[|y|^3\frac{\partial F}{\partial|y|}\right]_r^\infty - \int_r^\infty |y|^2\frac{\partial F}{\partial|y|}\,d|y|\right)\\
&= \frac{F}{r^2} + \frac{1}{r}\frac{\partial F}{\partial r} - \frac{2}{r^4}\int_r^\infty |y|F(y)\,d|y| - \frac{2F}{r^2} + \frac{1}{r^4}\left[|y|^3\frac{\partial F}{\partial|y|}\right]_r^\infty - \frac{1}{r^4}\left[|y|^2F\right]_r^\infty\\
&\qquad + \frac{2}{r^4}\int_r^\infty |y|F(y)\,d|y| = 0.
\end{aligned}$$

As every F harmonic in Ω, regular at infinity and such that $h_1(x)$ is zero, may be expressed as a sum of a harmonic function F' such that $|x|^3F'$ is bounded and a function of the form $h_0/|x|$ as we have solved the simple Molodenskiy problem for both forms of harmonic functions and as the problem is linear, we have in fact proved that the condition $h_1(x) = 0$ is not only necessary but also sufficient.

3. Stability — The Prague Method

We shall say that Molodenskiy's problem is *stable* if 'small' variations of the given function f inside the domain in which the problem is soluble result in 'small' variations of the admissible solution T. Here the adjective 'small' must be given reasonable meanings at both of its occurrences. Inequalities expressing stability are standard in modern treatises on elliptic boundary value problems—this is also the case covering our situation, but it is extremely difficult to find practical values for the constants entering these inequalities for actual problems. Guided by a knowledge about the form of such inequalities I have recently succeeded in finding a method by which actual constants can be found for the simple Molodenskiy problem.

The decisive step towards this determination is the introduction of an integral equation method for the solution of the simple Molodenskiy problem which I have called the Prague method.

The Prague method consists firstly to solve the Dirichlet problem for the function F by the classical integral-equation method: Principally we look for a continuous double layer distribution μ on ω corresponding to F such that

$$F(x) = \int_\omega \mu(y) \frac{\partial}{\partial n_y}\left(\frac{1}{|x-y|}\right) d\omega_y, \qquad x \in \Omega. \tag{6}$$

F has the continuous boundary values f if and only if μ satisfies the Fredholm integral equation

$$2\pi\mu(x) - \lambda \int_\omega \mu(y) \frac{\partial}{\partial n_y}\left(\frac{1}{|x-y|}\right) d\omega_y = f(x), \qquad x \in \omega \tag{7}$$

with $\lambda = -1$.

Instead of effectively computing F in the whole outer space Ω by (6), the idea of the Prague method is to find T directly from μ by an integral of a form not very different from the form of (6), of course using the formula (5).

Now the total mass of a double layer distribution is zero; therefore F can not be represented as in (6) if its mass does not vanish. But F can always be represented as

$$F = \frac{h_0}{|x|} + F' \tag{8}$$

where the mass of F' vanishes and h_0 is a suitable constant. Accordingly the corresponding homogeneous integral equation (7) has $\lambda = -1$ as an eigenvalue with the corresponding eigensolution $\mu_0 = 1$ and the associated homogeneous integral equation has an eigensolution say σ_0 corresponding to the conductor potential of ω. That is (7) has a solution if and only if $\int_\omega \sigma_0 f \, d\omega = 0$ and by the condition

$$\int_\omega \mu \, d\omega = \int_\omega \mu_0 \mu \, d\omega = 0 \tag{9}$$

this solution is uniquely determined and generates F by (6). Therefore our Dirichlet problem can be solved in the following way.

Write

$$f(x) = \frac{h_0}{|x|} + f', \qquad x \in \omega$$

and determine the constant h_0 so that (7) has a solution for f' on the right side instead of f:

$$0 = \int_\omega f'\sigma_0 \, d\omega = \int_\omega f\sigma_0 \, d\omega + h_0 \int_\omega \frac{\sigma_0}{|x|} \, d\omega.$$

This equation has a solution because $1/|x|$ is never orthogonal to σ_0. Then solve (7) and (9) for μ with f' instead of f. The corresponding potential F' is found by (6), and the solution of the Dirichlet problem F is given by (8) where h_0 and F' already have been determined.

From F we shall find T. If the problem has a solution it can be written as

$$T = \frac{h_0}{|x|} + T',$$

where T' corresponds to F' in (8), i.e., we have reduced the problem to determining T' corresponding to a potential F' which can be expressed from a double layer μ by (6). In the following we may and shall restrict ourselves to considering this problem without mentioning this restriction any more and consequently omitting the dots.

But first we must consider the potential T corresponding to the potential F generated by a unitary mass situated at some point y on ω. We have

$$F = \frac{1}{|x-y|}$$

and using (5) we find

$$\begin{aligned} T(x) &= -\left[\frac{1}{|z|^2} \int \frac{|z|}{|z-y|} \, d|z|\right]_{z=x}^{z=\infty} \\ &= -\left[\frac{1}{|z|^2}\left(|z-y| + \frac{(z,y)}{|z|}\left(\ln \frac{|z-y| + |z| - \frac{(z,y)}{|z|}}{2|z|} + \ln 2|z|\right) + C\right)\right]_{z=x}^{z=\infty} \\ &= \frac{|x-y|}{|x|^2} + \frac{(x,y)}{|x|^3}\left(\ln \frac{|x-y| + |x| - \frac{(x,y)}{|x|}}{2|x|} + \ln 2|x|\right) + \frac{C}{|x|^2}. \end{aligned}$$

Here the 'constant' C does not depend on $|x|$, i.e., it must be a function of y and $x/|x|$. Let us write the last line as

$$K(x,y) + \frac{C\left(\frac{x}{|x|}, y\right)}{|x|^2}.$$

From this follows—at least formally—that if F is given by (6) then the corresponding T is given by

$$T(x) = \int_\omega \mu(y) \frac{\partial}{\partial n_y} K(x,y)\, d\omega_y + \frac{1}{|x|^2} \int_\omega \mu(y) \frac{\partial}{\partial n_y} C\Big(\frac{x}{|x|}, y\Big)\, d\omega_y. \tag{10}$$

As long as x is in Ω, all the functions involved are very smooth so that the changing of order of integration and differentiation involved in the deduction of (10) are harmless. We shall later investigate the problem for $x \in \omega$. Now the question is whether T is a solution of our problem.

By direct calculation it can be proved that T satisfies Equation (3), but is T harmonic?

First consider the last term in (10):

$$\frac{1}{|x|^2} \int_\omega \mu(y) C\Big(\frac{x}{|x|}, y\Big)\, d\omega_y.$$

This is a function of x alone, and it is positively homogeneous of degree -2; that is it must be of the form $h_1(x)/|x|^3$ in order to be harmonic. This term is again the zero solution of Molodenskiy's problem and we need not be interested in it once more.

Next let us regard $K(x,y)$. It may be split into two parts:

$$\frac{|x-y|}{|x|^2} + \frac{(x,y)}{|x|^3} \ln \frac{|x-y| + |x| - \frac{(x,y)}{|x|}}{2|x|} \tag{11}$$

and

$$\frac{(x,y)}{|x|^3} \ln 2|x|. \tag{12}$$

(12) is certainly not harmonic, but a direct calculation shows that (11) is. (Here harmonic means harmonic with respect to x.)

The part of T in (10) coming from (12) is

$$T_2(x) = \frac{\ln 2|x|}{|x|^3} \int_\omega \mu(y)(n_y, x)\, d\omega_y,$$

(n_y is the unit vector in the direction of the normal of ω at y) which is harmonic only if it is zero, i.e., if

$$\int_\omega \mu(y)\, \alpha_y^i\, d\omega_y = 0 \qquad i = 1,\ 2,\ 3,\ \ldots \tag{13}$$

where α_y^i are the direction cosines of the normal to ω at the point y. (13) are the sufficient and necessary conditions for $h_1(x)$ to vanish in the expansion of F at infinity, i.e., again the conditions for the simple Molodenskiy problem to be soluble.

That is if μ satisfies (13) then

$$T(x) = \int_\omega \mu(y) \frac{\partial}{\partial n_y} K(x,y)\, d\omega_y \tag{14}$$

where now and in the following $K(x,y)$ shall mean that expression (11) should be a solution of Molodenskiy's problem.

It is now time to find the kernel $\frac{\partial}{\partial n_y} K(x,y)$ explicitly:

$$\frac{\partial}{\partial n_y} K(x,y) = \frac{(y-x,n_y)}{|x|^2|x-y|} + \frac{(x,n_y)}{|x|^3} \ln \frac{|x-y|+|x|-\frac{(x,y)}{|x|}}{2|x|} + \frac{(x,y)}{|x|^3} \frac{\frac{(y-x,n_y)}{|x-y|} - \frac{(x,n_y)}{|x|}}{|x-y|+|x|-\frac{(x,y)}{|x|}}.$$

By using simple geometry the reader can convince himself that

$$|x-y|+|x|-\frac{(x,y)}{|x|} = 2|x-y|\cos^2\frac{\sigma}{2}$$

where σ is the angle between the two half lines from x through O and through y so we may write:

$$\frac{\partial}{\partial n_y} K(x,y) = \frac{(y-x,n_y)}{|x|^2|x-y|} + \frac{(x,n_y)}{|x|^3} \ln\left(\frac{|x-y|}{|x|}\cos^2\frac{\sigma}{2}\right) + \frac{\frac{(y-x,n_y)}{|x-y|} - \frac{(x,n_y)}{|x|}}{2|x-y|\cos^2\frac{\sigma}{2}} \frac{(x,y)}{|x|^3}. \tag{15}$$

We observe that the star shape of ω is still essential for $\partial K/\partial n_y$ is not defined for x between O and y. We observe further that the two last terms are $O(|x|^{-3})$ for $x \to \infty$, while the first term is $O(|x|^{-2})$. It may be written

$$\frac{1}{|x|^2}\frac{(y-x,n_y)}{|x-y|} = -\frac{1}{|x|^2}\frac{(x,n_y)}{|x|} + O(|x|^{-3}).$$

But from the conditions (13) follows that the first term here is without influence, so that we have proved that if the simple Molodenskiy problem, has a solution then the Prague method automatically gives the admissible solution.

As we saw before that $T(x)$ as defined by (14) is very smooth (in fact it is analytic) for $x \in \Omega$, we shall now see how $T(x)$ behaves for x converging to a point $x_0 \in \omega$ from the outside. First suppose that x_0 is not an exceptional point of ω. By the exceptional points of ω we mean those where the tangent to ω passes through O.

The method we shall use is the well-known one used in potential theory for proving that the potential of a single layer is continuous by the passage of the surface supporting the single layer. The proof will be only sketched; for details the reader is referred to textbooks on classical potential theory.

Let $B_r(x_0)$ be the set of points x for which $|x - x_0| < r$ for $r > 0$ and set

$$\mu_1(y) = \begin{cases} 0 & \text{for } y \in B_r(x_0), \\ \mu(y) & \text{for } y \notin B_r(x_0) \end{cases}$$

and

$$\mu_2(y) = \begin{cases} \mu(y) & \text{for } y \in B_r(x_0), \\ 0 & \text{for } y \notin B_r(x_0) \end{cases}$$

and define $T_1(x)$ and $T_2(x)$ by

$$T_i(x) = \int_\omega \mu_i(x) \frac{\partial}{\partial n_y} K(x,y)\, d\omega_y \qquad \text{for } i = 1,\, 2.$$

The $\mu = \mu_1 + \mu_2$ and $T = T_1 + T_2$.

Now it is clear that for $x \in \Omega$

$$\lim_{x \to x_0} T_1(x) = T_1(x_0) \qquad \text{for every } r > 0.$$

On the other hand, for every $\epsilon > 0$ given we can choose $r > 0$ such that $|T_2(x)| < \epsilon$ for all $x \in B_r(x_0) \cap \overline{\Omega}$, because $(x - x_0)\frac{\partial}{\partial n_y} K(x,y)$ is bounded in $x, y \in B_r(x_0) \cap \overline{\Omega}$. From this follows the existence of $T(x_0)$ and that

$$\lim_{x \to x_0} T(x) = T(x_0) \qquad \text{for } x \in \Omega.$$

If x_0 is an exceptional point of ω then $T(x_0)$ is not defined. But if x converges along a straight line from the outside to x_0 then $1/\cos^2(\sigma/2)$ is bounded and therefore for each such line $\lim_{x \to x_0} T(x)$ will exist and its value will not depend on the choice of line along which x approximates x_0. Therefore we can define $T(x_0)$ to be this common limit.

It is not difficult to prove that $T(x)$ for $x \in \omega$ as a function on the surface ω is continuous at all non-exceptional points of ω.

As to the behavior of $\operatorname{grad} T$, it is a task of routine to prove that for $x_0 \in \omega$ a non-exceptional point $\lim_{x \to x_0} \operatorname{grad} T(x)$ exists and that if we define $\operatorname{grad} T(x_0)$ as this limit we have a vector field which is continuous on all subsets of $\overline{\Omega}$ not containing exceptional points, i.e., we have found for surfaces without exceptional points a solution to the simple Molodenskiy problem as it was defined in part 1). For surfaces with exceptional points we can only say that the solution is differentiable on such subsets of $\overline{\Omega}$ which do not contain exceptional points.

For points $x \in \Omega$ it is clear that $\operatorname{grad} T(x)$ may be calculated using (14) and changing the order of differentiation and integration. This is not so on ω: If $\mu(y)$ satisfies a Hölder-condition the expression

$$\int_\omega \mu(y) \operatorname{grad}_x \Big(\frac{\partial}{\partial n_y} K(x,y) \Big)\, d\omega_y$$

certainly exists for $x = x_0 \in \omega$ and is non-exceptional, but it is not continuous for $x \to x_0$ $(x \in \Omega)$. Certain jump relations exist similar to those satisfied by the gradients of potentials of a single layer. The derivative of T in the direction of the radius vector exists and has a well-defined limit at the boundary so that the boundary condition has a meaning even if μ is not Hölder-continuous. Summarizing we can say that the solution T of the simple Molodenskiy problem (if it exists) behaves outside the exceptional points very much as a potential of a single layer, whereas the solution of the Dirichlet problem behaves as a potential of a double layer, i.e., the solution of Molodenskiy's problem in smoother than the solution of Dirichlet problem. This important remark shall be made more explicit in the following. In these more precise statements we shall instead of continuous functions on ω use functions belonging to certain Lebesgue spaces $L^p(\omega)$ for $1 \leq p < \infty$. For the theory of these spaces, see for example [7] or [1].

Our main tool will be the following theorem (see [7], pp. 190–191 or [1], pp. 228–290):

> Let $(X, \mathcal{X}, m_X)$ and $(Y, \mathcal{Y}, m_Y)$ be σ-finite measure spaces, let p, r, μ_1, and μ_2 be real numbers such that
>
> $$1 \leq p \leq r, \qquad \frac{\mu_1}{p'} + \frac{\mu_2}{r} = 1, \qquad \frac{1}{p'} + \frac{1}{p} = 1$$
>
> and let ψ be an $\mathcal{X}, \mathcal{Y}$-measurable function. Suppose there are non-negative constants M_1 and M_2 such that
>
> $$\int_X \left|\psi(x,y)\right|^{\mu_1} dm_Y(y) \leq M_1 \qquad \text{for all } x \in X \tag{i}$$
>
> $$\int_Y \left|\psi(x,y)\right|^{\mu_2} dm_X(x) \leq M_2 \qquad \text{for all } y \in Y. \tag{ii}$$
>
> If the operator A is defined by
>
> $$A(f)(x) = \int_Y f(y)\psi(x,y)\, dm_Y(y), \qquad x \in X$$
>
> then
>
> $$A\colon L^p(Y) \to L^r(X),$$
>
> and
>
> $$\left\|A(f)\right\|_r \leq M_1^{1/p'} M_2^{1/r} \|f\|_p.$$

In our application of this theorem, Y is always the surface ω and X is a closed measurable subset of ω which does not contain exceptional points. If ω is without exceptional points then X may be ω itself but in applications it can be of interest to work with smaller subsets if one is interested in the local behavior of the solution of the simple Molodenskiy problem. The measure will always be the Lebesgue-measure (the area) on ω. The kernel $\psi(x,y)$ is our $\frac{\partial}{\partial n_y}K(x,y)$. Obviously from our definition of X follows that $1/\cos^2(\sigma/2)$ has an upper bound for $x \in X$ and $y \in \omega$. (i) and (ii) then become

$$\int_X \left|\frac{\partial}{\partial n_y}K(x,y)\right|^{\mu_1} d\omega_y \leq M_1 \qquad \text{for all } x \in X \tag{16}$$

and

$$\int_\omega \left|\frac{\partial}{\partial n_y}K(x,y)\right|^{\mu_2} d\omega_x \leq M_2 \qquad \text{for all } y \in \omega, \tag{17}$$

and it is clear that for $\mu_1 = \mu_2 = 2 - \epsilon$ the two integrals are in fact bounded if $\epsilon > 0$ but not if $\epsilon = 0$.

A simple manipulation of the relations between p, r, μ_1, and μ_2 cited in the theorem and $\mu_1 = \mu_2 = 2 - \epsilon$ gives

$$\frac{1}{p} - \frac{1}{r} = \frac{1-\epsilon}{2-\epsilon} \qquad \text{for} \quad 0 < \epsilon \leq 1. \tag{18}$$

We shall apply the result on the operator (confer (14))

$$A(\mu) = T\colon T(x) = \int_\omega \mu(y)\frac{\partial}{\partial n_y}K(x,y)\, d\omega_y.$$

If we put $\epsilon = 1$, the theorem gives:

Let the constants M_1 and M_2 be such that

$$\int_X \left|\frac{\partial}{\partial n_y}K(x,y)\right| d\omega_y \leq M_1 \qquad \text{for all } x \in X \tag{19}$$

and

$$\int_\omega \left|\frac{\partial}{\partial n_y}K(x,y)\right| d\omega_x \leq M_2 \qquad \text{for all } y \in \omega \tag{20}$$

then for any p such that $1 \leq p < \infty$

$$\left(\int_X |T(x)|^p\, d\omega\right)^{1/p} \leq M_1^{1-\frac{1}{p}} M_2^{\frac{1}{p}} \left(\int_\omega |\mu(x)|^p\, d\omega\right)^{1/p}.$$

A realistic estimation for the constants M_1 and M_2 is principally not difficult but would certainly demand patience.

The fact that T is more well-behaved than μ is expressed by the results of the theorem for $\epsilon > 1$ (but $\epsilon < 2$). As the reader may verify we can find in that way that $T \in L^p(X)$ for all $2 \leq p < \infty$ if μ is quadratic integrable, i.e., $\mu \in L^2(\omega)$ which certainly is true if $f \in L^2(\omega)$.

The following remark may not be superfluous.

Let $\delta\mu$ denote an increment to μ and let δT be the corresponding increment to T. As the operator $A\colon \mu \to T$ is linear, all our inequalities connecting norms of μ and of T are valid for the same norms of $\delta\mu$ and δT, respectively.

At last a few words should be mentioned about estimates for $\operatorname{grad} T$ at the boundary or more correctly for the limit of $\operatorname{grad} T$ as the point goes to the boundary from the outside.

In deducing such estimates we meet two difficulties: 1°) The integrals over ω must be complemented of terms caused by the jump relations mentioned (but not explicitly found) above. 2°) As the reader can easily find, the above cited theorem on integral operators between L^p-spaces cannot be applied here because the kernel now has become singular. The operator from μ to $\operatorname{grad} T$ (or perhaps more correctly to each component of $\operatorname{grad} T$) is now expressed by an integral operator of Calderon-Zygmund type (see e.g. [7] or [4]). The result is the following: For every $1 < p < \infty$ there is a constant C_p such that

$$\big\| |\operatorname{grad} T| \big\|_p \leq C_p \|\mu\|_p$$

where the norm at left is over X and that at right is over ω. This estimate is *not* true for $p = \infty$ if we do not impose a Hölder-condition on μ.

II. The Mushroom Problem

1. Reformulation of the Prague Method

In Section I a kernel was introduced:

$$K(x,y) = \frac{|x-y|}{|x|^2} + \frac{(x,y)}{|x|^3} \ln \frac{|x-y| + |x| - \frac{(x,y)}{|x|}}{2|x|} \tag{21}$$

which has the following properties:

1. $K(x,y)$ is defined for all pairs (x,y) of points in the three-dimensional Euclidian space $\mathbf{R}^3$ with the exception of those pairs for which there exist an α, $0 \leq \alpha \leq 1$, such that $x = \alpha y$.
2. For fixed y $K(x,y)$ is harmonic in x and regular at infinity.
3. For fixed x $K(x,y)$ is harmonic in y.
4. If $F(x)$ is harmonic in Ω, the exterior of a closed star-shaped surface ω, and is generated by a mass-distribution $\mu(y)$ in $\overline{\Omega}$, the interior of ω, and if μ satisfies the identities

$$\int\limits_{\overline{\Omega}} y_i \mu(y)\, dy = 0 \qquad \text{for } i = 1,\, 2,\, 3 \tag{22}$$

then $T(x)$ defined by

$$T(x) = \int\limits_{\overline{\Omega}} \mu(y) K(x,y)\, dy \tag{23}$$

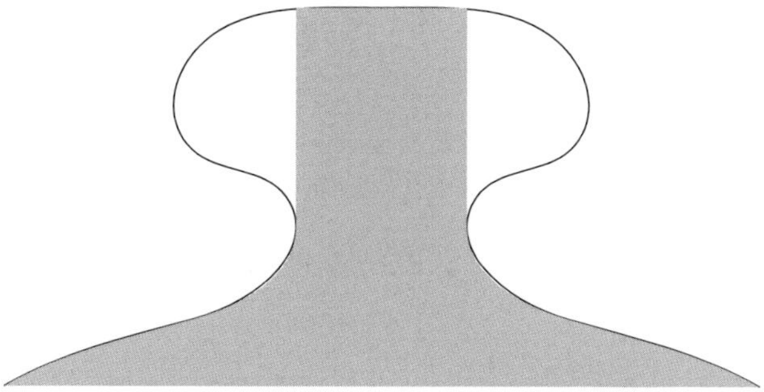

Fig. 8.1. Mushroom-shaped landscape

is the admissible solution of the simple Molodenskiy problem corresponding to the surface ω and the boundary values $f(x)$, where $f(x) = F(x)$ for $x \in \omega$.

5. If the identities (22) are satisfied, then the corresponding Molodenskiy problem has a solution.

At first sight it could make one wonder that the integral in (23) is independent of which of the different mass-distribution generating $F(x)$ has been chosen but this is a simple result of 3). (Use Green's formula!)

2. The Mushroom Problem and its Difficulty

We shall now modify the simple Molodenskiy problem to the effect that the surface ω is not necessarily star-shaped but may contain a 'mushroom-shaped landscape' as depicted in Figure 8.1.

If we try to use the Prague method, we can find T overall in Ω^- (by Ω^- we denote Ω with exception of the part of Ω situated under the cap) if $\mu(x)$ is not zero in the 'cap' outside the gray part, and we can in no way be sure that the potential we have found has a harmonic continuation in the part of Ω under the cap.

By following the reasoning in Section I it is not difficult now to prove the following:

The necessary and sufficient conditions for the existence of a solution of the mushroom problem are

1. that the given boundary values f are such that the identities (22) are satisfied, and
2. that the resulting T determined from F in Ω^- may be continuated harmonically until ω.

As the question as to the existence of a harmonic continuation of an approximately given potential is meaningless (confer some of my earlier letters to the study group), the question arises as to whether we can find an approximative solution to the problem.

3. Approximative Solution to the Mushroom Problem

I want to prove the following theorem:

1. If and only if the identities (22) are satisfied, it is possible to apply an arbitrarily small correction to the boundary values f so as to make the mushroom problem solvable.
2. The approximative solution of the mushroom problem depends *in* Ω^- continuously on f.

Let $H(\Omega^-)$ denote the normal vector space consisting of functions ϕ which are harmonic in Ω^-, regular at infinity, continuous in $\overline{\Omega}$, the closure of Ω^-, and have gradients which are continuous in $\overline{\Omega^-}$ with norm

$$\|\phi\| = \sup_{x\in\overline{\Omega^-}} \big(|\phi| + r|\operatorname{grad}\phi|\big).$$

A version of Runge's theorem says that the subset of $H(\Omega^-)$ which consists of potentials which are regular not only in Ω^- but in Ω is overall dense in $H(\Omega^-)$, i.e., in every neighborhood of a given $\phi_0 \in H(\Omega^-)$ there is a $\phi \in H(\Omega^-)$ which is regular in Ω.

Through the solution of Dirichlet's problem for Ω and the Prague method we have established a mapping

$$M\colon f \to T \in H(\Omega^-);$$

if we now can prove that the image by M of every neighborhood of any f contains a neighborhood in $H(\Omega^-)$, then part 1) of the theorem is proved.

In order to do this we must put a topology on the set of possible f. We choose the $C(\omega)$ topology, i.e., we take the space $C(\omega)$ of functions which are continuous on ω and define the norm

$$\|f\|_\omega = \sup_{y\in\omega}\big|f(y)\big|.$$

The mapping M^{-1} inverse to M is

$$M^{-1}\colon T \to f = r\frac{\partial T}{\partial r} + 2T + \frac{h_1(x)}{r^2},$$

where $h_1(x)/r^2$ is the first order term in the expression of T at infinity is continuous in the topologies we have defined. And now from the very definition of continuity follows what we wanted to prove.

But alas! We have met with a new difficulty. Instead the original T_0 corresponding to the given f_0, we see that we may choose another T_1 which is regular in Ω and which corresponds to an f_1 arbitrarily near to f_0 but we cannot be sure that h_1 for this T_2 is zero. But h_1 must be small so that also

$$f_2(y) = f_1(y) - \frac{h_1(y)}{|y|^2}$$

is near to f_0 and

$$T_2 = T_1 - \frac{h_1(x)}{|x|^2},$$

which corresponds to f_2 is an admissible solution and has the same domain of regularity as T_1 has and so the situation is saved.

Point 2) is simpler. We already know from Section I that T_0 depends continuously on f in Ω^-, as $|T_2 - T_0|$ can be made arbitrarily small T_2 must also depend continuously on f in Ω^-.

III. A Mathematical Formulation of Molodenskiy's Problem

1. Assumptions, *U*-Potentials, and *W*-Potentials

It is impossible to apply mathematics to the so-called 'real world.' We first have to make an idealized picture, a physical model, to which we may apply mathematics exactly and afterwards find out if this model is reasonable and if necessary find out how we can modify assumptions and results so as to arrive at a sufficiently good approximation to 'reality.' Therefore I shall start with a short description of that physical model which I have found most suitable for a mathematical treatment.

1) Stationarity. We assume that time does not enter into our problem, i.e., the Earth rotates as a solid with constant angular velocity ω around a fixed axis (the x_3-axis of our Cartesian coordinate system which rotates with the Earth) passing through the mass center of the Earth (the origin O of our Cartesian coordinate system).

2) Concentration of the masses. We assume the existence of a smooth surface ω (not to be confounded with the angular velocity) such that all gravitating masses are in $\widetilde{\Omega}+\omega$ ($\widetilde{\Omega}$ is the interior of ω and Ω is its exterior). We do not make any other assumptions with respect to the mass distribution than that its support is in $\widetilde{\Omega} + \omega$ and its mass center is O.

3) Knowledge. We assume known at every point x of ω a) the value $w(x)$ of the gravity potential W of the Earth (from spirit leveling and an intelligent guess of an additive constant) and b) the value $g(x)$ (observe that g is here a vector) of the gradient $\operatorname{grad} W$ of the same potential (from astronomy and gravity measurements). We shall suppose W and g to be sufficiently smooth functions.

4) Classicity. We shall work in three-dimensional space and assume that Euclidean geometry and Newtonian mechanics are valid.

5) A supplementary technical assumption. We shall assume that the gravity potential W satisfies the following condition (the *Marussi condition*): there exist two positive numbers h and a such that for every point in Ω with distance from ω less than h,

$$\left|\det \frac{\partial^2 W}{\partial x_i \partial x_j}\right| > a.$$

Remark. We assume no knowledge as to the Cartesian coordinates of the surface ω—in fact Molodenskiy's problem may be defined as the problem to find these coordinates from the assumptions 1–5.

From our assumption follows in the well-known way that the gravity potential can be written as

$$W = V + \tfrac{1}{2}\omega^2(x_1^2 + x_2^2),$$

where V is harmonic in Ω, regular at infinity having vanishing first order terms at infinity, and where W satisfies assumption 5.

Every function of this form will be called a W-potential (in Ω) and we shall not ascribe any value to it in $\widetilde{\Omega}$.

By a U-potential we shall mean a function similar to a W-potential, the only difference being that the domain of regularity for its harmonic part is so much larger that we can be sure that ω is contained in it and in a domain where $\left|\det \frac{\partial^2 W}{\partial x_i \partial x_j}\right| > a$ for some $a > 0$.

It is clear that every U-potential determines a W-potential which simply is the U-potential restricted to Ω. On the other hand from Runge's theorem follows a given W-potential that may be approximated arbitrarily well (in suitable topologies) by U-potentials.

2. Molodenskiy's Problem as a Problem of Coordinate Transformations

Curvilinear coordinates have the curious advantage that one can know, e.g., the coordinates of the points of a surface without knowing the shape and the position of the surface. We shall make use of this fact in the following.

By the *gravimetric coordinates* for a point in the domain of definition of a U-potential (or of a W-potential) we shall mean the components of grad U (or of grad W) at the point. It is clear from the Marussi condition that there exists a neighborhood around ω in which the Cartesian coordinates and the U-gravitational coordinates are in a one-to-one correspondence.

We may now formulate Molodenskiy's problem in the following way (here W means the actual gravity potential of the Earth):

Given the W-gravimetric coordinates of ω and the values w of W at ω find the Cartesian coordinates of ω.

As we have seen, if W is known in a neighborhood of ω the coordinate transformation is trivial, our problem is therefore to prove that it suffices to know the values of W on ω. On the other hand if w and the Cartesian coordinates of ω are known we can find W in Ω (Dirichlet's problem).

3. Mathematical Formulation of Molodenskiy's Problem

Molodenskiy's problem as defined above is highly non-linear, and in this respect it is similar to most conventional geodetic problems, a fact we often seem to forget.

The problem of adjustment of a geodetic network is not less non-linear. Traditionally this difficulty is evaded by starting the adjustment from rather well approximated coordinates and not iterating the process so that the non-linearity is not ascertained. The fact that today it would be much more rational and give more precise results to use a more dexterous adjustment procedure and iterate this process seems to prevail very slowly in practical geodesy.

In order to have a starting position we choose a U-potential U and a surface ω' which is situated in the domain of regularity of U and a one-to-one mapping of ω onto ω'. U shall be an approximation to the gravity potential W of the Earth—a possible choice is the normal potential. ω' shall be an approximation to ω; a possible choice is the surface consisting of points P for which the gravimetric coordinates with respect to U $(\operatorname{grad} U)_P$ are the same as the gravimetric coordinates with respect to W for the corresponding points Q of ω), $(\operatorname{grad} W)_Q$. In this case ω' is called the *gravimetric telluroid* (with respect to U). The traditionel choice would be the points P which have Marussi coordinates (with respect to U) identical to the Marussi coordinates (with respect to W) of Q. Then we would have the traditionel telluroid or, more correctly, the *Marussi telluroid* (with respect to U). In any case we call ω' the telluroid.

For every point $P \in \omega'$ we can compute U and $\operatorname{grad} U$, as we know the Cartesian coordinates x of P and for the corresponding point $Q \in \omega$ we know W and $\operatorname{grad} W$ according to our assumptions. That is, we can find

$$\delta W = W_Q - U_P \tag{24}$$

and

$$\delta g = (\operatorname{grad} W)_Q - (\operatorname{grad} U)_P. \tag{25}$$

where we call δW *the potential anomaly* and δg *the (vector) gravity anomaly.* If ω' is the gravimetric telluroid then $\delta g = 0$ and if it is the Marussi telluroid then $\delta W = 0$ and $\delta g = \Delta\left(\frac{\operatorname{grad} U}{|\operatorname{grad} U|}\right)_Q$, where Δg is the conventional (scalar) gravity anomaly.

Now let x be the (known) Cartesian coordinates of P and $x + \Delta x$ be the unknown Cartesian coordinates of Q and put $W = U + T$, where U is known and T unknown.

Then T must be harmonic outside ω, regular at infinity and have vanishing first order terms, and we can write (24) and (25) as

$$\begin{aligned} T(x + \Delta x) + U(x + \Delta x) - U(x) &= \delta W \\ (\operatorname{grad} T)_{x+\Delta x} + (\operatorname{grad} U)_{x+\Delta x} - (\operatorname{grad} U)_x &= \delta g. \end{aligned} \tag{26}$$

We linearize these equations, i.e., we rewrite them ignoring the following terms which are $o(\Delta x, T)$:

$$T(x + \Delta x) - T(x), \qquad (\operatorname{grad} T)_{x+\Delta x} - (\operatorname{grad} T)_x,$$

$$U(x+\Delta x) - \big(U(x) + \Delta x^{\mathrm{T}} (\operatorname{grad} U)_x\big), \qquad (\operatorname{grad} U)_{x+\Delta x} - \big((\operatorname{grad} U)_x - M\Delta x\big),$$

where $M = \left\{\frac{\partial^2 U}{\partial x_i \partial x_j}\right\}$ is the Marussi tensor evaluated at P, and we obtain

$$T + \Delta x^{\mathrm{T}} \operatorname{grad} U = \delta W \tag{27}$$
$$\operatorname{grad} T + M \Delta x = \delta g. \tag{28}$$

The system (27) and (28) consists of four scalar equations corresponding to each point of the telluroid. As three linear equations in the three components of Δx (28) are linearly independent according to our assumptions (the Marussi condition), we may solve them for Δx:

$$\Delta x = -M^{-1} \operatorname{grad} T + M^{-1} \delta g.$$

With this value of Δx, (27) becomes:

$$T - (\operatorname{grad} U)^{\mathrm{T}} M^{-1} \operatorname{grad} T = \delta W - M^{-1} \delta g.$$

Let us put

$$\tfrac{1}{2} f = \delta W - M^{-1} \delta g$$

and call it *the total anomaly* which is computable from known data, and we can write the last equation as

$$T - (\operatorname{grad} U)^{\mathrm{T}} M^{-1} \operatorname{grad} T = \tfrac{1}{2} f. \tag{29}$$

(29) gives one equation for every $P \in \omega'$, i.e., it is a boundary condition for T. As in the simple case of Molodenskiy's problem it is a first order linear partial differential operator.

We have now arrived at a boundary value problem which we will call *the linear Molodenskiy problem:*

> Find a function T which is harmonic outside ω', regular at infinity and which satisfies the boundary condition (29) at ω'.

A solution of the linear Molodenskiy problem for which the first order terms vanish is called *an admissible solution.*

4. The Linear Molodenskiy Problem

The equation (29) is a representation of the exact boundary conditions for the anomalous potential T but it is not very lucid from a geometrical point of view. Therefore we shall give it another form which is also exact but from which it is evident to see for example in which direction the derivative of T is taken. In order to do this we must use gravimetric coordinates instead of Cartesian coordinates.

The gravimetric U-coordinates are defined to be the components of the vector $u = \operatorname{grad} U$, and according to our assumptions these coordinates are in one-to-one correspondence with the Cartesian coordinates in some neighborhood of ω and so also of ω'. The Marussi tensor M and its inverse may be written

$$M = \left\{ \frac{\partial u_i}{\partial x_j} \right\}$$

and

$$M^{-1} = \left\{ \frac{\partial x_i}{\partial u_j} \right\}$$

so that (29) becomes

$$T - \sum u_i \frac{\partial T}{\partial u_i} = \frac{1}{2} f. \tag{30}$$

An even better form is obtained by the introduction of *quasi-spherical coordinates* (ρ, ϕ, λ) defined by

$$\begin{aligned} \frac{1}{\rho^2} \cos\phi \cos\lambda &= -u_1 = -\frac{\partial U}{\partial x_1} \\ \frac{1}{\rho^2} \cos\phi \sin\lambda &= -u_2 = -\frac{\partial U}{\partial x_2} \\ \frac{1}{\rho^2} \sin\phi &= -u_3 = -\frac{\partial U}{\partial x_3} \end{aligned} \tag{31}$$

where $\rho \geq 0$.

In quasi-spherical coordinates the boundary condition reads

$$\rho \frac{\partial T}{\partial \rho} + 2T = f. \tag{32}$$

The derivative $\partial/\partial\rho$ is taken for constant ϕ and λ, i.e., along an isozenithal. (32) has exactly the same form as the boundary condition has in the simple Molodenskiy problem; it is also evident from (31) that if $U = k/r$ then $\rho = r/k^{1/2}$.

If the Earth were non-rotating so that the potential $\omega^2(x_1^2 + x_2^2)/2$ of the centrifugal force vanished, then the correspondence between the Cartesian and the gravimetric coordinates would have been one-to-one overall outside ω' and so the quasi-spherical coordinates would also have been reasonable there. But the centrifugal potential spoils this bijectivity. If we multiply $\omega^2(x_1^2+x_2^2)/2$ by a smooth function of $|x|$ which is equal to one near the Earth and converges sufficiently fast to zero for $|x| \to \infty$ and uses this in the definition of U for the definition of the gravimetric and the quasi-spherical coordinates, then we have arrived at an exact definition of what by some geodesists is called the spherical mapping which underlies the theory of the so-called *spherical approximation to Molodenskiy's problem.*

The spherical mapping maps every point $P \in \Omega + \omega$ to the point Q so that the spherical coordinates of Q are identical to the quasi-spherical coordinates of P.

Naturally the expression for Laplace's operator in quasi-spherical coordinates (which is rather easily found) is different from its expression in spherical coordinates. The idea behind the spherical approximation is to ignore this difference and then solve a problem formally identical to the simple Molodenskiy problem if there is only one point of ω' corresponding to every set (ϕ, λ) of geographical coordinates. In this way one finds the values of T at ω' and then finds T in outer space (the true outer space). It is perhaps plausible that this T would be an approximation to the true T, and it would be promising to try to find sufficient conditions for an exact solution of the linear Molodenskiy problem using these ideas by successive approximation. I shall nevertheless use another method in my fourth letter.

5. Some Results from the Theory of Boundary Value Problems

I shall give here a concise account of the results from the theory of boundary value problems for elliptic partial differential equations relevant to the linear Molodenskiy problem. These results are compiled from [3] which is certainly not the simplest book for a geodesist to obtain for primary information about the modern theory of boundary value problems, but it seems to me to be almost indispensible for everyone who wants to penetrate seriously into Molodenskiy's problem.

Let ω be a closed surface in $\mathbf{R}^3$ diffeomorphic to a sphere, i.e., a sphere may be mapped by arbitrarily often differentiable functions onto ω and the inverse mapping is also arbitrarily often differentiable, and let Ω be the open part of $\mathbf{R}^3$ outside ω. By V we denote the set of functions such that for every $\phi \in V$ there exists an open subset Ω_ϕ of $\mathbf{R}^3$, such that $\Omega \subset \Omega_\phi$ and $\omega \subset \Omega_\phi$, where ϕ is harmonic and every $\phi \in V$ is regular at infinity.

For every $\phi \in V$ the function $\phi|_\omega$ (the function ϕ restricted to ω) and the vector field $(\operatorname{grad} \phi)|_\omega$ are well defined so that

$$\|\phi\|_1 = \left(\int_\omega (\phi^2 + |\operatorname{grad} \phi|^2)\, d\omega \right)^{1/2} \tag{33}$$

exists for every $\phi \in V$ and it is a norm. With this norm V becomes a pre-Hilbert space. The Hilbert space is H_1, and the completion of this pre-Hilbert space consists of functions ϕ which all are harmonic in Ω and regular at infinity and for which $\phi|_\omega$ has a meaning as a *trace*, see [3], p. 39, and (33) has a meaning for them also.

Let there be given on ω 1) an arbitrarily often differentiable scalar function a and 2) an arbitrarily often differentiable (three-dimensional) vector field l so that $|l| \neq 0$ everywhere on ω.

It is clear that for every $\phi \in H_1$

$$l^{\mathrm{T}} \operatorname{grad} \phi + a\phi \tag{34}$$

has a meaning (it would have been more correct to write $l^{\mathrm{T}}(\operatorname{grad}\phi)_\omega + a\phi|_\omega)$ as a function defined on ω and that

$$\|l^{\mathrm{T}} \operatorname{grad}\phi + a\phi\|_0 = \left(\int_\omega \left(l^{\mathrm{T}} \operatorname{grad}\phi + a\phi\right)^2 d\omega\right)^{1/2}$$

exists for all $\phi \in H_1$. Let the Hilbert space of all functions ψ on ω for which

$$\|\psi\|_0 = \left(\int_\omega \psi^2 \, d\omega\right)^{1/2} < \infty$$

be called H_0. Then we may express this fact as follows:

The operator

$$A\colon H_1 \to h_0\colon \phi \to l^{\mathrm{T}} \operatorname{grad}\phi + a\phi$$

is a linear continuous operator from H_1 to H_0.

The problem $f \in H_0$ given to find a $\phi \in H_1$ such that

$$A\phi = f$$

is called *the oblique derivative problem* (for the spaces H_1 and H_0).

If the vector field l on ω is such that there are no points on ω where l is tangent to ω, then the problem is said to be a *regular oblique derivative problem.* If the problem is not regular then there are difficulties of 'mushroom type' and they are not yet completely solved. Therefore we shall concentrate here on regular oblique derivative problems.

It is clear that if the boundary differential operator (34) corresponds to a regular oblique derivative problem, then by multiplying by a function with constant sign it may be expressible in the form

$$\frac{\partial\phi}{\partial n} + p^{\mathrm{T}} \operatorname{grad}_2 \phi + b\phi \tag{35}$$

where $\partial/\partial n$ is the derivative along the outward unit normal to ω and grad_2 is the two-dimensional gradient on ω, p is a two-dimensional tangential vector field on ω and b is a scalar function on ω. p and b are again arbitrarily often differentiable. If the boundary operator has been put into this form, then the boundary value problem is said to be *normalized.* We can then write the boundary condition as

$$A_n\phi = g,$$

where

$$A_n = \frac{\partial}{\partial n} + p^{\mathrm{T}} \operatorname{grad}_2 + b.$$

We shall now define the boundary problem adjoint to the given problem. Let ψ be an arbitrary function of H_1. Then from Green's formula and from

$$\int_\omega \mathrm{div}_2(\psi\phi p)\, d\omega = 0$$

and

$$\mathrm{div}_2(\psi\phi p) = \psi\phi\,\mathrm{div}_2\, p + p^{\mathrm{T}}(\psi\,\mathrm{grad}_2\,\phi + \phi\,\mathrm{grad}_2\,\psi),$$

it follows that

$$\int_\omega \psi\Big(\frac{\partial\phi}{\partial n} + p^{\mathrm{T}}\,\mathrm{grad}_2\,\phi + b\phi\Big)\, d\omega = \int_\omega \phi\Big(\frac{\partial\psi}{\partial n} - p^{\mathrm{T}}\,\mathrm{grad}_2\,\psi + (b - \mathrm{div}_2\, p)\psi\Big)\, d\omega. \tag{36}$$

If we define

$$A_n^*\colon H_1 \to H_0\colon \psi \to \frac{\partial\psi}{\partial n} - p^{\mathrm{T}}\,\mathrm{grad}_2\,\psi + (b - \mathrm{div}_2\, p)\psi,$$

then we can write (36) as

$$\int_\omega \psi A_n\phi\, d\omega = \int_\omega \phi A_n^*\psi\, d\omega, \qquad \text{for all } \phi, \psi \in H_1, \tag{37}$$

and we call A_n^* the boundary differential operator adjoint to A_n.

It is also clear that A_n^* defines a regular oblique derivative problem which will be called the boundary value problem adjoint to the given one and that A_n is the adjoint to A_n^*.

The subsets of H_1 consisting of functions $\phi \in H_1$ for which

$$A_n\phi = 0$$

will be called N and that for which

$$A_n^*\phi = 0$$

will be called N^*.

Clearly N and N^* are linear spaces which may be zero dimensional, i.e., consisting of $\phi = 0$ only.

By well-known reasoning it follows that a necessary condition for the linear problem

$$A_n\phi = g, \qquad \phi \in H_1, \quad g \in H_0, \quad g \text{ given}$$

to have a solution is

$$\int_\omega \psi g\, d\omega = 0, \qquad \text{for all } \psi \in N^*,$$

and analogously a necessary condition for the problem

$$A_n^*\phi = h, \qquad \phi \in H_1, \quad h \in H_0, \quad h \text{ given}$$

to have a solution is that

$$\int_\omega \psi h \, d\omega = 0, \qquad \text{for all } \psi \in N.$$

Now the first result is that these conditions are also sufficient; in fact:

If and only if g is orthogonal (with respect to the metric of H_0) to every solution $\psi \in H$, to the homogeneous equation

$$A_n^* \psi = 0$$

then the inhomogeneous equation

$$A_n \phi = g, \qquad \phi \in H_1, \quad g \in H_0, \quad g \text{ given} \tag{38}$$

has a unique solution ϕ_1 for which

$$\|\phi_1\|_1 = \min_{\phi:\, A_n\phi = g, \phi \in H_1} \|\phi\|_1$$

and every solution ϕ of (38) may be written as

$$\phi = \phi_1 + \phi_0, \qquad \phi_0 \in H_1, \quad A_n \phi_0 = 0.$$

There exists a constant C such that

$$\|\phi_1\| \leq C \|g\|_0. \tag{39}$$

The analogy between the equations of the type of (38) and linear algebraic equations

$$Bx = a,$$

where B is an $n \times m$ matrix:

$$B\colon \mathbf{R}^m \to \mathbf{R}^N : x \to Ax, \quad x \in \mathbf{R}^m, \quad a \in \mathbf{R}^n$$

is evident.

Here we can also define N and N^* such that $N \subset \mathbf{R}^m$ and $N^* \subset \mathbf{R}^n$, and results on solvability and unicity are similar in the two cases.

For the case of matrices we may define the index of B: $\operatorname{ind} B$ as

$$\operatorname{ind} B = \dim N - \dim N^*$$

where $\dim N$ ($\dim N^*$) is the number of dimensions of the linear vector space N (N^*) and we see that even if we can not in general say anything about $\dim N$ and $\dim N^*$ separately we always have

$$\operatorname{ind} B - n - m.$$

For the regular oblique derivative problem, the second result is now

$$N \text{ and } N^* \text{ are finite dimensional} \quad \text{and} \quad \operatorname{ind} A = \operatorname{ind} A^* = 0.$$

We may loosely express this last result by saying that A and A^* are analogue to square matrices.

Faced with these results from the theory of elliptic partial differential equations, a reasonable question from the reader would be if it would be possible to use other spaces than H_0 and H_1 for defining the regular oblique derivative problem and if the conditions for solvability and unicity could possibly be modified in that way.

In fact many other pairs of spaces H_0, H_1 are possible and give reasonable results. I have chosen H_0 and H_1 for simplicity and because I think that they are the most relevant spaces for the treatment of Molodenskiy's problem. It is a striking result of the theory that for every reasonable choice of the pair of spaces, the spaces N and N^* are identical as well, i.e., our conditions are definitive.

6. The Regular Linear Molodenskiy Problem in Light of the General Theory

It is easily seen that the simple Molodenskiy problem without exceptional points is a regular oblique derivative problem and the results we found in Section I are also in good agreement with the general results referred above. But as regards the simple Molodenskiy problem we were so fortunate to be able to prove that the dimensionality of N (and of N^*) is 3.

We also see that the linear Molodenskiy problem is an oblique derivative problem and that it is regular if and only if the isozenithals are never tangent to the surface ω'; in this case we shall call it the regular (linear) Molodenskiy problem.

In order to find some result concerning $\dim N = \dim N^*$ for the regular Molodenskiy problem, we shall first investigate the case of a non-rotating planet, i.e., the Molodenskiy problem where U is harmonic outside ω'. If U is harmonic then also $C^{\mathrm{T}} \operatorname{grad} U$ is harmonic for C any constant three-dimensional vector and so if $(T_0, \Delta x_0)$ is a solution of the linear Molodenskiy problem then $(T_0 + C^{\mathrm{T}} \operatorname{grad} U, \Delta x_0 - C)$ is also a solution. Confer (27) and (28); that is in this case $\dim N \geq 3$.

It is natural that it must be so: From our assumptions and observations we can not find the absolute position of the Earth in space but only its form and direction, and position relative to the mass center.

That a parallel translation of a solution of the non-linearizied Molodenskiy problem is again a solution is evident, which an examination of (26) can show.

We have $U = (M/|x|) + O(|x|^{-3})$, $M > 0$ and therefore

$$\operatorname{grad} U = -\frac{Mx}{|x|^3} + O(|x|^{-4}),$$

so that if T_0 is any solution of the linear Molodenskiy problem then we can find C uniquely so that

$$T_0 + C^{\mathrm{T}} \operatorname{grad} U$$

is an admissible solution. Therefore we may conclude:

In the regular Molodenskiy problem for a non-rotating planet $\dim N \geq 3$. The total anomaly has to satisfy $\dim N$ independent linear conditions for the problem to have a solution. If these conditions are satisfied, the admissible solution depends on $\dim N - 3$ arbitrary constants.

It is at first a little astonishing that the circumstances are different for the case of a rotating planet; it seems as though here a parallel translated solution should again be a solution so that $\dim N$ ought to be ≥ 3. The reason this is not the case is a purely formal one and has to do with the fact that our definition of regularity at infinity is not physically relevant here. It is true that the difference between two potentials of the centrifugal force corresponding to different axes but to the same angular velocity is harmonic in the whole space but not regular at infinity as it is not even bounded, and the same is valid for $\frac{\partial U}{\partial x_1}$ and $\frac{\partial U}{\partial x_2}$; only $\frac{\partial U}{\partial x_3}$ is regular at infinity. Therefore we find $\dim N \geq 1$. In this case we may conclude:

For the regular Molodenskiy problem (for a rotating planet) $\dim N \geq 1$. The total anomaly has to satisfy $\dim N + 2$ independent linear conditions for the problem to have *an admissible* solution. If these conditions are satisfied the admissible solution depends on $\dim N - 1$ arbitrary constants.

If we define the *degeneracy* δ, an integer, as

$$\delta = \begin{cases} \dim N - 3 & \text{for a non-rotating planet} \\ \dim N - 1 & \text{for a rotating planet,} \end{cases}$$

then we can say generally—for the rotating as well as the non- rotating planet:

For the regular Molodenskiy problem we have $\delta \geq 0$. The total anomaly has to satisfy $3 + \delta$ independent linear conditions for the problem to have an admissible solution. If these conditions are satisfied, the general admissible solution depends on δ arbitrary constants.

For the simple Molodenskiy problem $\delta = 0$, and it would be very satisfactory if one could prove that $\delta = 0$ in general. Unfortunately I have not been able to prove this. I believe that in exceptional cases $\delta > 0$ and in the fourth letter I shall give a sufficient condition for $\delta = 0$.

As we have seen, the conditions for the solvability depend on the solutions of the homogeneous adjoint problem. In practice it is very difficult to find these solutions, and often even their number.

This was the question of solvability and unicity. What does the general result say about stability?

Formally the stability is expressed by (39), where C depends on U and ω' in an unknown way. It is plausible however that C is 'large' if the problem is in the vicinity of a problem where $\delta > 0$. Therefore it seems to me that it is important to find out under which conditions $\delta > 0$ and to be able to say something about the nature of the corresponding eigensolutions.

7. Concluding Remarks

We have seen that results from the theory of partial differential equations can give us some information about the regular Molodenskiy problem but that there are still questions left. I shall try in my last Molodenskiy letter to partly solve these questions. I shall conclude Section III with a few remarks on the nonlinear Molodenskiy problem.

From a mathematical point of view it would be interesting to prove that the non-linear Molodenskiy problem could be solved by successive approximations by linear steps by using the W-potential found in one step to construct a U-potential for the next step. The possibility of having $\delta > 0$ makes however such a proof very difficult.

From a geodetic point of view however I think that a good theory for the linear Molodenskiy problem must be sufficient. Starting from the best available U-potential, a few linear approximations should suffice to make the numerical precision of the solution an order of magnitude better than the precision of the observations and this should be satisfactory. But in order to achieve this we must use the general regular Molodenskiy problem; the simple one is not enough.

IV. Application of the Prague Method on the Regular Molodenskiy Problem

1. Reformulation of the Prague Method Once Again

In the first letter we saw how the Prague method could be used to find the admissible solution of the simple Molodenskiy problem if a solution exists. Let

$$A_0\colon H_1 \to H_0\colon T \mapsto \sum_{i=1}^{3} x_i \frac{\partial T}{\partial x_i} + 2T = f, \tag{40}$$

then A_0 has no inverse, but if the problem has a solution the admissible solution is given by

$$T = Pf,$$

where

$$P\colon H_0 \to H_1\colon f \mapsto T = \int\limits_{\widetilde{\Omega}} \rho(y) K(x,y)\, dy,$$

where $\rho(y)$ is any mass distribution in $\widetilde{\Omega}$ generating the potential F such that $F|_\omega = f$.

As A_0 has no inverse, P can not be its inverse but we find

$$PA_0 = I, \qquad A_0 P f = -\frac{h_1(x)}{|x|^3},$$

where $h_1(x)/|x|^3$ is the first order term in the expansion at infinity for F.

Therefore if we define the operator A'_0 as

$$A'_0\colon H_1 \to H_0\colon T \mapsto \sum_1^2 x_i \frac{\partial T}{\partial x_i} + 2T + \frac{h_1(x)}{|x|^3}, \tag{41}$$

where $h_1(x)/|x|^3$ is the first order term of T, then

$$A'_0 P = P A'_0 = I,$$

i.e., the inverse of A'_0 is P.

The result of all this is that we may formulate the Prague method in this way: Instead of solving the original boundary value problem

$$A_0 T = f, \tag{42}$$

we solve the modified problem

$$A'_0 T = f, \tag{43}$$

which always has a unique solution. If and only if this solution is admissible, then the original problem (42) has a solution and the solution of (43) is the admissible solution of (42).

2. The Modified Regular Molodenskiy Problem and its Solution

In the regular Molodenskiy problem we work with an operator A such that

$$A\colon H_1 \to H_0\colon T \mapsto l^{\mathrm{T}} \operatorname{grad} T + 2T. \tag{44}$$

Let us modify this operator in the same way as we modified the operator A in the foregoing paragraph:

$$A'\colon H_1 \to H_0\colon T \mapsto l^{\mathrm{T}} \operatorname{grad} T + 2T + \frac{h_1(x)}{|x|^3}, \tag{45}$$

where $h_1(x)/|x|^3$ again is the first order term of T and let us together with the original regular Molodenskiy problem, the solution of

$$AT = f \tag{46}$$

regard the modified regular Molodenskiy problem, the solution of

$$A'T = f. \tag{47}$$

The index of A' is the same as the index of A viz. zero because the index of an operator is left unchanged by the addition of an operator with range of finite dimension. Therefore it is at least plausible that if our regular Molodenskiy problem is not too far from a simple Molodenskiy problem then A' is invertible, i.e., (47) has a unique solution. It is easy to prove that if (47) has a unique solution then if and only if this solution is admissible then the original problem (46) has a solution, and the solution of (47) is the admissible solution of (46). (This follows simply from the fact that for admissible T, the operators A and A' are identical.)

For the following it is necessary to assume that the surface ω is not only such that the Molodenskiy problem is regular but also that ω is star-shaped (with respect to the mass center) and without exceptional points.

We can now in a unique way write

$$A = A_0 + A_1$$

where A_0 is the differential operator defined in (40) for the corresponding simple Molodenskiy problem and A_1 is a 'small' differential operator of order 1:

$$A_1 T = l_1^{\mathrm{T}} \operatorname{grad} T.$$

Thus, we can write (46) as

$$A_0 T + A_1 T = f,$$

and (47) as

$$A_0' T + A_1 T = f. \tag{48}$$

It suggests itself to solve (48) by successive approximations, i.e., to find a sequence of potentials $T_0 = 0, T_1, T_2, T_3, \ldots$ where T_i is found from T_{i-1} by the i'th step $i = 1, 2, \ldots$ by solving the equation

$$A_0' T_i = f - A_1 T_{i-1}$$

or

$$T_i = Pf - PA_1 T_{i-1}.$$

This is exactly what we shall do, but it is perhaps more convenient to express it in another way.

Put

$$g = A_0' T,$$

then g is a function on ω and if $T \in H_1$ then $g \in H_0$.

From g we can find T by the Prague method:

$$T = Pg$$

so that we may write (48) as

$$g + A_1 P g = f \tag{49}$$

$$T = Pg. \tag{50}$$

We saw in the first letter that P is a bounded operator from H_0 to H_1, so that $A_1 P$ is bounded too from H_0 to H_0. We also saw that we can at least principally find the explicite bound for P—in practice it may be very difficult—these bounds found it is much easier through estimates for the coefficients l_1 to find the bound for the operator $A_1 P$.

Now if

$$\|A_1 P\| \equiv \sup_{\|g\|_0 = 1} \|A_1 P g\|_0 < 1 \tag{51}$$

then we can solve the equation (49) by a Neumann series:

$$g = (I + A_1P)^{-1}f = \sum_0^{\infty}(-1)^i(A_1P)^i f,$$

so that in this case (49) always has a unique solution g and the modified regular Molodenskiy problem (48) has the unique solution $T = Pg$ and as we saw above the original Molodenskiy problem then has a solution if and only if $T = Pg$ is admissible.

To the members of SSG 4:31 March 22, 1973

Dear Colleagues,

In terminating this sequence of letters on Molodenskiy's problem, which should be a rather subjective account of the situation of Molodenskiy's problem today, I shall try to clear up the connection of some of the ideas in the existing literature.

The astronomical variant of Molodenskiy's problem (Section III) goes back to [5], pp. 90–91, but by the introduction of approximations his equations were less transparent. (The reader should try to prove the invariance of Molodenskiy's equations with respect to a parallel translation for a nonrotating planet.) At the time when Molodenskiy's historical papers were written, boundary value problems for partial differential equations were solved using integral equations, and it is natural that very much transparency is lost by the transformation of a problem to integral equations. The boundary condition, see [5], p. 80, (V.4.5)

$$-(g - \gamma) = \left(\frac{\partial T}{\partial \nu} - \frac{T}{\gamma}\frac{\partial \gamma}{\partial \nu}\right)_h$$

is very similar to the correct one for the astronomical variant. It is easy to prove that it is made correct if

1. $\frac{\partial}{\partial \nu}$ means derivation along the isozenithal (and not along the vertical),
2. the left side is multiplied by $\cos\alpha$ where α is the angle between vertical and isozenithal.

This translation invariance of Molodenskiy's problem is—as it appears from my letter—essential for the understanding of it and for the results, as is also the assumption that the axis of rotation passes through the mass center. It could perhaps be interesting to find out how the theory should be modified if this assumption were dropped. But I must say that until somebody has convinced me to the contrary I believe that the distance of the mass center from the axis is small compared with the precision with which it might be determined by a thus modified Molodenskiy method—at least as long as time does not enter the problem. I should perhaps also say that I have not mentioned the two well-known admissibility conditions for the second order term of T because they do not play any essential role in the conditions for solvability and unicity.

The fact which makes the simple Molodenskiy problem simple is that the differential operator in (2) when applied on a harmonic function gives a harmonic function. This has been known and used by almost all authors who have written about Molodenskiy's problem. Moritz has used the fact to prove that the simple Molodenskiy problem is stable, see [6], p. 32. The idea to use it to derive a stronger result (i.e., that also the first derivatives of T at the surface depend continuously on f) and from that deduce a sufficient condition for the stability of the general Molodenskiy problem has been proposed in [2] but as I did not yet know the idea described in the fourth letter of modifying Molodenskiy's problem so that it becomes uniquely solvable, I did not succeed in carrying out my program and this disappointment has forced me to return to Molodenskiy's problem again and again during the last five years. I hope for myself and for my environments that the writing of these four letters will make it possible for me to forget everything about unsolved problems in connection with Molodenskiy's problem.

The definition of Molodenskiy's problem and the form for the boundary conditions also stem from [2]. The replacement of the classical gravity anomaly by more general forms of anomaly in the third letter has been done mainly in order to make possible the assumption of a C^∞-smooth boundary surface even if the input data is only in L_2.

At last I wish to thank the members of the study group for their patience and for the reaction I have received and the reactions to come in the future.

References

[1] L. W. Kantorowitsch and G. P. Akilow. *Funktionalanalysis in normierten Räumen.* Akademie-Verlag, Berlin, 1964.

[2] T. Krarup. *Molodenskij's Problem.* XIV General Assembly. IUGG, Lucerne, 1967.

[3] Jacques-Louis Lions and Enrico Magenes. *Non-Homogeneous Boundary Value Problems and Applications*, volume 181 of *Grundlehren der mathematischen Wissenschaften.* Springer, Berlin, 1972.

[4] S. G. Mikhlin. *Multidimensional Singular Integrals and Integral Equations.* Pergamon Press, Oxford, 1965.

[5] M. S. Molodenskii, V. F. Eremeev, and M. I. Yurkina. *Methods for Study of the External Gravitational Field and Figure of the Earth.* Translated from Russian by the Israel Program for Scientific Translations, Jerusalem, 1962.

[6] Helmut Moritz. *Series Solutions of Molodensky's Problem.* Deutsche Geodätische Kommission. C. H. Beck, München, 1971.

[7] G. O. Okikiolu. *Aspects of the Theory of Bounded Integral Operators in L^p-Spaces.* Academic Press, London and New York, 1971.

9

On the Spectrum of Geodetic Networks

Abstract

The spectrum of a geodetic network is defined as the eigenvalues of the matrix $A^{\mathrm{T}}A$ of normal equations corresponding to the observation equations $Ax = b$ in the coordinates.

We find the meaning of and some general properties for the eigenvectors of the symmetric matrices $A^{\mathrm{T}}A$ and AA^{T} and some results concerning the distribution of the eigenvalues when the network is relatively large.

The ultimate goal for these investigations is to get a deeper insight into the relations between netform and netquality. Unfortunately the results are rather fragmentary because we have only recently arrived at what we consider the core of the problem after years of research.

Introduction

So far there has been moderate interest in the spectral theory of geodetic networks. By the spectrum in a narrow sense we mean the eigenvalues. The interest has hitherto focused on the two functions of the spectrum: the trace and the condition number. But in questions concerning optimum netforms, optimum utilization of additional observations, and statistical and information-theoretical investigations the spectrum plays the main role. Besides this, the spectrum is invariant under similarity transformations, so in some sense this reflects the most useful information of geodetic networks.

In some respects the value of the smallest eigenvalue of the normal equations gives a good measure of the quality of the networks, and then the concentration of eigenvalues in the lower end of the spectrum says something about the number of complementary observations necessary to increase the netquality appreciably. However, as will be seen, the asymptotic properties of the spectrum will be of great interest in the above-mentioned problems.

The Canonical Form of the Adjustment Problem

Let the linearized observation equations be

$$Ax = b \tag{1}$$

where x is the n-dimensional vector of the increments of the coordinates and b the m-dimensional vector of observations which are supposed to be of equal weights (weight-normalized) and uncorrelated. In other words, the covariance matrix is equal to the unit matrix I.

It is always possible to find an orthogonal matrix V (in the coordinate space) and another U (in the space of measurements) such that with

$$y = Vx, \qquad c = Ub \tag{2}$$

the equation (1) can be written as

$$By = c, \tag{3}$$

where

$$B = UAV^{\mathrm{T}} \tag{4}$$

and such that B has the form

$$B = \left[\begin{array}{c} D \\ \hline \Theta \end{array}\right] \begin{array}{l} n \\ m-n \end{array} \tag{5}$$

where D is the diagonal matrix

$$D = \begin{bmatrix} \lambda_1 & & & \\ & \lambda_2 & & \\ & & \ddots & \\ & & & \lambda_n \end{bmatrix}, \tag{6}$$

and Θ is the $(m-n) \times n$ zero matrix. Further, all the λ's are ordered as

$$|\lambda_1| \geq |\lambda_2| \geq |\lambda_3| \geq \cdots \geq |\lambda_n| \geq 0.$$

Obviously we have

$$B^{\mathrm{T}}B = VA^{\mathrm{T}}U^{\mathrm{T}}UAV^{\mathrm{T}} = VA^{\mathrm{T}}AV^{\mathrm{T}} = D^2 \tag{7}$$

and

$$BB^{\mathrm{T}} = UAV^{\mathrm{T}}VA^{\mathrm{T}}U^{\mathrm{T}} = UAA^{\mathrm{T}}U^{\mathrm{T}} = \left[\begin{array}{c|c} D^2 & \Theta \\ \hline \Theta & \Theta \end{array}\right]. \tag{8}$$

Confer, e.g. [6], Chapter 3. Recall that an orthogonal matrix Q is characterized by the property $Q^{\mathrm{T}}Q = QQ^{\mathrm{T}} = I$. If we write

$$V^{\mathrm{T}} = \begin{bmatrix} \phi_1 & \phi_2 & \ldots & \phi_n \end{bmatrix}, \qquad U^{\mathrm{T}} = \begin{bmatrix} \psi_1 & \psi_2 & \cdots & \psi_n & \rho_1 & \rho_2 & \cdots & \rho_{m-n} \end{bmatrix}, \tag{9}$$

then we see that $\{\phi_i\}$ is an orthonormalized set of n-dimensional vectors, and $\{\psi_i\} \cup \{\rho_i\}$ is another of m-dimensional vectors, and $(2)^{\mathrm{T}}$ then shows that y may be considered the coefficients in the expansion of the vector x in the set $\{\phi_i\}$, and correspondingly c may be considered the coefficients in the expansion of b in $\{\psi_i\}$ and $\{\rho_i\}$. The sets $\{\phi_i\}$, $\{\psi_i\}$, and $\{\rho_i\}$ will be called the first, the second, and the third set of canonical vectors, although they are not uniquely determined.

The reader interested in a statistical description of the canonical form of the adjustment problem is referred to [10], Chapter I. Here the space spanned by the $\{\rho_i\}$'s is named the error space and that one spanned by the $\{\phi_i\}$'s is called the estimation space.

From (4) and (5) the following relations connecting the first two sets of canonical vectors result in the following

$$A\phi_i = \lambda_i \psi_i, \qquad i = 1, 2, \ldots, n \tag{10}$$

and from $(4)^{\mathrm{T}}$ and (5) follows:

$$A^{\mathrm{T}}\psi_i = \lambda_i \phi_i, \qquad i = 1, 2, \ldots, n \tag{11}$$

$$A^{\mathrm{T}}\rho_i = 0, \qquad i = 1, 2, \ldots, m - n. \tag{12}$$

As the orthogonal transformation by U of the observation vector b leaves the 'observations' c in (3) weight-normalized, we can solve the adjustment problem via the normal equations

$$B^{\mathrm{T}}By = B^{\mathrm{T}}c \tag{13}$$

or

$$\lambda_i^2 y_i = \lambda_i c_i, \qquad i = 1, 2, \ldots, n$$

or finally

$$y_i = \frac{c_i}{\lambda_i}, \qquad i = 1, 2, \ldots, p \tag{14}$$

where p is the number of λ_i's different from zero and the index i of the Latin letters denote the ith component of these vectors.

Already from these results we may draw some conclusions:

1. Regarding the first set of canonical vectors: The components of the coordinate vector x in the directions defined by the ϕ_i are better determined from the observations the larger the corresponding value for λ_i is; the components of ϕ_i with $\lambda_i = 0$ are completely undetermined. For this result to be interesting the coordinate increments must be measured on (approximately) the same scale (e.g., not as $\Delta\phi$ and $\Delta\lambda$) as we shall in fact suppose they are. Normally one is interested in all components of x being equally well determined; therefore we have to ask what can be done in order to have the spectrum strongly concentrated.

2. Regarding the second set, we can say that this can be used to find out which of the measurements should have been carried out so as to have greater weight by expanding the single measurements (that is the unit vectors in the observation space) in $\{\psi_i\}$ and $\{\rho_i\}$ and selecting those for which the coefficients corresponding to small $\lambda_i \neq 0$ are dominating. These coefficients may be found by inspecting the matrix U. This follows from the property $U^{\mathrm{T}}U = I$ which may be looked upon as giving the expansion of the unit vectors in $\{\psi_i\}$ and $\{\rho_i\}$.
3. Regarding the third set, we see that the components of the observations in the directions determined by $\{\rho_i\}$ bring no information about the coordinates; the same holds also true for the ψ_i for $i > p$, i.e., those vectors corresponding to eigenvalues $\lambda_i = 0$. It could perhaps be relevant to define the 'redundancy' of the ith measurement, say, as

$$\mathrm{red}_i = \sqrt{\sum_{j=p+1}^{m} u_{ji}^2}. \tag{15}$$

The Spectral Density of the Discrete Laplacian

Now we shall study some features of the spectral distribution of the untransformed normal equation matrix $A^{\mathrm{T}}A = R$. It is possible to carry through the calculations in explicit form if R, for example, has the following simple form:

$$R_n = \begin{bmatrix} 1 & -1 & & & & & \\ -1 & 2 & -1 & & & & \\ & -1 & 2 & -1 & & & \\ & & & \ddots & & & \\ & & & -1 & 2 & -1 \\ & & & & -1 & 1 \end{bmatrix}_{n\times n}. \tag{16}$$

These normal equations correspond to a leveling network with n nodal points each having only two neighbors, except the terminal points each having one. R_n also arises in the case of a straight traverse where only the logarithm of the side lengths is subject to measurement. Suppose all weights are unit weights and all neighboring distances are equal. The eigenvalues of R are given by

$$\lambda_i = 4\sin^2\frac{i\pi}{2n}, \qquad i = 0,\ 1,\ \ldots,\ n-1. \tag{17}$$

Note that throughout the rest of the paper we use a reverse ordering of the eigenvalues compared to that one in the previous section. We also omit the exponent of λ_i, as further on this could lead to misunderstandings. Obviously

$$0 \leq \lambda_i < 4. \tag{18}$$

What is then the distribution of the eigenvalues λ_i on this interval for $n \to \infty$?

Let $N(\lambda)$ denote the nondecreasing step function

$$N(\lambda) = \sum_{\lambda_i < \lambda} 1, \tag{19}$$

then $N(A)$ is the number of eigenvalues less than λ. $N(\lambda)$ is also called the spectral distribution function with discontinuity points at λ_i. Suppose for a fixed n corresponding to a subdivision of the interval $[a, b]$ we have $M + 1$ points with equidistance k:

$$a = x_0 < x_1 < x_2 < \cdots < x_M = b. \tag{20}$$

We want to find the number of eigenvalues in the interval $[x_i, x_{i+1}] = k$:

$$T_i(N) = N(x_{i+1}) - N(x_i) \geq 0. \tag{21}$$

Letting $n \to \infty$, we see that in the case of $M \to \infty$ the spectral density is determined by

$$\mu(i) = \lim_{k \to 0} \frac{T_i(N)}{k} = \frac{dN}{d\lambda}. \tag{22}$$

In our special case (17) we get

$$\lambda_i = 4 \sin^2 \frac{i\pi}{2n} \sim 4 \sin^2 x = y, \qquad 0 \leq y < 4 \tag{23}$$

and the normalized density function is given by

$$\mu(y) = \frac{dx}{dy} = \frac{1}{\pi\sqrt{y(4-y)}}. \tag{24}$$

For $y \to 0+$ and $y \to 4-$ we have $\mu(y) \to \infty$, i.e., the eigenvalues for $n \to \infty$ are naturally separated into two groups, but still they are overall dense on the interval $[0, 4]$, cf. Figure 9.1.

So if we for example want to improve the condition number of this matrix, it is clear that very little is gained by removing the smallest or greatest eigenvalue, i.e., the network is unstable.

In a corresponding 2-dimensional rectangular network of dimension $m \times n$ described by means of the Kronecker product, the normal equation matrix is given by

$$L = I_m \otimes R_n + R_m \otimes I_n \tag{25}$$

and now are the eigenvalues

$$\lambda_{ij} = 4\Big(\sin^2 \frac{i\pi}{2n} + \sin^2 \frac{j\pi}{2m}\Big), \qquad \begin{cases} i = 0, 1, \ldots, n-1 \\ j = 0, 1, \ldots, m-1. \end{cases} \tag{26}$$

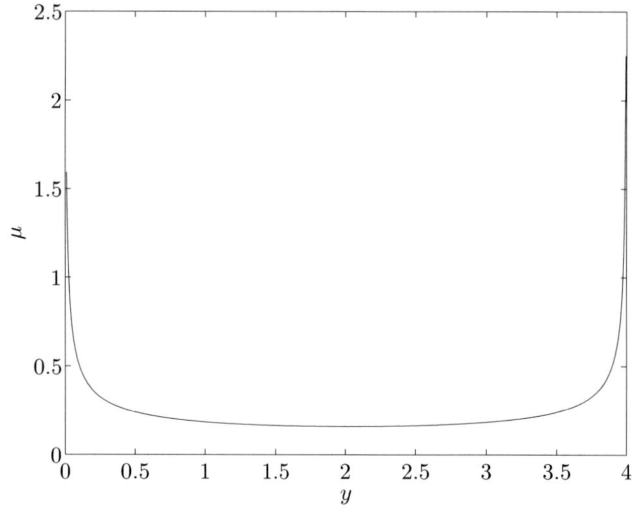

Fig. 9.1.

The spectral density is determined in the case of $m = n$ by the convolution

$$\lambda_{i.}(L) * \lambda_{.j}(L) \sim \mu(z) = \frac{1}{\pi^2}\int_0^z \frac{dy}{\sqrt{(y-4)(y-z)y(y-4+z)}}$$
$$= \frac{1}{2\pi^2} K\left(\sqrt{\frac{z(8-z)}{16}}\right), \qquad (27)$$

where $K(k)$ is the complete elliptic integral of the first kind, cf. Figure 9.2.

As it will be seen, there exists a remarkable difference in the spectral distributions for the cases $d = 1$, 2, and $d \geq 3$. By means of Young's inequality we may obtain a deep insight into the convolution of the spectrum in d-dimensions.

By definition the class of functions $f(x)$, which are measurable and for which $|f(x)|^p$ is integrable over $[a,b]$, is known as $L^p[a,b]$, $p > 0$. From integration theory we have the theorem: If $-\infty < a < b < +\infty$, then $f \in L^p[a,b]$, $p' < p$, which implies $f \in L^{p'}[a,b]$.

In the following we will use the pair of Hölder conjugate numbers (p, p') which are connected by

$$\frac{1}{p} + \frac{1}{p'} = 1,$$

and define the positive number $\|f\|_p$ by

$$\|f\|_p = \left(\int_a^b |f(x)|^p dx\right)^{1/p}.$$

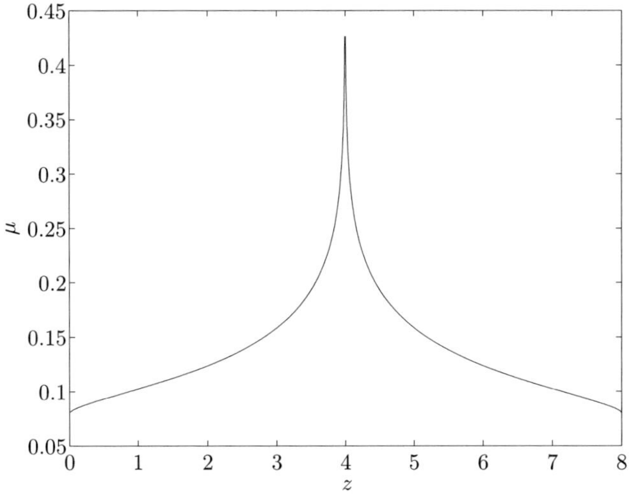

Fig. 9.2.

From [9], p. 192 we take the formulation of Young's Inequality: Let $f \in L^p(R^n)$, $g \in L^q(R^n)$, where $n \geq 1$ (an integer), $p, q \geq 1$, $\frac{1}{p} + \frac{1}{q} \geq 1$, and let

$$\frac{1}{r} = \frac{1}{p} + \frac{1}{q} - 1.$$

Then $(f * g) \in L^r(R^n)$, and

$$\|f * g\|_r \leq \|f\|_p \, \|g\|_q. \tag{28}$$

For the considered $f = g$ we have

$$f(x) = \frac{1}{\pi\sqrt{x(4-x)}}. \tag{29}$$

It is easily seen that $f \in L^p[0,4]$ for $p < 2$. Young's inequality now yields $r < \infty$, and consequently $f * g \in L^p$ for $p < \infty$ and is thus unbounded on $[0, 8]$. The convolution iterate

$$f^{(3)} = f * f * f = f^{(1)} * f^{(2)} \tag{30}$$

with $f^{(1)} \in L^p$ for $p = 2 - \epsilon$ and $f^{(2)} \in L^q$ for $q = N$ has the Hölder numbers $(2 - \epsilon, N)$. If the conditions for Young's inequality shall be fulfilled then

$$\frac{1}{2-\epsilon} + \frac{1}{N} \geq 1, \tag{31}$$

which leads to

$$1 > \epsilon > \frac{N-2}{N-1} \tag{32}$$

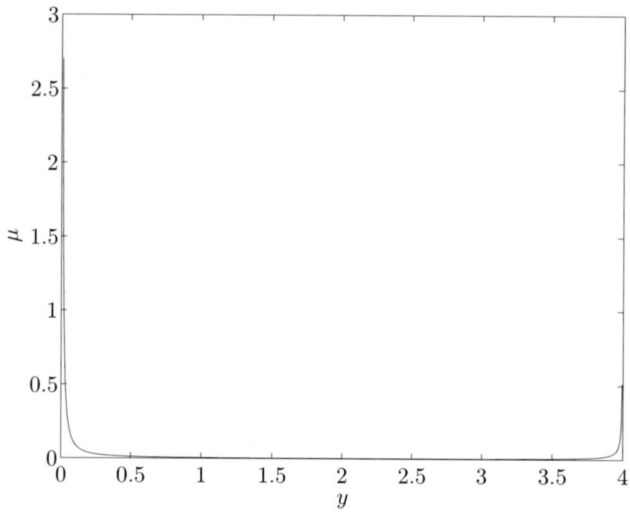

Fig. 9.3.

and consequently

$$p = 2 - \epsilon > 1 \qquad \text{and} \qquad q = N > 3. \tag{33}$$

This is valid as we only require $q < \infty$. Then $f^{(3)} \in L^\infty$ and is thus bounded.

(In statistics it is a well-known fact that $\lim_{n\to\infty} f^{(n)}$ is the normal distribution function when $f^{(1)}$ fulfils certain restrictions; one of which is

$$\int_0^4 f'(x)\,dx \ < \ \text{a constant.} \tag{34}$$

It is easily seen that this is not the case. Cf. [4], p. 164 ff.)

The matrix R_n^2 is related to observations of a straight traverse where only the angles are subject to measurements. So in this case we have

$$y = 4\sin^4 x \tag{35}$$

and

$$\mu(y) = \frac{1}{2\pi^4\sqrt{y^3(y+4-4\sqrt{y})}}. \tag{36}$$

Obviously the distribution function still has the same structure as in the former case, but now $\mu(y)$ is no more symmetric, cf. Figure 9.3. In the case of R_n^2 we are actually observing second order differences between the parameters contrary to first order differences in the case of R_n. So in this regular case we can conclude that angle observations in a certain sense decrease the larger eigenvalues compared to, e.g., length measurements.

So the fact that angle measurements in addition to first order differences of coordinates of neighboring stations contain second order differences of such coordinates and further exploiting that these second order differences are most dominating for rounds with a small number of directions the following could be proposed:

Conjecture *An increment of the rate of the number of angle measurements to the number of distance measurements will give an increment of the dispersion of the eigenvalues, i.e., make the concentration of them near the ends of the spectrum higher, and this tendency will be stronger as fewer stations are observed in each round.*

The authors must greatly deplore that up to now they can not can see how to prove this conjecture in the general form stated.

Finally we shall demonstrate how the spectral distribution for the discrete case can give valid results for the continuous case. We must then assume that the nodal points lie on a straight line with equidistance h. Then the eigenvalues of R_n are

$$\lambda_i = \frac{4}{h^2} \sin^2 \frac{hi\pi}{2M} = \frac{4}{h^2} \sin^2 \frac{i\pi}{2n}, \tag{37}$$

where $M = n \cdot h$. Now we keep M fixed and let $n \to \infty$:

$$\lim_{n\to\infty} \lambda_i = \left(\frac{i\pi}{M}\right)^2. \tag{38}$$

The number of eigenvalues less than $\Lambda = \left(i\pi/M\right)^2$ is

$$N(\Lambda) = \sum_{\lambda_i < \Lambda} 1 = i. \tag{39}$$

By means of $i = M\sqrt{\Lambda}/\pi$ we get

$$N(\Lambda) = \frac{M}{\pi}\sqrt{\Lambda}, \qquad \text{1-dim.} \tag{40}$$

cf. the result of Weyl

$$N(\Lambda) \sim \frac{|\Omega|}{2^d \sqrt{\pi^d}\,\Gamma(\frac{d}{2}+1)} \Lambda^{\frac{d}{2}}, \quad d\text{-dim.} \tag{41}$$

where the eigenvalues no longer are limited. We observe that $N(\Lambda)$ is proportional to $\Lambda^{\frac{d}{2}}$.

The Spectral Distribution Function $N(\lambda)$

It is comparatively easy to derive upper and lower bounds for the spectral distribution function $N(\lambda)$ in the special case treated in the previous section.

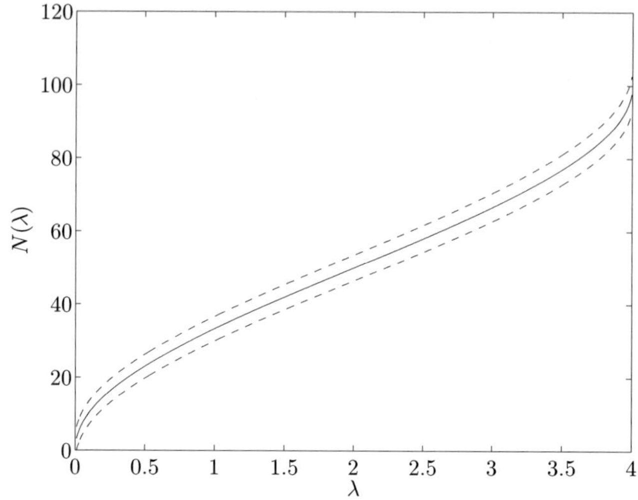

Fig. 9.4.

In the 1-dimensional case with

$$\lambda_i = 4\sin^2\frac{i\pi}{2n}, \qquad i = 0,\, 1,\, \ldots,\, n-1 \tag{42}$$

cf. equation (17) we get according to [3], Th. 2.2:

$$\frac{2n}{\pi}\arcsin\frac{\sqrt{\lambda}}{2} \le N(\lambda) \le \frac{2n}{\pi}\left\{1 + \frac{c}{n\sqrt{\lambda}}\right\}\arcsin\frac{\sqrt{\lambda}}{2}, \tag{43}$$

where c is a positive constant. The upper and lower bounds for $N(\lambda)$ are sketched in Figure 9.4.

In case we did not consider a free network, but one fixed at the terminals we would have

$$\lambda_i = 4\sin^2\frac{i\pi}{2(n-1)}, \qquad i = 1,\, 2,\, \ldots,\, n-2. \tag{44}$$

Now the bounds are given by Th. 2.1 in [3]:

$$\frac{2(n-1)}{\pi}\left\{1 - \frac{c}{(n-1)\sqrt{\lambda}}\right\}\arcsin\frac{\sqrt{\lambda}}{2} \le N(\lambda) \le \frac{2(n-1)}{\pi}\arcsin\frac{\sqrt{\lambda}}{2}. \tag{45}$$

The lower bound is shown with a dashed line in Figure 9.4.

Of course, this time the upper bound, apart from a minor change in the factor, is the lower bound in case of a free network. This is in agreement with Courant's minimax principle (enforcing constraints on an oscillating system at most increases all its eigenvalues) when recalling $N(\lambda)$ is reversely connected with this change of the eigenvalues.

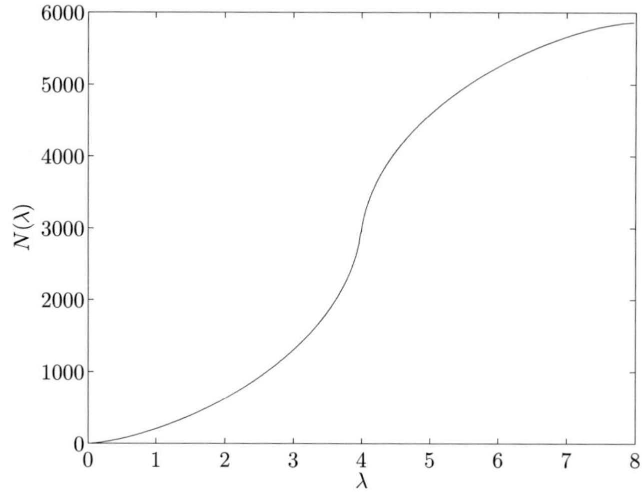

Fig. 9.5.

In the 2-dimensional case, however, it is not plain to find the explicit upper and lower bounds for $N(\lambda)$. In case of a free boundary we have, e.g.,

$$N(\lambda \leq 4) \geq \frac{2mn\sqrt{\lambda}}{\pi^2} \int_0^1 \frac{\arcsin\sqrt{\frac{\lambda}{4} - \frac{\lambda}{4}x^2}}{\sqrt{1 - \frac{\lambda}{4}x^2}} dx. \tag{46}$$

But because of Equation (22) and Figure 9.2 we must have a picture somewhat as sketched in Figure 9.5. The function corresponding to the central line will be denoted $\lambda mn \cdot A(\lambda)$.

As we especially are interested in small eigenvalues and their density, it is appropriate to point out the following upper bounds valid for free networks, cf. [3], p. 477:

Dimension	$N(\lambda) \leq$		
1	$\frac{n\sqrt{\lambda}}{\pi}(1 + O(\lambda))$		
2	$\frac{mn\lambda}{4\pi}(1 + O(\lambda))$	for $\lambda \to 0$,	(47)
3	$\frac{lmn\lambda^{3/2}}{6\pi^2}(1 + O(\lambda))$		
4	$\frac{klmn\lambda^2}{32\pi^2}(1 + O(\lambda))$		

where k, l, m, and n denote the number of points in various directions.

The corresponding lower bounds are obtained by multiplying by $\left(1-\frac{c}{a\sqrt{\lambda}}\right)$ where $a = \min(n, m, \ldots)$.

All bounds given till now have dealt with the discrete Laplacian on a rectangular grid with mesh width 1 in d-dimensional Euclidian spaces. Of special interest are bounds for 2-dimensional square grids with an arbitrary boundary. For networks with a fixed boundary we get, cf. [3], Th. 3.1:

$$A(\lambda)\lambda\Big(1-\frac{c_2}{\delta\sqrt{\lambda}}\Big)\big(|\Omega|-c_0\delta\big) \le N(\lambda) \le A(\lambda)\lambda\Big(1+\frac{c_1}{\delta\sqrt{\lambda}}\Big)\big(|\Omega|+c_0\delta\big), \quad (48)$$

where c_1 and c_2 are positive constants, $|\Omega|$ denotes the number of interior points of the grid, δ is the minimum length over all edges of the grid and then necessarily ≥ 1, and finally c_0 is a positive constant independent of δ but dependent of the 'pathology' of the boundary. A modified result is valid for free networks.

In a given network it may be difficult to obtain the best possible values for the constants c_0, c_1, and c_2. Therefore an easy estimate is always obtainable by circum- and inscribing rectangles with the smallest or largest number of interior points, respectively. But this may lead to bad estimates. For example in case of a circle we can never get a better upper bound than one which is twice as large as the lower one.

Essentially, however, the bounds are rather close for small eigenvalues. In a certain sense the larger ones only determine the scale of the whole problem. Confer the definition of the spectral condition number.

As earlier mentioned, the constants c_0, c_1, and c_2 depend on how frayed the boundary is. Whether an improved analysis can yield results so good that 'one can hear the shape of the boundary,' cf. [5] is still an open question.

On the Smoothness and Roughness of the Eigenvectors of the Normal Equation Matrix

In this section we shall discuss properties of a geodetic network which ensure that the eigenvectors belonging to small eigenvalues of the normal equations are smooth.

Let the observation equations be

$$b = Ax. \quad (49)$$

They give the change b of the observable quantities when the coordinates are changed by the amount x.

[8] calls a network relatively large when the distance of any pair of points between which observations have been taken is small compared to the size of the network.

Now if x is a smooth network distortion, then the distances and angles of local point configurations are small. Since only such local point configurations have been observed, it follows: In a relatively large network a smooth network distortion causes a small change of the observations.

For any eigenvector x, $x^{\mathrm{T}}x = 1$, we have

$$\lambda = x^{\mathrm{T}} A^{\mathrm{T}} P A x = b^{\mathrm{T}} P b. \tag{50}$$

If we agree to measure the size of b by the norm $\|b\| = \sqrt{b^{\mathrm{T}} P b}$ we immediately have the consequence: In a relatively large network a smooth eigenvector belongs to a small eigenvalue.

However, we also want to demonstrate the opposite: A small eigenvalue must belong to a smooth eigenvector. For this we require another network property also stated in [8]. A network is locally stable if the observations cannot be changed considerably without changing size and shape of local configurations appreciably.

For such a network, we can revert the above reasoning: A small A entails a small b which, by local stability, entails a smooth x. We summarize:

In a relatively large and locally stable network small eigenvalues of the normal equation matrix belong to smoothly varying eigenvectors, and vice versa.

The verbal formulation of the stated principle may not satisfy everybody. Instead of formalizing it in general terms, we illustrate by means of a specific example.

Consider a relatively large leveling network. A leveling network is automatically locally stable. Let S denote the set of pairs of points between which height differences have been measured. For any $(i, j) \in S$, i and j are necessarily close together (definition of relatively large network). Suppose a unit weight matrix. We have

$$b_{ij} = \delta H_j - \delta H_i, \qquad (i, j) \in S, \tag{51}$$

δH_i denoting the height distortion at i. We further have for any eigenvalue λ and corresponding eigenvector δH:

$$\lambda = \sum_{(i,j) \in S} \left(\delta H_j - \delta H_i\right)^2. \tag{52}$$

This demonstrates that a small λ must belong to a smooth height distortion with δH as eigenvector.

Green's Formula for Trigonometric Networks

For a trigonometric network, the analogue to Green's formula is formally

$$x^{\mathrm{T}}(A^{\mathrm{T}} A)x = (Ax)^{\mathrm{T}}(Ax). \tag{53}$$

This formula is less trivial than it looks at first glance.

In order to find the meaning of (53) we shall concentrate here on free networks.

The vector x has an immediate interpretation as a discrete vector field: x consists of a set of two dimensional vectors x_i (the coordinate increments), one for each station in the network: $x = [\, x_1^{\mathrm{T}} \;\; x_2^{\mathrm{T}} \;\; \dots \;\; x_n^{\mathrm{T}} \,]^{\mathrm{T}}$.

An observation equation corresponding to an observation of a distance or of an absolute direction between the stations i and j has the form

$$s_k \frac{e_k^{\mathrm{T}} x_i - e_k^{\mathrm{T}} x_j}{r_{ij}} = b_k, \qquad k = 1, 2, \dots, m \tag{54}$$

where k is the observation number, s_k has something to do with the weight, r_{ij} is the distance between the stations i and j, and e_k is a two-dimensional unit vector. In the case of distance observations, e is parallel to the direction station$_i$–station$_j$; in the case of a direction observation e is orthogonal to that direction.

If every observation in the network is of one of these two kinds, (53) may be written

$$x^{\mathrm{T}}(A^{\mathrm{T}}A)x = \sum_{k=1}^{m} s_k^2 \left(\frac{e_k^{\mathrm{T}} x_i - e_k^{\mathrm{T}} x_j}{r_{ij}} \right)^2,$$

and if ϕ is a normalized eigenvector for $A^{\mathrm{T}}A$ with eigenvalue λ we deduce

$$\sum_{k=1}^{m} s_k^2 \left(\frac{e_k^{\mathrm{T}} \phi_i - e_k^{\mathrm{T}} \phi_j}{r_{ij}} \right)^2 = \lambda. \tag{55}$$

$(e_k^{\mathrm{T}} x_i - e_k^{\mathrm{T}} x_j)/r_{ij}$ is a kind of 1. order vectorial difference quotient and (55) is thus a complete analogue to formula (52) in the case of a leveling network.

If we have in station i a horizon of directions to p other stations with observations of equal weight s, we have the observation equations

$$s\left(\frac{e_k^{\mathrm{T}} x_i - e_k^{\mathrm{T}} x_k}{r_{ik}} - \alpha_0 \right) = b_k, \qquad k = 1, 2, \dots, p$$

where α_0 is the initial direction.

It is not difficult to see that the contribution to the normal equations of such a horizon will be a positive definite quadratic form in the difference

$$D_k - D_l = \frac{e_k^{\mathrm{T}} x_i - e_k^{\mathrm{T}} x_k}{r_{ik}} - \frac{e_l^{\mathrm{T}} x_i - e_l^{\mathrm{T}} x_l}{r_{il}}, \qquad k, l = 1, 2, \dots, p$$

between difference quotients. In a realistic network, the contributions of this type from the different horizons have to be added at the right hand member of (55). The result is

$$\frac{s^2}{p} \sum_{k<l} (D_k - D_l)^2.$$

If the network is not free, every outer condition gives occasion to a 'boundary term' in addition to the terms mentioned above.

Difference and Differential Equations Corresponding to Networks With Various Types of Observation

In a triangulation network we assign to each station a two-dimensional vector describing the increments of coordinates of the station. All the elements of these vectors are contained in the x vector of equation (1). Let us imagine this discrete vector field prolonged so as to cover the whole network up to a boundary a little outside it. The prolongation shall take place such that the vector field becomes continuous and the gradient shall fulfill certain restrictions which we shall not describe in detail.

The coefficient matrix $A^{\mathrm{T}}A$ of the normal equations can be viewed as the result of a certain difference operator acting on this continuous vector field. The difference operator must act just between the stations of the given geodetic network. Obviously, this difference operator must depend on the sort of measurement taken.

Although this difference operator is already known for some classes of observations satisfying certain correlation patterns, cf. [2], such operators are not yet known for, e.g., pure distance or pure angular networks.

The introduction of the continuous prolongation of the discrete vector field also introduces infinitely many unbounded eigenvalues which in the discrete case are bounded, and their number here equals the dimension of the coefficient matrix. But in some sense, which may be defined in terms of concepts found in [1], the eigenvalues and -functions are good approximations to the corresponding eigensolutions of the discrete network. Especially for small eigenvalues the approximation is good and should thus indicate something about the weakness of the network. So there are good reasons for trying to obtain the continuous analogous operator corresponding to any form for geodetic observations and combinations of such.

In [2] one finds the following class of observations treated: Leveling networks, pure angular networks with known ratio of the sides, and networks with distance and absolute directions. For regularly shaped (isotropic) networks they all boil down to the discrete Laplacian, so the following partial differential equation is the analogue to the eigensolution problem of the normal equations:

$$\Delta u + \lambda u = 0 \tag{56}$$

with Neumann or Dirichlet boundary conditions depending on whether it is a free network or one fixed along the boundary.

A so-called relative leveling network, i.e., one in which the scale cannot be carried over from one station to another, satisfies

$$\Delta^2 u - \lambda u = 0. \tag{57}$$

Further, by more advanced methods which shall not be presented here, we get for distance measurements in a free equilateral triangular network

$$
\begin{aligned}
&3\frac{\partial^2 u_1}{\partial x^2} + \frac{\partial^2 u_1}{\partial y^2} + 2\frac{\partial^2 u_2}{\partial x \partial y} + \lambda u_1 = 0 \\
&2\frac{\partial^2 u_1}{\partial x \partial y} + \frac{\partial^2 u_2}{\partial x^2} + 3\frac{\partial^2 u_2}{\partial y^2} + \lambda u_2 = 0
\end{aligned}
\tag{58}
$$

with boundary conditions

$$
\begin{aligned}
&n_1\left(3\frac{\partial u_1}{\partial x} + \frac{\partial u_2}{\partial y}\right) + n_2\left(\frac{\partial u_1}{\partial y} + \frac{\partial u_2}{\partial x}\right) = 0 \\
&n_1\left(\frac{\partial u_1}{\partial y} + \frac{\partial u_2}{\partial x}\right) + n_2\left(\frac{\partial u_1}{\partial x} + 3\frac{\partial u_2}{\partial y}\right) = 0.
\end{aligned}
\tag{59}
$$

Here, u_1 and u_2 denote the components of the two-dimensional vector u, and n_1 and n_2 denote the components of the vector normal to the boundary. (In fact, equations (56) and (57) are scalar equations.) Equations (58) may be rewritten

$$3\Delta u + 2 \text{ curl curl } u + \lambda u = 0. \tag{60}$$

By the way, these equations are closely related to the equations of equilibrium for longitudinal deformations of thin plates, cf. [7], §13.

In a similar way, we get for a network of the same shape but now only with absolute direction measurements

$$\Delta u - 2 \text{ curl curl } u + \lambda u = 0. \tag{61}$$

Finally, the combination of the last two networks yields

$$2\Delta u + \lambda u = 0 \tag{62}$$

in accordance with (56) as we deal with observations of this class.

It shall be mentioned that the boundary value problem (58)–(59) can be solved explicitly within a circle in terms of Bessel functions. For small values of λ the eigenfunctions are nearly harmonic.

This continuous analogy to realistic, i.e., discrete networks may be viewed as a limit case of relatively large networks, namely networks which are infinitely large compared to the distances between stations with connecting observations.

For partial differential equations of the type we meet with here, many important results are known especially concerning the asymptotic behavior; therefore, we believe that by using them it will be possible to get a more quantitative formulation of results and conjectures referred to in this paper.

The different equations were found this summer, so we could not, as we would have liked, have built up the whole paper upon this single method. The deduction of the differential equations, which relies on Green's formula, both in its usual form and in its discrete form referred in the foregoing section, will be published as soon as it appears to us to be sufficiently clear.

Acknowledgment

The authors like to express their gratitude to Professor P. Meissl for valuable discussions on strength analysis of networks.

References

[1] Jean-Pierre Aubin. *Approximation of Elliptic Boundary-Value Problems.* Wiley-Interscience, New York, 1972.

[2] K. Borre and P. Meissl. *Strength Analysis of Leveling-Type Networks. An Application of Random Walk Theory.* Meddelelse 50. Geodætisk Institut, København, 1974.

[3] R. L. Burden and G. W. Hedstrom. The distribution of the eigenvalues of the discrete Laplacian. *BIT*, 12:475–488, 1972.

[4] A. J. Chintschin. *Mathematische Grundlagen der statistischen Mechanik.* Bibliographisches Institut, Mannheim, 1964.

[5] Marc Kac. Can one hear the shape of a drum? *American Mathematical Monthly*, 73:1–23, 1966.

[6] C. Lanczos. *Linear Differential Operators.* D. Van Nostrand, New York, 1961.

[7] L. P. Landau and E. M. Lifschitz. *Elastizitätstheorie.* Akademie-Verlag, Berlin, 1970.

[8] P. Meissl. Investigations of random error propagation in geodetic networks. In *Proceedings of the Symposium:* Satellite and Terrestrial Triangulation, pages 301–319, Graz, 1972.

[9] G. O. Okikiolu. *Aspects of the Theory of Bounded Integral Operators in L^p-Spaces.* Academic Press, London and New York, 1971.

[10] Henry Scheffé. *The Analysis of Variance.* John Wiley and Sons, New York, 1959.

10

Mathematical Geodesy

Considering the intimate relations which have always existed between geodesy and mathematics the recent notion of mathematical geodesy could seem rather superfluous. I think however that this notion indicates a new style for the relations between the two sciences and I shall try in this lecture to describe this new style in a manner which you may find rather subjective.

The notion itself 'mathematical geodesy' goes back to Martin Hotine who launched it as the title of a formidable book of his in 1969, and it is worth while to mention his name as it was his untiring struggle which has made it acceptable for a geodesist to work in the field of mathematical geodesy.

But the concept was born in 1951 with Marussi's beautiful paper Fondamenti di Geodesia Intrinseca where—in the language of the famous Italian school of differential geometry—the geodesy is edified in a radically new way. Marussi accepts only coordinates which, in principle, may be observed directly i.e. the two astronomical coordinates and the value of the potential (or rather the potential plus a fixed but unknown constant) as the third coordinate. Observations of distances and angles between points at the surface of the Earth are then used to find the metric fundamental tensor in the intrinsic coordinates. When this fundamental tensor is found the intrinsic coordinate system has a practical meaning and it is e.g. possible to calculate coordinates for points at the surface of the Earth in a global Cartesian system.

The really radical in Marussi's approach was not so much that it gave geodesy a three-dimensional foundation—although this for the contemporaries seemed to be the conspicuous innovation—it was rather that it departed with a geodetic tradition, or bad habit, which may be called that of inconsistent approximations.

This old style of application of mathematics in geodesy, which all of us know too well, may be described as follows: The treatment of a geodetic problem is split up into several steps and at each step one makes special simplifications and approximations in order to apply that mathematical method which the author, or tradition, has decided to use for the carrying through of that step. Use of this style has not only the effect that it makes the reading of geodetic texts unreasonably tiring even for geodesists who have been indoctrinated in this 'method' and makes our texts almost unreadable for mathematicians, but even if the approximations introduced at the single steps are clear and relatively harmless—as they sometimes are—it is usually impossible to trace the influence of the different approximations on the result and finally to say anything about the reliability of the result. Regarding the immense quantity of geodetical papers we see that it is in fact easy to arrive at numerical results of the observations but it is very difficult to arrive at a reliable estimate for the reliability of these numerical results.

It is the great merit of Marussi's that he has introduced a new style into geodetic research, i.e. that of translating the original geodetic problem into a mathematical problem (forming a mathematical model) and then to treat this mathematical problem so to say under the jurisdiction of mathematics. If one succeeds in solving the problem in its mathematical form one only has to find out how relevant the mathematical model is by investigating the possible pertubations caused by the geodetical factors ignored in the mathematical model, and this should be relatively simple because of the existence of an exact and consistent mathematical solution. I find that it is reasonable to use—and to reserve the use of—the notion of mathematical geodesy to such a method for approaching geodetic problems which is quite parallel to the philosophy of mathematical physics.

With this definition, which is parallel to that of mathematical physics, mathematical geodesy becomes a mathematical discipline, i.e. its form is mathematical but its content is geodetical.

I regard Marussi's demand for consistency of geodetic theories so important that the question whether he really has succeeded in giving geodesy a consistent mathematical foundation becomes of minor interest. Personally I think that he has not yet succeeded, but I do not go into that here.

After that of Marussi there have been other approaches to intrinsic geodesy. The most interesting of those is perhaps that of E. Grafarend and N. Grossmann. In contrast to Marussi they express geodesy in the language of the French school of differential geometry (Elie Cartan). They start from the fact that there is at every point a privileged direction, that of the plumb line, therefore they introduce in order to describe the geometry in three-dimensional Euclidean space a geometry in which all plumb lines are parallel. Naturally this describing geometry is not Euclidean and even not Riemannian (being not torsion-free). Although this approach could seem a little sophisticated it is not impossible that they can erect on this foundation a theory which is equivalent to that of Marussi but with a simpler mathematical formalism.

Almost at the same time as Marussi founded his intrinsic geodesy Molodenskiy published his important papers on physical geodesy.

The most famous part of Molodenskiy's investigations is the definition of Molodenskiy's problem and the attempt to solve it. I believe that the most permanent result of his work is his analysis of the concept of the geoid, from which results that it is principally impossible to determine the geoid from geodetic measurements and that the concept of the geoid is unnecessary, so when we continue to speak about the geoid it is a historical relict exactly as when we speak about the daily motion of the sun: it spares us from using more sophisticated expressions and everybody knows what we mean. It was a pitty that Molodenskiy not such as Marussi had an established mathematical tradition upon which to found his work—such a tradition in the field of the solution of complicated boundary value problems existed hardly at that time, therefore Molodenskiy's problem has been formulated in an unhappy way. It is well-known that one of the secrets of science is to ask the right questions: Our questions shall not only be relevant, there shall also exist an answer to them. Nevertheless Molodenskiy's problem has fascinated us for tens of years and we have wasted a lot of time on it. To give a better formulation of it is an important task which is in store for us.

We have also a Danish approach to mathematical geodesy, which we call 'integrated geodesy.'

Integrated geodesy is founded upon least-squares adjustment or rather upon a generalization of this concept arrived at by means of functional analysis.

We normally use Cartesian coordinates. We do not mean that the fact that these coordinates are not directly observable should be any reason for not using them provided that we can prove that they are principally computable.

The idea of integrated geodesy may shortly be described by describing the following two steps:

1. For every geodetical measurement it is possible to write down the correct linearized observation equation i.e. an equation which expresses the differential of the value of the observation (the correction) by the differentials of the Cartesian coordinates (the corrections to the reference coordinates) and by linear functionals of the disturbing potential of the Earth (generalized corrections to the reference potential). There may also enter some auxiliary parameters such as orientation constants etc. The difference from the classical observation equations is that no geodetic reduction is applied to the observations but that the reference potential and the disturbing potential enter the equations. Principally these observation equations are simple to deduce, but in practice we have seen that without a clever technique the formulas will become rather impenetrable. We have therefore spent some time bringing about such a technique and we think we have succeeded in finding one which is both handy and interesting.
2. The second step is naturally the adjustment.

The mathematical theory for such a generalized least-squares adjustment is almost trivial. An expression which is the sum of the weighted square sum of the corrections to the observations (as in classical adjustment) and a quadratic form in the disturbing potential shall be minimized in order to define a solution. This solution can be found numerically. Now the serious problem is that the solution depends of which quadratic form we choose. This problem is a principal one and is closely connected to the fact that the value of the potential in one point cannot be determined from any set of measurements in finitely many other points. Our common sense tells us that if these measurements are dense enough near the point in question then we can find the value by interpolation but there is no objective method to define what the best value is—and there will never be! It was an obvious idea to try to solve the problem by statistical methods similar to time series analysis. We have tried this and even if somebody still may hope for a solution from statistics I now feel convinced that this will be impossible. For the time being we must be satisfied with a reasonable numerical result which is consistent with the observations, but we have good prospects of a method which by using topographical and geological results will enable us to say something about the reliability of the result.

One could argue that it is a drawback for our method as compared with conventional methods that the result is not unique. To that I can only say that when conventional methods give unique results it is because in them there has already been made a choice normally tacitely and perhaps even subconsciously. I find it important to focus on such 'nightsides' of scientific methods and try to find the possible influence of such choices on the results.

I think it is in continuation of good geodetic tradition to work with such an adjustment model which involves all forms of geodetic observations even if there may be practical reasons for realizing the computations in sections each of which involves parts of the measurements, only in this way we can arrive at a theory for the correct joining together of partial results.

As I indicated before I do not see the greatest importance of mathematical geodesy in the new methods of calculation it furnishes but didactically in the possibility it gives us to present geodesy in a logical and coherent form and practically in that it should give us good methods for treating the important problems of reliability and optimisation.

If we shall discuss the aptitude of integrated geodesy for solving problems of optimisation we are faced with the deep and fundamental dualism of applied and numerical mathematics: that of the continuous and the discrete aspects.

I shall illustrate this dualism by two examples.

The first example is from the theory of elasticity.

The theory of elasticity treats the behaviour of solids under the influence of forces. Classically the theory investigates the continuous deformations under the influence of exteriour forces and the laws are expressed by partial differential equations. But this macroscopic behaviour can be explained from molecular forces and so we are led to the discrete aspect: the elastical behaviour of the solid can be expressed by a gigantic system of equations involving each of the molecules. On the other hand if one shall solve numerical problems in elasticity theory one usually has to solve the differential equations by discrete approximation methods using finite differences or finite element methods.

The other example concerns Molodenskiy's problem. Molodenskiy himself formulated this problem under the continuous aspect while other geodetical boundary value problems such as astrogeodetic levelling is traditionally treated under the discrete aspect. As Bjerhammar has pointed out a discrete approach to Molodenskiy's problem is not only possible but even more realistic, and no matter how the problem is formulated in order to find a numerical solution we must apply some finite i.e. discrete method, but the different formulations may lead to different discrete numerical methods, and if we want to formulate general laws e.g. about the influence on the result at one point from errors of the anomalies at more or less distant points then the continuous formulation is likely to be best suited.

Considerations similar to these have recently led us to the conviction that laws concerning the 'macroscopic' behaviour of geodetic networks under the influence of systematic and accidental errors are most easily found by using a continuous approach and we have therefore begun to work at what we call 'the elasticity theory of geodetic networks.' This theory makes a rapid progress so that the paper about it distributed at this meeting gives only an impression of what it is about, an impression which is already out of date today. It is my hope that it will give important help for the planning of geometrical networks but that it also as an organic part of integrated geodesy will be well suited for solution of planning problems for geodetic projects where physical as well as geometric observations are involved.

My personal experience from working with mathematical geodesy is that it is much easier to make new ideas—and perhaps even good ideas?—in that field than it is to work out these ideas into details so that they can have practical influence. This task is very work consuming and cannot be afforded by one man or even one institute. But most people seem to be more interested in presenting new ideas than in transforming good ideas into reality.

11

Foundation of a Theory of Elasticity for Geodetic Networks

1

Every geodesist has when assessing geodetic networks intuitively used analogies from the theory of elasticity, but so far nobody seems to have succeeded in making such an analogy rigorous. The authors of the present paper believe that the establishing of a theory of elasticity for geodetic networks will be important for the analysis of strength of relatively extended geodetic networks, i.e. networks for which the maximum net width is small with respect to the extension of the net.

The main tools in elasticity theory are the system of partial differential equations and boundary conditions for the displacements. A classical way to deduce these equations is to write down the expression for the elastic energy, which is a quadratic form in the displacements, and then using methods from the calculus of variations to find the conditions which the displacements must satisfy in order to minimize the energy: these conditions are exactly the equations we wanted.

In the theory for geodetic networks we also have a quadratic form in the displacements which we want to minimize: the weighted square sum of the residuals, and the conditions on the displacements for this minimization are expressed by the normal equations. Now our main idea is to consider these normal equations as expressing a discrete approximation to an ideal 'continuous network' problem described by a system of partial differential equations and corresponding boundary value conditions. Then we find it reasonable to expect that these equations may give us the same global information about the network as the partial differential equations in elasticity theory give about the macroscopic behaviour of solids.

There exist many methods for discrete approximation of problems expressed by partial differential equations, the most modern and perhaps the most powerful of those is 'the method of finite elements.' After eight years of dreaming about a geodetic theory of elasticity one of the authors to this paper was struck by the idea to use the method of finite elements in the opposed direction and some of the results attained in this way are presented in the following.

Naturally we are not the only geodesists who have speculated about finding a continuous model for networks, influences from [4], [6], and [2] have contributed to keeping our dream alive.

The present paper consists of two parts. In the first of them Sections 2 to 6 the partial differential equations and the boundary conditions governing continuous networks are deduced. A certain acquaintance with the calculus of variations and with the theory of elasticity will be helpful for the reader, on the other hand no knowledge about the method of finite elements will be necessary. The second part is written in order to help the reader to understand the ideas. Here the general method is applied to one-dimensional networks, where the main ideas are illustrated without distracting technical details which are inevitable in the two-dimensional case which naturally is the only geodetically relevant one, and which we hope to cover in a following paper, where we shall demonstrate that the time of dreaming about a geodetic theory of elasticity is definitely over.

2

We shall now concentrate on a single triangle (the finite element) of the network and find the contributions to normals from observations directly related to this triangle.

In order to take into consideration the commonly used geometric geodetic observations but also to give a certain symmetry to the resulting mathematical model we shall consider the following four types of observations:

1. Observations of the (logarithm of the) distance between two vertices of the network [*absolute distances*],
2. Observations of the logarithm of the distance from one vertex of the network to another vertex apart from an additive constant α, which is supposed to be the same for all observations of this type [*relative distances*] from the same vertex,
3. Observations of grid directions (in radians) for sides of triangles in the network [*absolute directions*],
4. Observations of grid directions from one vertex of the network to another vertex apart from an additive constant β, which is supposed to be the same for all observations of this type [*relative directions*] from the same vertex.

The observations of the types 1 and 4 are natural geodetic observations but as it may be of interest to study the difference between the behaviour of relative and absolute observations the inclusion of observations of the types 2 and 3 can be of some practical interest.

One advantage by concentrating on a single triangle is that we can choose a simple and systematic system for indexing of the quantities entering in our calculations.

We will number the vertices of the triangle by 1, 2, and 3 and index quantities referring to the vertex i ($i = 1, 2, 3$) by i e.g. we denote the (Cartesian) coordinate of the vertex 2 by x_2 and y_2 and the additive constants α and β connected with the relative observations in the vertex 2 by α_2 and β_2. The components of the displacement vector for the vertex i will be denoted by u_i and v_i.

Correspondingly quantities naturally pertaining to oriented sides of the triangle will be equipped with double indices, e.g. $x_{23} = x_3 - x_2$, the x-component of the oriented triangle side from vertex 2 to vertex 3.

If $i = 2$ then $x_{i-1,i+1}$ refers to the side opposite to vertex 2 (with a well determined direction). In summation formulas it is convenient to use the same expression also for $i = 1$ or 3. Therefore $i = 0$ respectively $i = 4$ shall mean the same as $i = 3$ respectively $i = 1$.

Relative observations refer naturally to oriented sides and one of the vertices, we will agree to let the first of the two indices refer to the vertex in which the observation has taken place.

The equations of observation which express the corrections to the observations by the displacements and the observed values can now be written as:

1. For absolute observations:

$$w = p_{ij}u_{ij} + q_{ij}v_{ij} - f_{ij} \tag{1}$$

where w is the correction to the observed value f_{ij}, $u_{ij} = u_i - u_j$, $v_{ij} = v_i - v_j$ where (u_i, v_i) is the displacement vector of the vertex i and

$$p_{ij} = \frac{x_{ij}}{r_{ij}^2}, \qquad q_{ij} = \frac{y_{ij}}{r_{ij}^2} \tag{2}$$

for observations of distances and

$$p_{ij} = -\frac{y_{ij}}{r_{ij}^2}, \qquad q_{ij} = \frac{x_{ij}}{r_{ij}^2} \tag{3}$$

for observations of directions. Here $r_{ij}^2 = x_{ij}^2 + y_{ij}^2$.

2. For relative observations:

$$w = p_{ij}u_{ij} + q_{ij}v_{ij} + \alpha_i - f_{ij} \tag{4}$$

for observations of distances (p and q are here again given by (2)) and

$$w = p_{ij}u_{ij} + q_{ij}v_{ij} + \beta_i - f_{ij} \tag{5}$$

for observations of directions (p and q from (3)).

We are interested in the contributions from such equations to the normals or rather to the weighted square sum of the corrections which we will call *the energy form.* The whole contribution from all the observations of relative distances in the triangle $1, 2, 3$ is for example:

$$\sum_{i,j} e_{ij}\big(p_{ij}u_{ij} + q_{ij}v_{ij} + \alpha_i - f_{ij}\big)^2 \qquad i, j = 1,\ 2,\ 3, \quad i \neq j \tag{6}$$

where e_{ij} are the weights.

3

The energy form to which we have found the contribution from a single triangle shall now be considered as a discrete approximation to a continuous quadratic form.

First we must prolong the displacement vectors (u, v) which are defined only at the vertices of the network, to a continuous vector field defined on the whole area covered by the network. Such a prolongation is uniquely defined by the demands that 1) u and v shall be continuous functions 2) which at the vertices coincide with the discrete displacements and 3) which in the interior of every triangle of the network vary linearly. From this definition follows that the partial derivatives $\partial u/\partial x$, $\partial u/\partial y$, $\partial v/\partial x$, $\partial v/\partial y$ are constant in the interior of each triangle and that

$$\begin{aligned} u_{ij} &= x_{ij}\frac{\partial u}{\partial x} + y_{ij}\frac{\partial u}{\partial y} \\ v_{ij} &= x_{ij}\frac{\partial v}{\partial x} + y_{ij}\frac{\partial v}{\partial y}. \end{aligned} \tag{7}$$

The additive constants α and β which also are defined only at the vertices are prolonged in the same way and if the values of the prolonged functions α and β at the barycentre of the triangle $1, 2, 3$ are denoted by α_0 and β_0 then we have

$$\begin{aligned} \alpha_i &= \alpha_0 + \tfrac{1}{3}\big(x_{i-1,i} + x_{i+1,i}\big)\frac{\partial \alpha}{\partial x} + \tfrac{1}{3}\big(y_{i-1,i} + y_{i+1,i}\big)\frac{\partial \alpha}{\partial y} \\ \beta_i &= \beta_0 + \tfrac{1}{3}\big(x_{i-1,i} + x_{i+1,i}\big)\frac{\partial \beta}{\partial x} + \tfrac{1}{3}\big(y_{i-1,i} + y_{i+1,i}\big)\frac{\partial \beta}{\partial y}. \end{aligned} \qquad i = 1,\ 2,\ 3 \tag{8}$$

Now we can substitute the expression (7) and (10) into the forms such as (6) and so we have a quadratic form in the partial derivatives of first order of the displacements of α and β and in α_0, β_0, and in the observed values with coefficients depending of the weights and the coordinate differences x_{ij}, y_{ij}. Before we write down this quadratic form explicitly we must make some small changes.

We want to write that contribution to the energy form on which we concentrate as an integral over the triangle 123. The partial derivatives of u, v, α, and β are constant and the mean values of α and β are α_0 and β_0 respectively, therefore we may omit the zero indices in α_0 and β_0 without spoiling the approximation the result of the integration is almost only that of multiplication of the value of the quadratic form by the area of the triangle. This multiplication can be corrected for by dividing every weight by the area of the triangle; the weight per unit of area is also the reasonable unit for weight in a continuous network.

We also have to suppose the observed value f_{ij} of the four different types to be prolonged in a reasonable way. This will be described later on, until then we shall concentrate on the 'left hand side' and formally treat f as a vector.

The vector field

$$W = \left[\frac{\partial u}{\partial x}, \frac{\partial u}{\partial y}, \frac{\partial v}{\partial x}, \frac{\partial v}{\partial y}, \frac{\partial \alpha}{\partial x}, \frac{\partial \alpha}{\partial y}, \alpha, \frac{\partial \beta}{\partial x}, \frac{\partial \beta}{\partial y}, \beta, f\right]^{\mathrm{T}}$$

is now well defined for the whole area Ω covered by the network, and we can now define the energy form E for the continuous network by the integral

$$E = \tfrac{1}{2} \sum_k \iint_\Omega W^{\mathrm{T}} M_k W \, dx \, dy, \tag{9}$$

where M_k for $k = 1, 2, 3, 4$ are the matrices defining the contributions to the energy form corresponding to the four different types of observations. M_k is constant in the interior of each triangle of the network, and E is likely to be a good continuous approximation to the discrete energy form for the original discrete network.

The factor 1/2 stems from the fact that all observations referring to interior sides of the network are included twice in the integral once for each of the triangles into which the side enters. Therefore observations referring to boundary sides of the network should enter with double weight into the formulas.

M_k is an 11 by 11 matrix and for each value of k a considerable part of its entries are zero. For typographical reasons it is most practical to partition the matrix in the following way before writing it out explicitly:

$$M = \begin{bmatrix} \underset{4\times 4}{M_{ww}} & \underset{4\times 3}{M_{w\alpha}} & \underset{4\times 3}{M_{w\beta}} & \underset{4\times 1}{M_{wf}} \\ \underset{3\times 4}{M_{w\alpha}^{\mathrm{T}}} & \underset{3\times 3}{M_{\alpha\alpha}} & \underset{3\times 3}{M_{\alpha\beta}} & \underset{3\times 1}{M_{\alpha f}} \\ \underset{3\times 4}{M_{w\beta}^{\mathrm{T}}} & \underset{3\times 3}{M_{\alpha\beta}^{\mathrm{T}}} & \underset{3\times 3}{M_{\beta\beta}} & \underset{3\times 1}{M_{\beta f}} \\ \underset{1\times 4}{M_{wf}^{\mathrm{T}}} & \underset{1\times 3}{M_{\alpha f}^{\mathrm{T}}} & \underset{1\times 3}{M_{\beta f}^{\mathrm{T}}} & \underset{1\times 1}{M_{ff}} \end{bmatrix}. \tag{10}$$

For all values of k, M_{ww} is given by

$$\begin{bmatrix} \sum e_{ij}p_{ij}^2x_{ij}^2 & \sum e_{ij}p_{ij}^2x_{ij}y_{ij} & \sum e_{ij}p_{ij}q_{ij}x_{ij}^2 & \sum e_{ij}p_{ij}q_{ij}x_{ij}y_{ij} \\ \sum e_{ij}p_{ij}^2x_{ij}y_{ij} & \sum e_{ij}p_{ij}^2y_{ij}^2 & \sum e_{ij}p_{ij}q_{ij}x_{ij}y_{ij} & \sum e_{ij}p_{ij}q_{ij}y_{ij}^2 \\ \sum e_{ij}p_{ij}q_{ij}x_{ij}^2 & \sum e_{ij}p_{ij}q_{ij}x_{ij}y_{ij} & \sum e_{ij}q_{ij}^2x_{ij}^2 & \sum e_{ij}q_{ij}^2x_{ij}y_{ij} \\ \sum e_{ij}p_{ij}q_{ij}x_{ij}y_{ij} & \sum e_{ij}p_{ij}q_{ij}y_{ij}^2 & \sum e_{ij}q_{ij}^2x_{ij}y_{ij} & \sum e_{ij}q_{ij}^2y_{ij}^2 \end{bmatrix}.$$

The summation is to be taken over $i \neq j$ but for the absolute types (i.e. for $k = 1$ and 3) there is only one observation for each side i.e. the summation is only over $i < j$.

All the submatrices containing an α-index vanish for $k \neq 2$ as those with a β vanish for $k \neq 4$, $M_{\alpha\beta}$ vanishes always as none of the observations we consider concerns both α and β constants.

$M_{w\alpha} = M_{w\beta}$ have the following form with the corresponding meaning of p and q ((2) or (3)):

$$M_{w\alpha} = -\begin{bmatrix} \frac{1}{3}\sum e_{ij}p_{ij}x_{ij}x_i' & \frac{1}{3}\sum e_{ij}p_{ij}x_{ij}y_i' & \sum e_{ij}p_{ij}x_{ij} \\ \frac{1}{3}\sum e_{ij}p_{ij}y_{ij}x_i' & \frac{1}{3}\sum e_{ij}p_{ij}y_{ij}y_i' & \sum e_{ij}p_{ij}y_{ij} \\ \frac{1}{3}\sum e_{ij}q_{ij}x_{ij}x_i' & \frac{1}{3}\sum e_{ij}q_{ij}x_{ij}y_i' & \sum e_{ij}q_{ij}x_{ij} \\ \frac{1}{3}\sum e_{ij}q_{ij}y_{ij}x_i' & \frac{1}{3}\sum e_{ij}q_{ij}y_{ij}y_i' & \sum e_{ij}q_{ij}y_{ij} \end{bmatrix}$$

where $x_i' = x_{i-1,i} + x_{i+1,i}$ and $y_i' = y_{i-1,i} + y_{i+1,i}$.

$M_{\alpha\alpha} = M_{\beta\beta}$ has the following appearance:

$$M_{\alpha\alpha} = \begin{bmatrix} \frac{1}{9}\sum e_{ij}(x_i')^2 & \frac{1}{9}\sum e_{ij}x_i'y_i' & \frac{1}{3}\sum e_{ij}x_i' \\ \frac{1}{9}\sum e_{ij}x_i'y_i' & \frac{1}{9}\sum e_{ij}(y_i')^2 & \frac{1}{3}\sum e_{ij}y_i' \\ \frac{1}{3}\sum e_{ij}x_i' & \frac{1}{3}\sum e_{ij}y_i' & \sum e_{ij} \end{bmatrix}.$$

In order to make possible a prolongation of the values f_{ij} of the observations and to obtain a convenient expression for the 'forces' we must look for a more invariant representation for f_{ij}.

Let us start with type 1. As we have here three values f_{ij} for every triangle—one value connected with the direction of every side of the triangle—it could perhaps be obvious to try to express them from a symmetric tensor.

In fact if we write

$$f_{ij} = \frac{1}{r_{ij}^2}\begin{bmatrix} x_{ij} & y_{ij} \end{bmatrix}\begin{bmatrix} g_1 & g_3 \\ g_3 & g_2 \end{bmatrix}\begin{bmatrix} x_{ij} \\ y_{ij} \end{bmatrix}, \qquad i,j = 1,2;\ 2,3;\ 3,1; \qquad (11)$$

then we have three linear equations for the determination of the three unknowns g_1, g_2, g_3. Furthermore the determinant of this system will vanish only if the area of the triangle vanishes. Therefore we can find the symmetric tensor

$$G = \begin{bmatrix} g_1 & g_3 \\ g_3 & g_2 \end{bmatrix}$$

for every triangle of the network. We assign this value of G to the barycentre of the triangle and prolong it to all the area covered by the network continuously and piecewise linearly in the following way: To the vertices we assign the mean value of the G's corresponding to the adjacent triangles (eventually with weights proportional to the respective triangles). If we then connect each barycentre with the vertices of its triangle by straight lines we have a new network of triangles for which G has been defined for all its vertices and we can in exactly one way prolong G to a continuous symmetric tensor field which varies linearly in each of the small triangles. This tensor field is obviously differentiable almost overall with uniformly bounded derivatives, and this is all we want.

We shall now find the matrix elements of the energy form involving g_1, g_2, and g_3.

The contribution from a finite element to E is (for type 1):

$$\Delta E = \sum_{12,23,31} e_{ij} \Bigg(p_{ij} x_{ij} \frac{\partial u}{\partial x} + p_{ij} y_{ij} \frac{\partial u}{\partial y} + q_{ij} x_{ij} \frac{\partial v}{\partial x} + q_{ij} x_{ij} \frac{\partial v}{\partial y} \\ - \frac{x_{ij}^2}{r_{ij}^2} g_1 - \frac{2 x_{ij} y_{ij}}{r_{ij}^2} g_3 - \frac{y_{ij}^2}{r_{ij}^2} g_2 \Bigg)^2$$

and we find

$$M_{wg} = - \begin{bmatrix} \sum e_{ij} p_{ij} x_{ij} \frac{x_{ij}^2}{r_{ij}^2} & 2 \sum e_{ij} p_{ij} x_{ij} \frac{x_{ij} y_{ij}}{r_{ij}^2} & \sum e_{ij} p_{ij} x_{ij} \frac{y_{ij}^2}{r_{ij}^2} \\ \sum e_{ij} p_{ij} y_{ij} \frac{x_{ij}^2}{r_{ij}^2} & 2 \sum e_{ij} p_{ij} y_{ij} \frac{x_{ij} y_{ij}}{r_{ij}^2} & \sum e_{ij} p_{ij} y_{ij} \frac{y_{ij}^2}{r_{ij}^2} \\ \sum e_{ij} q_{ij} x_{ij} \frac{x_{ij}^2}{r_{ij}^2} & 2 \sum e_{ij} q_{ij} x_{ij} \frac{x_{ij} y_{ij}}{r_{ij}^2} & \sum e_{ij} q_{ij} x_{ij} \frac{y_{ij}^2}{r_{ij}^2} \\ \sum e_{ij} q_{ij} y_{ij} \frac{x_{ij}^2}{r_{ij}^2} & 2 \sum e_{ij} q_{ij} y_{ij} \frac{x_{ij} y_{ij}}{r_{ij}^2} & \sum e_{ij} q_{ij} y_{ij} \frac{y_{ij}^2}{r_{ij}^2} \end{bmatrix} \tag{12}$$

a result which is valid also for type 3. The reader should notice that all the entries of this matrix are identical to (or twice the value of) entries occurring in M_{ww}.

Now to the types 2 and 4. Here we have six observations f_{ij} which fall into two groups: the positive group $i, j = 1, 2;\ 2, 3;\ 3, 1$ and the negative group $i, j = 2, 1;\ 3, 2;\ 1, 3$ and accordingly we define two symmetric tensors G^+ and G^-, G^+ being defined from the positive group by (11) and G^- by the same formula for $i, j = 2,1;\ 3,2;\ 1,3$. We may write ΔE as

$$\Delta E = \sum_{+} e_{ij}\Bigg(p_{ij}x_{ij}\frac{\partial u}{\partial x} + p_{ij}y_{ij}\frac{\partial u}{\partial y} + q_{ij}x_{ij}\frac{\partial v}{\partial x} + q_{ij}x_{ij}\frac{\partial v}{\partial y} \\ - \tfrac{1}{3}\big(x_{i,i-1} + x_{i,i+1}\big)\frac{\partial \alpha}{\partial x} - \tfrac{1}{3}\big(y_{i,i-1} + y_{i,i+1}\big)\frac{\partial \alpha}{\partial y} \\ - \alpha - \frac{x_{ij}^2}{r_{ij}^2}g_1^+ - \frac{2x_{ij}y_{ij}}{r_{ij}^2}g_3^+ - \frac{y_{ij}^2}{r_{ij}^2}g_2^+\Bigg)^2 \\ + \sum_{-} \cdots .$$

Here M_{wg^+} and M_{wg^-} are of the same form as (12) with summation over the positive respectively the negative group.

We find $M_{\alpha g^+}$ equals

$$\begin{bmatrix} \frac{1}{3}\sum_{+} e_{ij}\big(x_{ij} + x_{ik}\big)\frac{x_{ij}^2}{r_{ij}^2} & \frac{2}{3}\sum_{+} e_{ij}\big(x_{ij} + x_{ik}\big)\frac{x_{ij}y_{ij}}{r_{ij}^2} & \frac{1}{3}\sum_{+} e_{ij}\big(x_{ij} + x_{ik}\big)\frac{y_{ij}^2}{r_{ij}^2} \\ \frac{1}{3}\sum_{+} e_{ij}\big(y_{ij} + x_{ik}\big)\frac{x_{ij}^2}{r_{ij}^2} & \frac{2}{3}\sum_{+} e_{ij}\big(y_{ij} + y_{ik}\big)\frac{x_{ij}y_{ij}}{r_{ij}^2} & \frac{1}{3}\sum_{+} e_{ij}\big(y_{ij} + x_{ik}\big)\frac{y_{ij}^2}{r_{ij}^2} \\ \sum_{+} e_{ij}\frac{x_{ij}^2}{r_{ij}^2} & 2\sum_{+} e_{ij}\frac{x_{ij}y_{ij}}{r_{ij}^2} & \sum_{+} e_{ij}\frac{y_{ij}^2}{r_{ij}^2} \end{bmatrix}$$

where $k = i - 1$. For $M_{\alpha g^-}$ we have the same expression with summation over the negative groups and $k = i + 1$.

4

According to the method of least squares the solution of our continuous network problem is defined as that field of displacement vectors (u, v) and 'additives' α and β for which the energy form E attains its minimum for the given values of observations f. This problem will be treated here as a problem of variational calculus but it is of the same type as problems encountered in the least-squares methods used for the determination of the gravity field of the Earth, i.e. least-squares methods in Hilbert space and it could be treated in almost the same way.

In the variational calculus it is generally much simpler to find necessary conditions for a minimum than to find sufficient ones. So also here, and in this paper we shall find only necessary conditions. The sufficient conditions will be given at another occasion, we shall here only mention that it seems to be less difficult to prove the existence of a solution in the geodetic theory of elasticity than it is in classical elasticity theory [3], pp. 88–100.

In deducing the necessary conditions we meet with a technical difficulty. We have to apply Green's theorem on the energy integral but coefficients of the quadratic form in the integral i.e. the entries of the matrix M_k are not generally, as they should be, continuously differentiable, they are even not continuous under the passage of a side of the network. Only if the network is *homogeneous* i.e. $\sum M_k$ is constant all over the network we may apply Green's theorem without applying first a smoothing procedure on the coefficients. Therefore we shall describe here an interpolation method for the matrices $\sum M_k$, which is simple but certainly not the 'best' one. As we cannot warrant that the reader will find this method interesting we believe the best thing he could do (at least at the first reading) would be to jump directly to Section 5.

First we fix the value the matrix shall attain at the barycentre of a triangle to be the value we have found for the matrix in Section 3. Then we define that its value at any vertex of the network shall be the weighted mean of its original values in the adjacent triangles with weights the areas of the respective triangles.

The next step is then to interpolate between the vertices along the side using one polynomial for each side. The values of this polynomial at the two vertices corresponding to it shall be those just defined and the differential quotient of the polynomial shall vanish at the vertices.

If we regard the side in question to be the part of an axis between the points (the vertices) -1 and $+1$ and the value at -1 should be a and that at $+1$ should be b then the polynomial

$$P(x) = \frac{a+b}{2} + \frac{b-a}{4}\left(3x - x^3\right)$$

will do the job.

The values thus defined along all the sides of the network, all that is left is to interpolate to points in the interior of the triangles but different from the barycentre. Each such point defines a half line through the barycentre of the triangle to the interior which it belongs and this half line will meet the circumference of this triangle at one point where the value has already been fixed. All we have to do is to interpolate between this point and the barycentre in the same way as above.

It is not difficult (although some reflection may be needed) to persuade oneself that this method of interpolation has the following three properties which are essential for the application we shall make of it:

1. The first derivatives (with respect to x and y) of the interpolated values are continuous.
2. For each point of the network each entry of the matrix at that point depends linearly upon the values of the same entry of the original matrices—from which follows that symmetry properties of the original matrices will be preserved by the interpolation.
3. Properties of definiteness of the original matrices or of submatrices of those will be preserved by the interpolation.

We also observe that if the network is homogeneous then nothing is changed by using this method of interpolation.

5

In order to avoid typographical excesses we shall further on limit ourselves to an important special type of networks. Our method is sufficiently general to make it possible for the reader to deduce the general formulas.

To make the meaning clear we start with an example.

The simplest type of a geodetic network is perhaps that consisting of congruent equilateral triangles. Suppose we have made all four types of observations in such a network and that all observations of the same type is of the same weight. Let the side length be l and let one angle between the x-axis and one of the sides be ϕ. We will find first the submatrix M_{ww} of the matrix M corresponding to relative observations of directions with weight e per unit of area.

For the first element in the first column we have

$$
\begin{aligned}
m_{11} &= \sum_{ij} e \frac{x_{ij}^2 y_{ij}^2}{r_{ij}^4} = e \sum_{1}^{6} \cos^2\Big(\frac{k\pi}{3} + \phi\Big) \sin^2\Big(\frac{k\pi}{3} + \phi\Big) \\
&= \frac{e}{4} \sum_{1}^{6} \sin^2\Big(\frac{2k\pi}{3} + 2\phi\Big) = \frac{e}{8} \sum_{1}^{6} \Big(1 - \cos\Big(\frac{4k\pi}{3} + 4\phi\Big)\Big) = \frac{3e}{4}.
\end{aligned}
$$

In the same way we find

$$
m_{44} = \frac{3e}{4}
$$

and

$$
m_{14} = m_{23} = m_{32} = m_{41} = -\frac{3e}{4}.
$$

Similarly

$$
\begin{aligned}
m_{22} &= \sum_{ij} e \frac{x_{ij}^4}{r_{ij}^4} = e \sum_{1}^{6} \cos^4\Big(\frac{k\pi}{3} + \phi\Big) = \frac{e}{4} \sum_{1}^{6} \Big(1 + \cos\Big(\frac{2k\pi}{3} + 2\phi\Big)\Big)^2 \\
&= \frac{e}{4} \sum_{1}^{6} \Big(1 + 2\cos\Big(\frac{2k\pi}{3} + 2\phi\Big) + \cos^2\Big(\frac{2k\pi}{3} + 2\phi\Big)\Big) \\
&= \frac{e}{4} \Big(\sum_{1}^{6} 1 + 2 \sum_{1}^{6} \cos\Big(\frac{2k\pi}{3} + 2\phi\Big) + \frac{1}{2} \sum_{1}^{6} \Big(1 + \cos\Big(\frac{4k\pi}{3} + 4\phi\Big)\Big)\Big) \\
&= \frac{3e}{2} + \frac{3e}{4} = \frac{9e}{4}.
\end{aligned}
$$

In the same way we find

$$
m_{33} = \frac{9e}{4}.
$$

Moreover

$$
m_{12} = m_{21} = m_{34} = m_{43} = m_{13} = m_{24} = m_{31} = m_{42} = 0.
$$

Evidently absolute observations of directions will give the same M_{ww}-matrix if the weight per area unit is $2e$.

We write down this result together with the corresponding one for observations of distances:

$$M_{ww} = e \begin{bmatrix} \frac{3}{4} & 0 & 0 & -\frac{3}{4} \\ 0 & \frac{9}{4} & -\frac{3}{4} & 0 \\ 0 & -\frac{3}{4} & \frac{9}{4} & 0 \\ -\frac{3}{4} & 0 & 0 & \frac{3}{4} \end{bmatrix} \qquad \text{for directions,}$$

and

$$M_{ww} = e \begin{bmatrix} \frac{9}{4} & 0 & 0 & \frac{3}{4} \\ 0 & \frac{3}{4} & \frac{3}{4} & 0 \\ 0 & \frac{3}{4} & \frac{3}{4} & 0 \\ \frac{3}{4} & 0 & 0 & \frac{9}{4} \end{bmatrix} \qquad \text{for distances.}$$

With the same method we find the other submatrices of

$$M_{w\alpha} = e \begin{bmatrix} 0 & 0 & -3 \\ 0 & 0 & 0 \\ 0 & 0 & 0 \\ 0 & 0 & -3 \end{bmatrix} \qquad \text{for type-2 observations,}$$

$$M_{w\beta} = e \begin{bmatrix} 0 & 0 & 0 \\ 0 & 0 & 3 \\ 0 & 0 & -3 \\ 0 & 0 & 0 \end{bmatrix} \qquad \text{for type-4 observations,}$$

$$M_{\alpha\alpha} = M_{\beta\beta} = e \begin{bmatrix} \frac{l^2}{3} & 0 & 0 \\ 0 & \frac{l^2}{3} & 0 \\ 0 & 0 & 3 \end{bmatrix} \qquad \text{for type-2 or type-4 observations,}$$

$$M_{wg} = -e \begin{bmatrix} \frac{9}{4} & 0 & \frac{3}{4} \\ 0 & \frac{3}{2} & 0 \\ 0 & \frac{3}{2} & 0 \\ \frac{3}{4} & 0 & \frac{9}{4} \end{bmatrix} \qquad \text{for type-1 observations,}$$

$$M_{wg} = -e \begin{bmatrix} 0 & -\frac{3}{2} & 0 \\ -\frac{3}{4} & 0 & -\frac{9}{4} \\ \frac{9}{4} & 0 & \frac{3}{4} \\ 0 & \frac{3}{2} & 0 \end{bmatrix} \qquad \text{for type-3 observations.}$$

M_{wg^+} for type 2 is half M_{wg} for type 1, and M_{wg^-} for type 4 is half M_{wg} for type 3.

Finally we have

$$M_{\alpha g^+} = M_{\beta g^-}$$
$$= \begin{bmatrix} \frac{el}{4}\sqrt{3}\cos(3\phi+\pi/6) & \frac{el}{2}\sqrt{3}\sin(3\phi+\pi/6) & -\frac{el}{4}\sqrt{3}\cos(3\phi+\pi/6) \\ \frac{el}{4}\sqrt{3}\sin(3\phi+\pi/6) & -\frac{el}{2}\sqrt{3}\cos(3\phi+\pi/6) & -\frac{el}{4}\sqrt{3}\sin(3\phi+\pi/6) \\ \frac{3e}{4} & 0 & \frac{3e}{4} \end{bmatrix}$$

while $M_{\alpha g^-}$ and $M_{\beta g^+}$ are the same except for the opposite sign in the two first rows.

It is evident that networks of this type are homogeneous and we also see that the matrices M_k are independent of ϕ, i.e. they are invariant with respect to orthogonal transformations. Networks possessing this invariance property are said to be *isotropic*. (The expressions 'homogeneous' and 'isotropic' are borrowed from classical elasticity theory.)

The matrices we have found for this simple network are typical for the slightly more general isotropic networks. For general isotropic networks the parts of the integrands of the energy integral corresponding to the four types of observations are respectively:

$$\begin{aligned} E_1\Big[&\frac{\partial u}{\partial x}\Big(3\frac{\partial u}{\partial x}+\frac{\partial v}{\partial y}-3g_1-g_2\Big)+\frac{\partial u}{\partial y}\Big(\frac{\partial u}{\partial y}+\frac{\partial v}{\partial x}-2g_3\Big)+\frac{\partial v}{\partial x}\Big(\frac{\partial u}{\partial y}+\frac{\partial v}{\partial x}-2g_3\Big)\\ &+\frac{\partial v}{\partial y}\Big(\frac{\partial u}{\partial x}+3\frac{\partial v}{\partial y}-g_1-3g_2\Big)-g_1\Big(3\frac{\partial u}{\partial x}+\frac{\partial v}{\partial y}\Big)-2g_3\Big(3\frac{\partial u}{\partial y}+\frac{\partial v}{\partial x}\Big)\\ &-g_2\Big(\frac{\partial u}{\partial x}+3\frac{\partial v}{\partial y}\Big)+\text{terms quadratic in } g\Big]. \end{aligned} \tag{13}$$

Also in the three following formulas we shall ignore the terms quadratic in g because their explicit form is irrelevant.

$$\begin{aligned} E_2\Big[&\frac{\partial u}{\partial x}\Big(3\frac{\partial u}{\partial x}+\frac{\partial v}{\partial y}-4\alpha-3g_1-g_2\Big)+\frac{\partial u}{\partial y}\Big(\frac{\partial u}{\partial y}+\frac{\partial v}{\partial x}-2g_3\Big)+\frac{\partial v}{\partial x}\Big(\frac{\partial u}{\partial y}+\frac{\partial v}{\partial x}-2g_3\Big)\\ &+\frac{\partial v}{\partial y}\Big(\frac{\partial u}{\partial x}+3\frac{\partial v}{\partial y}-4\alpha-g_1-3g_2\Big)-\alpha\Big(4\frac{\partial u}{\partial x}+4\frac{\partial v}{\partial y}-8\alpha-4g_1-4g_2\Big)\\ &+\tfrac{4}{3}l\frac{\partial\alpha}{\partial x}\Big[l\frac{\partial\alpha}{\partial x}+\tfrac{\sqrt{3}}{4}\Big((\tilde g_1-\tilde g_2)\cos(3\phi+\pi/6)+2\tilde g_3\sin(3\phi+\pi/6)\Big)\Big]\\ &+\tfrac{4}{3}l\frac{\partial\alpha}{\partial y}\Big[l\frac{\partial\alpha}{\partial y}+\tfrac{\sqrt{3}}{4}\Big((\tilde g_1-\tilde g_2)\sin(3\phi+\pi/6)-2\tilde g_3\cos(3\phi+\pi/6)\Big)\Big]\\ &-g_1\Big(3\frac{\partial u}{\partial x}+\frac{\partial v}{\partial y}-4\alpha\Big)-2g_3\Big(3\frac{\partial u}{\partial y}+\frac{\partial v}{\partial x}\Big)-g_2\Big(\frac{\partial u}{\partial x}+3\frac{\partial v}{\partial y}-4\alpha\Big) \end{aligned}$$

$$+ \tfrac{\sqrt{3}}{3} l(\tilde{g}_1 - \tilde{g}_2)\Big(\frac{\partial \alpha}{\partial x}\cos(3\phi + \pi/6) + \frac{\partial \alpha}{\partial y}\sin(3\phi + \pi/6)\Big)$$

$$+ 2\tfrac{\sqrt{3}}{3} l\tilde{g}_3\Big(\frac{\partial \alpha}{\partial x}\sin(3\phi + \pi/6) - \frac{\partial \alpha}{\partial y}\cos(3\phi + \pi/6)\Big) + \cdots\Bigg] \tag{14}$$

where we have used the symbols

$$\begin{aligned} g_i &= \tfrac{1}{2}(g_i^+ + g_i^-); \\ \tilde{g}_i &= \tfrac{1}{2}(g_i^+ - g_i^-); \end{aligned} \qquad i = 1, 2, 3.$$

$$\begin{aligned} E_3\Bigg[&\frac{\partial u}{\partial x}\Big(\frac{\partial u}{\partial x} - \frac{\partial v}{\partial y} + 2g_3\Big) + \frac{\partial u}{\partial y}\Big(3\frac{\partial u}{\partial y} - \frac{\partial v}{\partial x} + g_1 + 3g_2\Big) - \frac{\partial v}{\partial x}\Big(\frac{\partial u}{\partial y} - 3\frac{\partial v}{\partial x} + 3g_1 + g_2\Big) \\ &- \frac{\partial v}{\partial y}\Big(\frac{\partial u}{\partial x} - \frac{\partial v}{\partial y} + 2g_3\Big) + g_1\Big(\frac{\partial u}{\partial y} - 3\frac{\partial v}{\partial x}\Big) \\ &+ 2g_3\Big(\frac{\partial u}{\partial x} - \frac{\partial v}{\partial y}\Big) + g_2\Big(3\frac{\partial u}{\partial y} - \frac{\partial v}{\partial x}\Big) + \cdots\Bigg]. \end{aligned} \tag{15}$$

$$\begin{aligned} E_4\Bigg[&\frac{\partial u}{\partial x}\Big(\frac{\partial u}{\partial x} - \frac{\partial v}{\partial y} + 2g_3\Big) + \frac{\partial u}{\partial y}\Big(3\frac{\partial u}{\partial y} - \frac{\partial v}{\partial x} - 4\beta + g_1 + 3g_2\Big) \\ &- \frac{\partial v}{\partial x}\Big(\frac{\partial u}{\partial y} - 3\frac{\partial v}{\partial x} - 4\beta + 3g_1 + g_2\Big) - \frac{\partial v}{\partial y}\Big(\frac{\partial u}{\partial x} - \frac{\partial v}{\partial y} + 2g_3\Big) \\ &- \beta\Big(4\frac{\partial u}{\partial x} - 4\frac{\partial v}{\partial y} - 8\beta + 4g_1 + 4g_2\Big) \\ &+ \tfrac{4}{3}l\frac{\partial \beta}{\partial x}\Big[l\frac{\partial \beta}{\partial x} + \tfrac{\sqrt{3}}{4}\Big((\tilde{g}_1 - \tilde{g}_2)\cos(3\phi + \pi/6) + 2\tilde{g}_3\sin(3\phi + \pi/6)\Big)\Big] \\ &+ \tfrac{4}{3}l\frac{\partial \beta}{\partial y}\Big[l\frac{\partial \beta}{\partial y} + \tfrac{\sqrt{3}}{4}\Big((\tilde{g}_1 - \tilde{g}_2)\sin(3\phi + \pi/6) - 2\tilde{g}_3\cos(3\phi + \pi/6)\Big)\Big] \\ &+ g_1\Big(\frac{\partial u}{\partial y} - 3\frac{\partial v}{\partial x} - 4\beta\Big) + 2g_3\Big(\frac{\partial u}{\partial x} - \frac{\partial v}{\partial y}\Big) + g_2\Big(3\frac{\partial u}{\partial y} - \frac{\partial v}{\partial x} - 4\beta\Big) \\ &+ \tfrac{\sqrt{3}}{3}l(\tilde{g}_1 - \tilde{g}_2)\Big(\frac{\partial \beta}{\partial x}\cos(3\phi + \pi/6) + \frac{\partial \beta}{\partial y}\sin(3\phi + \pi/6)\Big) \\ &+ 2\tfrac{\sqrt{3}}{3}l\tilde{g}_3\Big(\frac{\partial \beta}{\partial x}\sin(3\phi + \pi/6) - \frac{\partial \beta}{\partial y}\cos(3\phi + \pi/6)\Big) + \cdots\Bigg]. \end{aligned} \tag{16}$$

E_k is for given networks proportional to weight per unit of area of observations of type k. As an isotropic network is not necessarily a homogeneous one E_k may vary from point to point in the area.

6

It might be useful for the understanding of the following first to recall the classical connection between variational problems and the Neumann problem.

Let Ω be a domain in the plane bounded by a smooth closed curve ω. Given a continuous function f defined on Ω we want to find a function ϕ on Ω such that the integral

$$I(\phi) = \iint_{\Omega} \left[\left(\frac{\partial \phi}{\partial x}\right)^2 + \left(\frac{\partial \phi}{\partial y}\right)^2 - 2\phi f \right] dx\, dy \tag{17}$$

attains a minimum.

A necessary condition for $I(\phi)$ to be minimal is that for every differentiable function ψ on Ω

$$\delta I = \left(\frac{\partial}{\partial \epsilon} I(\phi + \epsilon\psi)\right)_{\epsilon=0} = 0,$$

or

$$\delta I = 2\iint_{\Omega} \left[\frac{\partial \phi}{\partial x}\frac{\partial \psi}{\partial x} + \frac{\partial \phi}{\partial y}\frac{\partial \psi}{\partial y} - \psi f \right] dx\, dy = 0. \tag{18}$$

According to Green's theorem (18) may be written

$$\delta I = -2\iint_{\Omega} \psi(\Delta\phi + f)\, dx\, dy + 2\int_{\omega} \psi \frac{\partial \phi}{\partial n}\, ds = 0, \tag{19}$$

where ds is the differential of the arc length on ω.

A form of reasoning well-known in the calculus of variations implies that for (19) to be satisfied for all differentiable ψ it is necessary that factors to ψ in the two integrands of (19) vanish identically, i.e. ϕ must satisfy the following partial differential equation

$$\Delta\phi + f = 0 \qquad \text{on } \Omega, \tag{20}$$

and the following boundary condition

$$\frac{\partial \phi}{\partial n} = 0 \qquad \text{on } \omega. \tag{21}$$

In more advanced treatises it is proved that in fact for certain well-defined function classes the boundary-value problem (20), (21) is equivalent with the original minimum problem for the integral in (17), [1], pp. 2–4.

If we will minimize the integral over Ω of an expression such as (13) we are faced with three new problems:

1. We shall minimize not over one function ϕ but over two functions u and v,
2. The factor to the given function f_1 is not 2ϕ but a linear combination $2(4\partial u/\partial x + 4\partial v/\partial y)$ of the differential quotients of the unknown functions,
3. There are also terms with g.

The last of these complications is easily solved, the term g is relevant only for the value of the minimum but not for the determination of the functions for which it is attained. We may therefore delete the term which is quadratic in g. Neither the two first points are really serious as we shall see now.

Instead of (17) we have now the integral

$$I(u,v) = \iint_{\Omega} E_1 \left[\frac{\partial u}{\partial x}\left(3\frac{\partial u}{\partial x} + \frac{\partial v}{\partial y} - 3g_1 - g_2\right) + \frac{\partial u}{\partial y}\left(\frac{\partial u}{\partial y} + \frac{\partial v}{\partial x} - 2g_3\right) \right.$$
$$+ \frac{\partial v}{\partial x}\left(\frac{\partial u}{\partial y} + \frac{\partial v}{\partial x} - 2g_3\right) + \frac{\partial v}{\partial y}\left(\frac{\partial u}{\partial x} + 3\frac{\partial v}{\partial y} - g_1 - 3g_2\right)$$
$$\left. - g_1\left(3\frac{\partial u}{\partial x} + \frac{\partial v}{\partial y}\right) - 2g_3\left(\frac{\partial u}{\partial y} + \frac{\partial v}{\partial x}\right) - g_2\left(\frac{\partial u}{\partial x} + 3\frac{\partial v}{\partial y}\right) \right] dx\, dy. \tag{22}$$

Exactly as above we see that a necessary condition for $I(u,v)$ to be minimal is that for every pair of differentiable functions u' and v' on Ω we have

$$\delta I = \left(\frac{\partial}{\partial \epsilon} I(u + \epsilon u', v + \epsilon v')\right)_{\epsilon=0} = 0,$$

or

$$\delta I = \iint_{\Omega} E_1 \left[\frac{\partial u'}{\partial x}\left(3\frac{\partial u}{\partial x} + \frac{\partial v}{\partial y} - 3g_1 - g_2\right) \right.$$
$$+ \frac{\partial u'}{\partial y}\left(\frac{\partial u}{\partial y} + \frac{\partial v}{\partial x} - 2g_3\right) + \frac{\partial v'}{\partial x}\left(\frac{\partial u}{\partial y} + \frac{\partial v}{\partial x} - 2g_3\right)$$
$$\left. + \frac{\partial v'}{\partial y}\left(\frac{\partial u}{\partial x} + 3\frac{\partial v}{\partial y} - g_1 - 3g_2\right) \right] dx\, dy = 0. \tag{23}$$

Here we may apply Green's theorem to obtain

$$\delta I = -\iint_{\Omega} \left[u'\left(\frac{\partial}{\partial x}\left(E_1\left(3\frac{\partial u}{\partial x} + \frac{\partial v}{\partial y} - 3g_1 - g_2\right)\right) + \frac{\partial}{\partial y}\left(E_1\left(\frac{\partial u}{\partial y} + \frac{\partial v}{\partial x} - 2g_3\right)\right)\right) \right.$$
$$+ v'\left(\frac{\partial}{\partial x}\left(E_1\left(\frac{\partial u}{\partial y} + \frac{\partial v}{\partial x} - 2g_3\right)\right) \right.$$
$$\left.\left. + \frac{\partial}{\partial y}\left(E_1\left(\frac{\partial u}{\partial x} + 3\frac{\partial v}{\partial y} - g_1 - 3g_2\right)\right)\right)\right] dx\, dy$$
$$+ 2\int_{\omega} E_1 \left[u'\left(n_x\left(3\frac{\partial u}{\partial x} + \frac{\partial v}{\partial y} - 3g_1 - g_2\right) + n_y\left(\frac{\partial u}{\partial y} + \frac{\partial v}{\partial x} - 2g_3\right)\right) \right.$$
$$\left. + v'\left(n_x\left(\frac{\partial u}{\partial y} + \frac{\partial v}{\partial x} - 2g_3\right) + n_y\left(\frac{\partial u}{\partial x} + 3\frac{\partial v}{\partial y} - g_1 - 3g_2\right)\right) \right] ds = 0, \tag{24}$$

but this is possible only if the factors to u' and v' in both of the integrals vanish. (n_x and n_y are the x- and y-components of the unit normal vector to ω.) The resulting partial differential equations and the boundary conditions are easily deduced, but we shall do it here only for homogeneous networks, i.e. for E_1 = constant:

$$\left.\begin{aligned} 3\frac{\partial^2 u}{\partial x^2} + \frac{\partial^2 u}{\partial y^2} + 2\frac{\partial^2 v}{\partial x \partial y} &= 3\frac{\partial g_1}{\partial x} + \frac{\partial g_2}{\partial x} + 2\frac{\partial g_3}{\partial y} \\ 2\frac{\partial^2 u}{\partial x \partial y} + \frac{\partial^2 v}{\partial x^2} + 3\frac{\partial^2 v}{\partial^2 y} &= 2\frac{\partial g_3}{\partial x} + \frac{\partial g_1}{\partial y} + 3\frac{\partial g_2}{\partial y} \end{aligned}\right\} \text{ in } \Omega \tag{25}$$

and

$$\left.\begin{aligned} n_x\Big(3\frac{\partial u}{\partial x} + \frac{\partial v}{\partial y}\Big) + n_y\Big(\frac{\partial u}{\partial y} + \frac{\partial v}{\partial x}\Big) &= n_x(3g_1 + g_2) + 2n_y g_3 \\ n_x\Big(\frac{\partial u}{\partial y} + \frac{\partial v}{\partial x}\Big) + n_y\Big(\frac{\partial u}{\partial x} + 3\frac{\partial v}{\partial y}\Big) &= 2n_x g_3 + n_y(g_1 + 3g_2) \end{aligned}\right\} \text{ on } \omega. \tag{26}$$

If we in the networks have all four types of observations we only have to add the expressions (13)–(16) and apply Green's theorem as above. The resulting differential equations may be written as

$$\begin{aligned} A\frac{\partial^2 u}{\partial x^2} + B\frac{\partial^2 u}{\partial y^2} + (A-B)\frac{\partial^2 v}{\partial x \partial y} - a\frac{\partial \alpha}{\partial x} - b\frac{\partial \beta}{\partial y} &= F_u \\ (A-B)\frac{\partial^2 u}{\partial x \partial y} + B\frac{\partial^2 v}{\partial x^2} + A\frac{\partial^2 v}{\partial^2 y} - a\frac{\partial \alpha}{\partial y} + b\frac{\partial \beta}{\partial x} &= F_v \\ a\frac{\partial u}{\partial x} + a\frac{\partial v}{\partial y} + \frac{al^2}{3}\Delta\alpha - 2a\alpha &= F_\alpha \\ b\frac{\partial u}{\partial y} - b\frac{\partial v}{\partial x} + \frac{bl^2}{3}\Delta\beta - 2b\beta &= F_\beta \end{aligned} \tag{27}$$

where

$$\begin{aligned} A &= \tfrac{9}{4}(E_1 + E_2) + \tfrac{3}{4}(E_3 + E_4) \\ A &= \tfrac{3}{4}(E_1 + E_2) + \tfrac{9}{4}(E_3 + E_4) \\ a &= 3E_2 \\ b &= 3E_4. \end{aligned}$$

The forces F_u, F_v, F_α, and F_β depend on the observations and may be found as in the special case (25). But for the study of random errors in geometric networks we are normally interested only in the left hand side of the equations which together with the boundary conditions play a role corresponding to the role played by the left hand side of the normal equations. But for the study of systematic errors and of the interplay between geometric and physical data they are extremely important.

The boundary conditions are

$$
\begin{aligned}
n_x\Big(A\frac{\partial u}{\partial x}+\frac{A-B}{2}\frac{\partial v}{\partial y}-a\alpha\Big)+n_y\Big(B\frac{\partial u}{\partial y}+\frac{A-B}{2}\frac{\partial v}{\partial x}-b\beta\Big)&=f_u\\
n_x\Big(\frac{A-B}{2}\frac{\partial u}{\partial y}+B\frac{\partial v}{\partial x}-a\alpha\Big)+n_y\Big(\frac{A-B}{2}\frac{\partial u}{\partial x}+A\frac{\partial v}{\partial y}+b\beta\Big)&=f_v\\
n_x\frac{\partial \alpha}{\partial x}+n_y\frac{\partial \alpha}{\partial y}&=f_\alpha\\
n_x\frac{\partial \beta}{\partial x}+n_y\frac{\partial \beta}{\partial y}&=f_\beta .
\end{aligned}
\tag{28}
$$

It seems to be so that the boundary forces f_u, f_v, f_α, and f_β may be supposed vanish if the boundary is shifted a half side length away from the physical boundary of the network. But as we have not yet found a rigorous proof for this strong conjecture we only mention it here and refer to the following treatment of the one-dimensional case where the correctness of the conjecture is obvious.

7

This last section shall try to describe how we got the idea of establishing a continuous analogue of the geodetic problem. When once established, then it is so apparent!

For the sake of simplicity consider a 1-dimensional case of absolute distance measurements. The straight line from 0 to L is subdivided by points P_1, P_2, P_3, ..., P_{n-1}, P_n. The length of each subinterval between consecutive points P_{i-1} and P_i is measured and the logarithm to the length is given with equal weight 1. Hence the observation equations are

$$\ln(x_{i-1,i}) = s_{i-1,i} \qquad i = 2,\, 3,\, \ldots,\, n. \tag{29}$$

The linearized observation equations are then

$$\frac{u_i - u_{i-1}}{x_{i-1,i}} = s_{i-1,i}. \tag{30}$$

Finally we put $x_{i-1,i}s_{i-1,i} = f_{i-1,i}$ and connect the observation $f_{i-1,i} \equiv f_{i-1}$ with the coordinate $i-1$:

$$
\left.
\begin{aligned}
u_2 - u_1 &= f_1\\
u_3 - u_2 &= f_2\\
&\vdots\\
u_n - u_{n-1} &= f_{n-1}
\end{aligned}
\right\}\quad n-1 \text{ equations}
\tag{31}
$$

or in matrix notation

$$Au = f \tag{32}$$

where

$$A = \begin{bmatrix} -1 & 1 & & & & \\ & -1 & 1 & & & \\ & & -1 & 1 & & \\ & & & \ddots & & \\ & & & & -1 & 1 \end{bmatrix}_{(n-1)\times n}. \tag{33}$$

The normals are obtained in the usual way:

$$Nu = A^{\mathrm{T}} Au = A^{\mathrm{T}} f \tag{34}$$

or

$$\begin{bmatrix} 1 & -1 & & & \\ -1 & 2 & -1 & & \\ & & \ddots & & \\ & & -1 & 2 & -1 \\ & & & -1 & 1 \end{bmatrix} \begin{bmatrix} u_1 \\ u_2 \\ \vdots \\ u_{n-1} \\ u_n \end{bmatrix} = \begin{bmatrix} -1 & & & \\ 1 & -1 & & \\ & \ddots & & \\ & & 1 & -1 \\ & & & 1 \end{bmatrix} \begin{bmatrix} f_1 \\ f_2 \\ \vdots \\ f_{n-2} \\ f_{n-1} \end{bmatrix}. \tag{35}$$

The n-dimensional matrix equation (35) can be looked upon as a submatrix equation of an infinite matrix equation with the general term

$$-u(j-1) + 2u(j) - u(j+1) = f(j-1) - f(j). \tag{36}$$

The transition from the infinite matrices to the finite ones demands some modifications near the indices 1 and n. We write fully out the equations concerned:

$$-u_0 + 2u_1 - u_2 = f_0 - f_1$$

and

$$-u_{n-1} + 2u_n - u_{n+1} = f_{n-1} - f_n.$$

It is easily seen that we have subtracted

$$-u_0 + u_1 = f_0 \qquad \text{and} \qquad u_n - u_{n+1} = -f_n \tag{37}$$

respectively, from the infinite matrices in order to get the normals (35).

Now we apply the method of images to determine the values of u_0 and u_{n+1}. By placing the "mirrors" at $x = 0$ and $x = n$ we get

$$u_0 = u_1 \qquad \text{and} \qquad u_n = u_{n+1}. \tag{38}$$

From (36) then follows

$$f_0 = f_n = 0. \tag{39}$$

The eigenvalue problem defined by the homogeneous difference equation (36) and the boundary conditions (37), or equivalently the eigenvalue problem of the matrix N, has the eigenvalues

$$\lambda_i = 4\sin^2\frac{i\pi}{2n} \qquad i = 0, 1, \ldots, n-1 \tag{40}$$

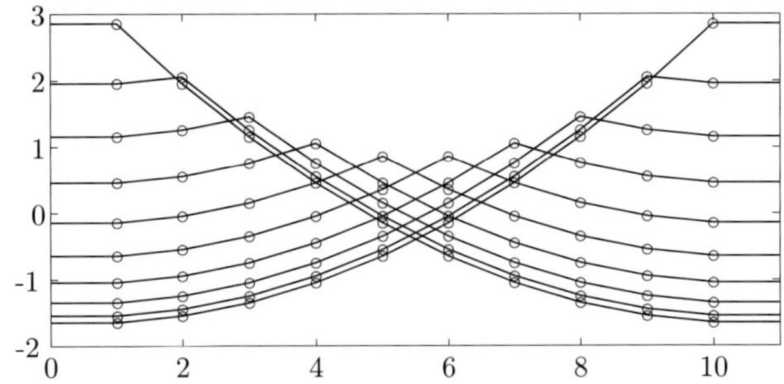

Fig. 11.1. The Green Function. Note that $G'(0,\xi) = 0$ and $G'(L,\xi) = 0$

and corresponding eigenvectors

$$\phi_i(j) = \cos\frac{2j-1}{2n}i\pi \qquad \begin{array}{l} i = 0, 1, \ldots, n-1 \\ j = 1, 2, \ldots, n. \end{array} \tag{41}$$

Normalization of the ϕ's yields

$$\psi_i(j) = \begin{cases} \frac{1}{\sqrt{n}} & \\ \sqrt{\frac{2}{n}}\cos\frac{2j-1}{2n}i\pi & \begin{array}{l} i = 1, 2, \ldots, n-1 \\ j = 1, 2, \ldots, n. \end{array} \end{cases} \tag{42}$$

In our simple example it has been possible to find an explicit expression for the pseudo-inverse of N:

$$N_{ij}^{+} = \frac{1}{2n}\Big(\frac{n-1}{3}(2n+5-6j) + (j-1)(j-2) + i(i-1)\Big) \qquad \text{for } j \geq i. \tag{43}$$

Notice that the row/column sums are zero, i.e. the vectors are orthogonal to a constant, corresponding to the fact that every vector orthogonal to a constant vector is estimable.

The continuous analogue to the eigenvalue problem for the difference equation (36) is

$$u'' + \lambda u = 0 \tag{44}$$

with the boundary conditions

$$u'(0) = u'(L) = 0. \tag{45}$$

Notice that these conditions are attached to $x = 0$ and $x = L$ which are points half a mesh width outside the physical network in analogy to the discrete case.

This problem has non-trivial solutions for the eigenvalues

$$\lambda_i = \Big(\frac{i\pi}{L}\Big)^2 \qquad i = 0, 1, \ldots \tag{46}$$

and the functions

$$u(x) = \cos\frac{i\pi}{L}x \tag{47}$$

are the corresponding eigenfunctions.

Obviously the set of eigenvectors/functions (42), (47) of the discrete and continuous problems are very similar. The same holds true for the pseudoinverse N_{ij}^{+} (43) and for the Green function of the problem (44), (45). The generalized Green function is

$$G(x,\xi) = \begin{cases} \frac{x^2+\xi^2}{2} + \frac{L}{3} - \xi & \text{for } x \leq \xi \\ \frac{x^2+\xi^2}{2} + \frac{L}{3} - x & \text{for } x \geq \xi. \end{cases} \tag{48}$$

In Figure 11.1 we have graphed the Green function with arguments $x = 1, 2, \ldots, 10$. Besides the circles are indicating the entries of N^{+} such that point No i has coordinate i. The absolute error of approximation $\delta(n) = N^{+} - G$ is $\delta(n) = -\frac{1}{12n}$ where $n = L$. The relative error is $\approx -\frac{1}{4n^2}$. An astonishing good approximation.

Let $\phi_{pq} = x_p - x_q$, $p < q$. Then the variance of this difference is given by

$$\sigma^2(\tilde{\phi}_{pq}) = G(p,p) + G(q,q) - G(p,q) - G(q,p) = q - p \tag{49}$$

i.e. the variance of a difference of lengths grows proportionally to this. The result is in good agreement with the result known from the discrete case.

It turned out that it was easy to determine the generalized Green function for absolute distance measurements while already in the case of 1-dimensional relative distance measurements we are faced with a generalized Green function expressed in not quite simple terms, cf. [5].

This and many other problems need thorough investigation before the model can be given the adequate geodetic interpretation. About this in forthcoming publications.

References

[1] Jean-Pierre Aubin. *Approximation of Elliptic Boundary-Value Problems.* Wiley-Interscience, New York, 1972.

[2] H. M. Dufour. Générations et applications des tableaux de variances des systèmes de moindres carrés. *Bulletin Géodésique*, 98:309–339, 1970.

[3] Gaetano Fichera. *Linear Elliptic Differential Systems and Eigenvalue Problems*, volume 8 of *Lecture Notes in Mathematics.* Springer, Berlin, 1965.

[4] G. Förster. Statik trigonometrischer Netze. *Mitteilungen des Reichsamts für Landesaufnahme*, 1934/35:13–25, 1935.

[5] W. S. Loud. Generalized inverses and generalized Green's functions. *Journal of Applied Mathematics*, 14:342–369, 1966.

[6] Peter Meissl. Über die Fehlerfortpflanzung in gewissen regelmäßigen flächig ausgebreiteten Nivellementnetzen. *Zeitschrift für Vermessungswesen*, 95:103–109, 1970.

12

Integrated Geodesy

Introduction

Though geodesy, as the theory of shape and size of the earth, may seem to be a purely geometrical science, the measurements performed depend not only on the relative position of the stations involved, but also on a physical entity, the potential W of the Earth. So, in order to determine the position of a station, some information concerning W is necessary. If, on the other hand, we know $W(x, y, z)$ as a function of position, then measurements of, say, the direction and magnitude of the gradient would determine the coordinates of a point and so it becomes an object for geodesy to determine W.

The classical way to find W is to split it up in a normal part U and an anomalous part T through the equation (all the references in the Introduction refer to [3])

$$T = W - U \tag{2-137}$$

where T is determined so as to satisfy the Laplacian $\Delta T = 0$ outside some surface, say, the geoid. On this surface a boundary condition found

$$\Delta g = -\frac{\partial T}{\partial h} + \frac{1}{\gamma}\frac{\partial \gamma}{\partial h}T \tag{2-147c}$$

presenting a third boundary value problem of potential theory, which can be solved in spherical approximation by Stokes formula

$$T = \frac{R}{4\pi}\iint_{\sigma} \Delta g S(\psi)\, d\sigma. \tag{2-163a}$$

The disadvantage of this method is that the geoid, being the equipotential surface $W =$ constant of the mean sea level, inside the Earth depends on the density ρ because of Poisson's formula

$$\Delta W = -4\pi k\rho + 2\omega^2. \tag{2-6}$$

In order to find the gravity anomaly Δg and at the same time satisfy that T shall be regular outside the geoid, we have to compute the gravity g_G on the geoid, that is, we have to reduce gravity measured at ground level, and at the same time remove the masses outside the geoid. In order to do that we have to make assumptions concerning the density ρ, and this is unsatisfactory at least from a theoretical point of view even though the practical influence of these assumptions usually is very small.

For this reason it is of basic importance, that Molodenskiy was able to go through essentially the same development as Stokes, without assuming anything about the density, by introducing the telluroid and letting this surface take the place of the geoid. The solution to his problem presents itself in spherical approximation as

$$T = \frac{R}{4\pi} \iint_\sigma \Delta g S(\psi)\, d\sigma + \frac{R}{4\pi} \iint_\sigma G_1 S(\psi)\, d\sigma \tag{8-50}$$

where the second term on the right side expresses the influence of the topographic effect, and Δg is calculated as the difference of measured gravity at ground level and normal gravity at the telluroid. Both Stokes' and Molodenskiy's method, however, presupposes gravity known overall on the surface of the Earth, and as this is impossible, ways of interpolating the gravity anomalies have to be found in both cases. This exposes the weakness of (8-50) as the telluroid as a whole follows the surface of the Earth, and so the interpolated gravity anomalies become strongly correlated with the height, whereas the geoid represents the intuitive realization of a level surface.

Now the question naturally arises if it were not possible to select an equipotential surface $W =$ constant, which did not intersect the Earth, reduce measured gravity upwards and carry out the same calculations as Stokes and Molodenskiy—thereby getting the advantages of both methods i.e. a smooth interpolation and no assumptions concerning the mass inside the Earth. The solution T could then be added to the normal potential U to get W outside this surface. W could then be split up according to

$$W = V + \Phi \tag{2-5}$$

thereby determining a regular potential V as Φ is given analytically as

$$\Phi = \tfrac{1}{2}\omega^2 \left(x^2 + y^2\right). \tag{2-3}$$

Now we could use, that if for two regular potentials V and V' the following equation holds

$$V(x) = V'(X), \qquad x \in \Omega$$

for an open set Ω, then it holds everywhere in the region of regularity of V and V', and so the potential W could be analytically continued "downwards" to the surface of the Earth.

Actually, however, due to inaccuracy of computation and interpolation, the above equation cannot be fulfilled exactly, but even if there exists an $\epsilon > 0$ so that

$$\left|V(x) - V'(x)\right| < \epsilon, \qquad x \in \Omega$$

potential theory shows us that there exists no $\delta > 0$ so that the "downward" analytically continued potential differs less than δ from the actual potential of the Earth, in other words, we risk to get arbitrarily great errors if we try to continue such a solution downwards. Now, this remarkable feature of the potential functions not only impairs the practical value of the preceding example—it also suggests, that if we look for an approximation of a potential regular in Ω, then we may have several choices, even if we as approximating functions restrict ourselves to potentials regular in $\Omega' \Supset \Omega$. In [4] is proved the existence of a solution to such an interpolation problem, using as Ω' the exterior of a sphere situated inside the Earth, and as Ω the exterior of the Earth, and this important theorem enables us to regard the determination of T as a problem of interpolation rather than a boundary value problem.

One consequence of this different point of view is immediately appealing: Whereas we in order to solve the boundary value problem have to know gravity all over the world, or at least so dense that we can interpolate with sufficient accuracy, we now can seek local solutions of T, which again will enable us to compute absolute coordinates of the geodetic stations, based only upon geodetic measurements performed at the area in question.

As mentioned before almost any measurement, even a seemingly purely geometrical one as an azimuth measurement, attaches a bond to the potential function at the station where the measurement takes place. So the result of the measurement expresses an interplay between the position of the stations involved and the potential field. This interplay can be expressed in differential equations (so called observation equations), which relate linear functionals of the disturbing potential to variations in coordinates. Now, as the disturbing potential is determined so as to be very "small," we cannot afford to falsify the information we get about it through the measurements by using some assumption concerning the density ρ; so we cannot use any form of projection of the stations and the measurements down to an ellipsoid. This is the main reason why, in this paper, the observation equations have been carried out in three dimensions.

The problem of interpolating the disturbing potential T outside the surface of the Earth, by means of potentials regular down to a "Bjerhammar" sphere, demands us to give these potentials a mathematical structure, by which we can be able to measure the fitness of the approximation.

In [4] is found, that the class of approximating functions could conveniently be structured as a Hilbert space with reproducing kernel, see [6]. Various reasons for this choice present themselves in his paper, and we shall only mention a few here:

1. Due to the fact that the Laplacian is a linear differential operator, the class of potentials regular outside a sphere forms a vector space over the reals.
2. By introducing a norm in the space it is not only possible to decide if a given function is an element of this class, but it also, in the case where several solutions to a problem are available, introduces the possibility to choose the "smoothest" function in the given topology, by applying the condition of least norm.
3. This, however, requires that the space is complete.
4. The above mentioned linear functionals on T can be identified with elements in the dual space of a Hilbert space.
5. The last reason we will mention here is, that the reproducing kernel, due to the fact that it determines the norm in the dual space, can be used to find the approximating potential, as is shown in Section III.

This reason exposes the strength and the weakness of the method of interpolation. Given a norm in a Hilbert space we can calculate the reproducing kernel (if it exists), which again delivers the interpolating function, that is, the problem of interpolation becomes inextricably involved with the problem of choosing the norm. This problem is not treated in this paper, but even if we in some way or another found an optimal norm, we could not be sure that the reproducing kernel existed and even if it existed it might not be suitable for use in computations.

As a consequence of this line of reasoning it looks more promising to try to find a kernel which gives the "best" approximation, bearing in mind that it has to have a simple, closed expression for the sake of computation. Some work has been done in this field, but it is sufficient here to refer to [8].

I

Let us by W denote the Earth's potential outside the attracting masses and let U be some reference potential, determined so that

$$T = W - U \tag{1}$$

is regular outside the surface at the Earth. If we replace the surface of the Earth by a surface α with finite curvature all over, and confine ourselves to determine T outside α, we can make use of an important variant of Runge's Theorem, which states that the set of potentials regular outside a ball B (situated inside $\alpha, B \cap \alpha = \emptyset$) is dense relative to the set of potentials regular outside α. As ball it is common to use the Bjerhammar sphere $B_{0,B}$ with radius B, surface σ and complement Σ.

The following theorem follows from the fact that the Laplacian is a linear differential operator.

Theorem 12.1 *The class of functions harmonic on the open set Σ is a linear vector space over the reals.*

Now, let R be the distance of the volume element $d\Sigma$ from the center of the ball $B_{0,B}$, then the following theorem holds:

Theorem 12.2 *Let H_B be the class of potential functions regular in Σ for which*

$$\frac{1}{2\pi}\int\limits_{\Sigma}\left(\frac{1}{B}-\frac{1}{R}\right)\left(\operatorname{grad}\phi\right)^2 d\Sigma \tag{2}$$

exists. The class H_B is a separable Hilbert space with a reproducing kernel, in which the scalar product $\langle\phi,\psi\rangle_H$ is given by

$$\langle\phi,\psi\rangle_H = \frac{1}{2\pi}\int\limits_{\Sigma}\left(\frac{1}{B}-\frac{1}{R}\right)\operatorname{grad}\phi\operatorname{grad}\psi\, d\Sigma. \tag{3}$$

First we shall prove that H_B is a Hilbert space. The only non-trivial part is to prove, that any Cauchy sequence in H_B is convergent relative to H_B. To a given $\delta > 0$ let $\Sigma_\delta \subset \Sigma$ be an open subset defined by

$$x \in \Sigma_\delta \quad \Leftrightarrow \quad B_{x,\sigma} \cap \sigma = \emptyset.$$

Now let ϕ_n be a Cauchy sequence, that is

$$\forall\, \epsilon>0 \quad \exists\, N\in\grave{N} \quad \forall\, n,m:$$
$$\left(n,m>N \Rightarrow \|\phi_n-\phi_m\|_H^2 = \frac{1}{2\pi}\int\limits_{\Sigma}\left(\frac{1}{B}-\frac{1}{R}\right)\left(\operatorname{grad}(\phi_n-\phi_m)\right)^2 d\Sigma < \epsilon^2\right).$$

Green's third formula, [7], Chapter III (34), used for the (not harmonic) function $\left(\phi_n(x)-\phi_m(x)\right)^2$, $\phi_n, \phi_m \in H_B$ gives

$$4\pi\left(\phi_n(x)-\phi_m(x)\right)^2 = \iint\limits_{\delta B_{x,\delta}}\left(\tfrac{2}{r}(\phi_n-\phi_m)\frac{\partial(\phi_n-\phi_m)}{\partial n} - (\phi_n-\phi_m)^2\frac{\partial\frac{1}{r}}{\partial n}\right) d\delta B$$
$$- \iiint\limits_{B_{x,\delta}} \tfrac{1}{r}\Delta(\phi_n-\phi_m)^2\, d\Sigma$$

as

$$\left(\phi_n(x)-\phi_m(x)\right)^2 \geq 0, \iint\limits_{\delta B_{x,\delta}}(\phi_n-\phi_m)^2\frac{\partial\frac{1}{r}}{\partial n} \geq 0, \text{ and } \iiint\limits_{B_{x,\delta}}\tfrac{1}{r}\Delta(\phi_n-\phi_m)^2\, dB \geq 0$$

we have

$$4\pi\left(\phi_n(x)-\phi_m(x)\right)^2 \leq \iint\limits_{\delta B_{x,\delta}}\tfrac{2}{r}(\phi_n-\phi)\frac{\partial(\phi_n-\phi_m)}{\partial n}\, d\delta B$$

or

$$4\pi\big(\phi_n(x)-\phi_m(x)\big)^2 \le \tfrac{2}{\delta}\iint\limits_{\delta B_{x,\delta}} (\phi_n-\phi)\frac{\partial(\phi_n-\phi_m)}{\partial n}\, d\delta B \tag{4}$$

using the fact, that we on $\delta B_{x,\delta}$ have $r=\text{constant}=\delta$.

Green's first identity

$$\iiint\limits_{B_{x,\delta}} u\Delta v\, dB + \iiint\limits_{B_{x,\delta}} \operatorname{grad} u \operatorname{grad} v\, dB = \iint\limits_{\delta B_{x,\delta}} u\frac{\partial v}{\partial n}\, d\delta B$$

with $u=v=\phi_n-\phi_m$ gives

$$\tfrac{2}{\delta}\iint\limits_{\delta B_{x,\delta}} (\phi_n-\phi_m)\frac{\partial(\phi_n-\phi_m)}{\partial n}\, d\delta B = \tfrac{2}{\delta}\iiint\limits_{\delta B_{x,\delta}} \big(\operatorname{grad}(\phi_n-\phi_m)\big)^2\, dB$$

and (4) becomes

$$2\pi\delta\big(\phi_n(x)-\phi_m(x)\big)^2 \le \iiint\limits_{B_{x,\delta}} \big(\operatorname{grad}(\phi_n-\phi_m)\big)^2\, dB$$

as any $x\in\Sigma_\delta$ at least is the distance δ from σ, a scalar k can be found (i.e. $k<\frac{1}{B}-\frac{1}{B+\delta}$) so that

$$\begin{aligned}\delta k 2\pi\big(\phi_n(x)-\phi_m(x)\big)^2 &\le k\iiint\limits_{B_{x,\delta}} \big(\operatorname{grad}(\phi_n-\phi_m)\big)^2\, dB\\ &\le \int\limits_{\Sigma}\Big(\frac{1}{B}-\frac{1}{R}\Big)\big(\operatorname{grad}(\phi_n-\phi_m)\big)^2\, d\Sigma \end{aligned}\tag{5}$$

so that $\phi_n(x)$, $x\in\Sigma_\delta$ is uniformly convergent. Consequently there exists a function ϕ harmonic in Σ_δ (Harnack's Theorem) for which

$$\phi(x)=\lim_{n\to\infty}\phi_n(x), \qquad x\in\Sigma_\delta \tag{6}$$

in the sense of pointwise convergence, i.e.

$$\forall\, x\in\Sigma_\delta \quad \forall\,\eta>0 \quad \exists\, N\ \big(n>N \Rightarrow |\phi(x)-\phi_n(x)|<\eta\big). \tag{7}$$

As Σ is open and δ can be made arbitrary small we can replace $\forall x\in\Sigma_\delta$ with $\forall x\in\Sigma$ getting the sharper result that there exists a harmonic function $\phi(x)$, $x\in\Sigma$ for which

$$\phi(x)=\lim_{n\to\infty}\phi_n(x), \qquad x\in\Sigma.$$

We shall now prove, that ϕ is an element of H_B, that is has a finite norm. First we observe that the sequence $\|\phi_n\|_H$ is convergent owing to the fact, that $\|\phi_n\|_H - \|\phi_m\|_H \leq \|\phi_n - \phi_m\|_H$. Let $\lambda = \lim_{n\to\infty} \|\phi_n\|_H$ and let us take a closer look at the equation

$$\|\phi\|^2_{\Sigma_\delta} = \frac{1}{2\pi} \int\limits_{\Sigma_\delta} \Big(\frac{1}{B+\delta} - \frac{1}{R}\Big) (\operatorname{grad}\phi)^2 \, d\Sigma_\delta . \tag{8}$$

We repeat Green's second identity

$$\int\limits_{\Sigma_\delta} u\Delta v - v\Delta u \, d\Sigma_\delta = \int\limits_{\sigma'} \Big(u\frac{\partial v}{\partial n} - v\frac{\partial u}{\partial n}\Big)\, d\sigma'$$

where σ' is the surface of Σ_δ.

Put $u = \rho$, $v = \phi\psi$ where ρ, ϕ, ψ satisfy Laplace's equation in a domain including Σ_δ and ϕ, ψ are regular at infinity:

$$2\int\limits_{\Sigma_\delta} \rho \operatorname{grad}\phi \operatorname{grad}\psi \, d\Sigma_\delta = \int\limits_{\sigma'} \rho\Big(\phi\frac{\partial \psi}{\partial n} + \psi\frac{\partial \phi}{\partial n}\Big)\, d\sigma' - \int\limits_{\sigma'} \frac{\partial \rho}{\partial n}\phi\psi \, d\sigma' \tag{9}$$

putting $\rho = \frac{1}{B+\delta} - \frac{1}{R}$ we obtain

$$2\int\limits_{\Sigma_\delta} \Big(\frac{1}{B+\delta} - \frac{1}{R}\Big) \operatorname{grad}\phi \operatorname{grad}\psi \, d\Sigma_\delta = \frac{1}{(B+\delta)^2} \int\limits_{\sigma'} \phi\psi \, d\sigma'$$

or

$$\frac{1}{4\pi(B+\delta)^2} \int\limits_{\sigma'} \phi\psi \, d\sigma' = \frac{1}{2\pi} \int\limits_{\Sigma_\delta} \Big(\frac{1}{B+\delta} - \frac{1}{R}\Big) \operatorname{grad}\phi \operatorname{grad}\psi \, d\Sigma_\delta \tag{10}$$

using (10) we can now write (8):

$$\|\phi\|^2_{\Sigma_\delta} = \frac{1}{4\pi(B+\delta)^2} \int\limits_{\sigma'} \phi^2 \, d\sigma' . \tag{11}$$

The triangular inequality now gives:

$$\|\phi\|_{\Sigma_\delta} \leq \|\phi - \phi_n\|_{\Sigma_\delta} + \|\phi_n\|_{\Sigma_\delta}$$

now we use (7) on (11) and observe that $\|\phi_n\|_{\Sigma_\delta} \leq \|\phi_n\|_H$:

$$\|\phi\|_{\Sigma_\delta} \leq \eta + \|\phi_n\|_H$$

and we have for each $\delta > 0$ that $\|\phi\|_{\Sigma_\delta}$ is finite. Now follows the existence of

$$\|\phi\|^2_H = \frac{1}{2\pi} \int\limits_{\Sigma} \Big(\frac{1}{B} - \frac{1}{R}\Big) (\operatorname{grad}\phi)^2 \, d\Sigma \leq \lambda$$

from a well-known theorem concerning monotone convergence of Lebesgue integrals (B. Levi's monotone convergence theorem). The limit function ϕ is then a member of our class H_B and as a consequence H_B is a Hilbert space. Following [6], Theorem III.2, H_B has a reproducing kernel if $\phi(x)$ is a linear functional for every fixed $x \in \Sigma$. It is enough here to show that it is bounded.

(5) yields

$$2\pi k\big(\phi(x)\big)^2 \le \int\limits_{\Sigma}\Big(\frac{1}{B}-\frac{1}{R}\Big)(\operatorname{grad}\phi)^2\,d\Sigma$$

or

$$\frac{\phi(x)}{\|\phi\|_H} \le \sqrt{\frac{1}{k(x)}} \tag{12}$$

where k depends solely on the distance from x to σ. Now from (12) and the fact that $\phi \in H_B$ is continuous follows, that H_B is separable, see [6], Theorem III.9, i.e. there exists a complete orthonormal system for H_B. This means that every $\phi \in H_B$ may be represented by a series expansion

$$\phi(x) = \sum_{v=1}^{\infty}\langle\phi,\phi_v\rangle_H\,\phi_v(x), \qquad x \in \Sigma \tag{13}$$

this, however, is to be understood in the sense of convergence in the Hilbert space metric

$$\lim_{N\to\infty}\left\|\phi(x) - \sum_{v=1}^{N}\langle\phi,\phi_v\rangle_H\,\phi_v(x)\right\|_H = 0$$

in addition, see [6], Theorem III.4, for every subset $\Sigma_\delta \subset \Sigma$ the series (13) is uniformly convergent.

The kernel function $K_B(x,y)$, $x,y \in \Sigma$ is uniquely determined, see [6], Theorem III.1 and is for every fixed $x \in \Sigma$ a regular potential function of y and vice versa. As the space H_B is separable $K_B(x,y)$ can be written as a convergent series

$$K_B(x,y) = \sum_{v=1}^{\infty}\phi_v(x)\phi_v(y) \qquad x,y \in \Sigma \tag{14}$$

for any complete orthonormal system $\{\phi_v\}$, $v = 1, 2, \ldots$.

We have now finished the proof of Theorem 12.2. Of course it could have been made the other way around: By 1) introducing the spherical harmonics suitably normalized, 2) prove that they constitute a complete orthonormal system, and 3) that the series (14) is convergent. This has been done by Krarup, getting the reproducing kernel

$$K(x,y) = \frac{\|x\|^2\|y\|^2 - B^4}{\big(B^4 - 2B^2\|x\|\|y\|\cos\omega + \|x\|^2\|y\|^2\big)^{3/2}}$$

where ω is the angle between the vectors x and y.

We are in the following going to use two important facts concerning a reproducing kernel, first the reproducing quality:

$$\phi(x) = \big\langle K(x,y), \phi(y) \big\rangle_H \qquad \forall \phi \in H_B, \quad x, y \in \Sigma \tag{15}$$

which is well known, and second that it can be used to find a representation of certain bounded linear functionals. Let ev_x (the evaluation at x) be defined by

$$ev_x \in H_B^{\mathrm{T}}, \qquad ev_x \colon \phi \to \phi(x), \qquad \forall \phi \in H_B, \quad x \in \Sigma$$

from (15) follows

$$\phi(x) = \big\langle K(x,\cdot), \phi \big\rangle_H \tag{16}$$

so that $ev_x \in H_B^{\mathrm{T}}$ may be represented by the element $K(x,\cdot) \in H_B$. That ev_x is bounded follows by applying Cauchy-Schwarz' inequality on (16) using the same technique as in

Theorem 12.3 *Any directional derivative $D_{x,\mathbf{e}}$ taken at a fixed point x in the direction $\mathbf{e}$ is a bounded linear functional $D_{x,\mathbf{e}} \in H_B^{\mathrm{T}}$ and may be represented by the element $D_{x,\mathbf{e}}\big(K(x,\cdot)\big)$ where the index x means that the directional derivative is to be applied on $K(x,y)$ considered as a function of x for every fixed $y \in \Sigma$.*

Following (16) we have for every fixed $h \neq 0$

$$\frac{\phi(x) - \phi(x + h\mathbf{e})}{h} = \left\langle \frac{K(x,\cdot) - K(x + h\mathbf{e},\cdot)}{h}, \phi \right\rangle_H, \qquad h \neq 0 \tag{17}$$

or, in a condensed notation

$$\Delta_{x,h}\,\phi = \big\langle \Delta_{x,h} K(x,\cdot), \phi \big\rangle_H$$

using Cauchy-Schwarz' inequality we get

$$|\Delta_{x,h}\,\phi| \leq \|\Delta_{x,h} K(x,\cdot)\|_H \|\phi\|_H$$

but

$$\big\|\Delta_{x,h} K(x,\cdot)\big\|_H^2 = \left\langle \frac{K(x,\cdot) - K(x + h\mathbf{e},\cdot)}{h}, \frac{K(x,\cdot) - K(x + h\mathbf{e},\cdot)}{h} \right\rangle_H$$

using the reproducing quality of $K(x,y)$ we get

$$\big\|\Delta_{x,h} K(x,\cdot)\big\|_H^2 = \frac{K(x + h\mathbf{e}, x + h\mathbf{e}) - 2K(x, x + h\mathbf{e}) + K(x,x)}{h^2}$$

which can be arranged as

$$\begin{aligned} &\Delta_{x,h}\Delta_{y,h} K(x,y) \\ &\quad = \frac{\big(K(x + h\mathbf{e}, x + h\mathbf{e}) - K(x, x + h\mathbf{e})\big) - \big(K(x, x + h\mathbf{e}) - K(x,x)\big)}{h^2}. \end{aligned}$$

Now, for every fixed $h \neq 0$ the above expression is finite. Using that $K(x,y)$ is a potential regarded as a function of one of the variables when the other is fixed, and that the derivative of a potential is a potential we can to every $r > 0$ find M so that

$$|h| < r \quad \Rightarrow \quad \Delta_{x,h}\Delta_{y,h}K(x,y) < M$$

by taking the limit in (17) the rest of Theorem 12.3 follows.

II

We will start this section by establishing the mathematical model chosen in this paper to fit the physical reality and the preceding theory.

Regarding the former it comes natural to refer the Earth and thereby the geodetic stations and the reference potential U to the 3-dimensional Euclidean vector space with a basis e_1, e_2, e_3 and origin at the Earth's center of gravity. e_3 is chosen as a parallel to the meridian plane of Greenwich, e_1 coincides with the Earth's axis of rotation. With special reference to polar migration the choice of e_1 formally requires a fixed moment of time but as that problem is of no immediate consequence for the actual derivation of the observation equations we will for the rest of this section regard it as non-existing. Relative to this "global" coordinate system each geodetic station P will have attached reference coordinates X_R (written as a 3×1 matrix) and values U_{X_R}, $(\operatorname{grad} U)_{X_R}$ and $\left(\frac{\partial^2 U}{\partial x_i \partial x_j}\right)_{X_R}$.

The preceding theory demands observation equations relating measurements to linear operators of the anomalous potential T and variations in coordinates and as a consequence hereof it has proven a facility at each station to introduce local Cartesian coordinate systems after an idea developed in [2].

Now, suppose that we have carried out angular measurements at the station Q to the station P. In the moment of observation the theodolite defines a "local" Cartesian coordinate system with origin at Q, first axis along $(\operatorname{grad} W)_Q$ and third axis situated in the plane defined by the first axis and the center of the cross hairs of the theodolite, i.e. in the direction of the measured angle α_m.

The idea now is to approximate this local system by a "local reference system" with origin at the reference coordinate Y_R of Q, first axis along $(\operatorname{grad} U)_{Y_R}$ and third axis turned the angle α_m relative to some fixed approximation of the zero-point of the circle. If we, in this system, make a simulation of the measurement by moving the cross hairs along a unit sphere in the plane defined by the first and the third axes until the almucantar goes through the point X_R defined by the reference coordinates of P, we can introduce the two variables K, H which both are angles defined along the lines of the cross hairs relative to its center. K is defined positive along the almucantar towards the second axis, H is defined negative along the vertical towards the first axis.

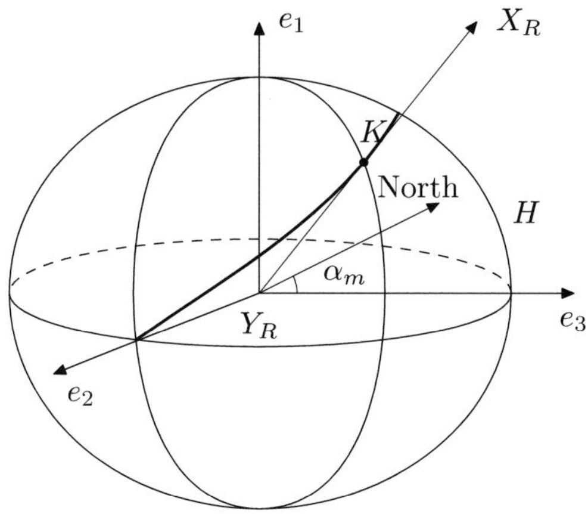

Fig. 12.1. Definition of the variables H and K for given measurement station Y_R and target station X_R

Thus, if the point X_R has the local direction cosines (ξ_R, η_R, ζ_R), H and K are given by

$$\begin{aligned} \xi_R &= \cos K \sin H \\ \eta_R &= \sin K \\ \zeta_R &= \cos K \cos H \end{aligned} \tag{18}$$

or, in the reference coordinates relative to the local system (small x), with $R = \|X_R - Y_R\|$:

$$\begin{aligned} x_{1R} &= R \cos K \sin H \\ x_{2R} &= R \sin K \\ x_{3R} &= R \cos K \cos H. \end{aligned} \tag{19}$$

The local systems introduced this way may at first sight seem a little tedious but there are among others three reasons underlying our choice:

1. Regardless of the position of the station and the object there are induced no singularities, and so we can use the systems to deduce both terrestrial and astronomical observation equations.
2. Corrections for instrumental errors can be applied at the intuitive correct place i.e. at H and K.
3. The systems apply directly to horizontal wire passages of stars, the problem of convergence between the almucantar of latitude and the great circle induced by the horizontal wire never arising.

To do the observation equations will now be to establish a bond between the variations leading

(a) the local reference system onto the local system (by modifications applied to the reference coordinates and to the directions of the axes), and

(b) the changes thereby induced in the quantities H and K.

The derivation of (a) might be done by means of ordinary vector calculus, but we think the reader will agree, that we can cut a short way by utilizing a property by the Euclidean space, observing that any two of the coordinate systems introduced so far transform into each other by a so called rigid motion i.e. a motion composed by a translation and a rotation. Especially, the rotations in this case being orthogonal transformations in a Euclidean space, the transformations form the Euclidean Group of the space. According to [9] the members of this group may be represented as 4×4 matrices

$$\mathcal{F} = \begin{bmatrix} 1 & 0 \\ Y & A \end{bmatrix}$$

where Y is a 3×1 matrix and A is a 3×3 orthogonal matrix. For our purpose it may be suggestive to interpret Y as the global coordinates of the local system and A as the matrix of transformation which rotates the global coordinate axes onto the local ones. Using this interpretation we will call $\mathcal{F}$ a frame and it is easily seen that there is a one-to-one correspondence between frames and coordinate systems, for instance the frame corresponding to the global system is the 4×4 unit matrix E.

Let P and Q have the global coordinates X and Y, and let P have the local coordinates x relative to a local system situated at Q. The equation of transformation is

$$X - Y = Ax \tag{20}$$

which is equal to writing

$$\begin{bmatrix} 1 \\ X \end{bmatrix} = \begin{bmatrix} 1 & 0 \\ Y & A \end{bmatrix} \begin{bmatrix} 1 \\ x \end{bmatrix} \tag{21}$$

thus establishing a one-to-one correspondence between the coordinates of a point relative to a coordinate system and relative to a frame by extending the vector

$$X \rightarrow \begin{bmatrix} 1 \\ X \end{bmatrix}. \tag{22}$$

If we denote the extended vectors as $\left[\begin{smallmatrix} 1 \\ X \end{smallmatrix}\right]$, $\left[\begin{smallmatrix} 1 \\ x \end{smallmatrix}\right]$ by $\widetilde{X}$, $\widetilde{x}$ we can write (21) simply as

$$\widetilde{X} = \mathcal{F}\widetilde{x} \tag{23}$$

which has the solution

$$\widetilde{x} = \mathcal{F}^{-1}\widetilde{X} \tag{24}$$

where

$$\mathcal{F}^{-1} = \begin{bmatrix} 1 & 0 \\ -A^{-1}Y & A^{-1} \end{bmatrix}. \tag{25}$$

(24) yields the local coordinates of a point as a function of its global coordinates and the frame in question. The above formulas apply also for a direction when we interpret it as a difference between points, i.e.

$$\widetilde{X} - \widetilde{Y} = \begin{bmatrix} 0 \\ X - Y \end{bmatrix}. \tag{26}$$

In order to compare measurements in a local frame $\mathcal{F}$ to coordinates in a local reference frame $\mathcal{F}_R$, we seek the variation $\delta\widetilde{x}$ in $\widetilde{x}_R$ when we apply the modifications

$$\widetilde{X}_R \rightarrow \widetilde{X}_R + d\widetilde{X} \tag{27}$$

$$\widetilde{Y}_R \rightarrow \widetilde{Y}_R + d\widetilde{Y} \tag{28}$$

to the reference coordinates X_R, Y_R of P, Q and simultaneously change the reference frame according to

$$\mathcal{F}_R \rightarrow \mathcal{F}_R + d\mathcal{F} \tag{29}$$

where the variation of $\mathcal{F}_R$ may be seen as the result of the following three points

1. The first column in $\mathcal{F}_R$ is changed according to (28).
2. The first coordinate axis is rotated onto a parallel to $(\operatorname{grad} W)_Q$ or, expressed by U and T, to the gradient of

$$W = U_{Y_R} + dY^{\mathrm{T}}(\operatorname{grad} U)_{Y_R} + T \tag{30}$$

   where we can suppose T to be taken at Q.
3. The third axis (and thereby the second too) is rotated according to a correction $d\alpha_0$ to the fixed approximation to the zero-direction.

As a consequence of the way we have chosen the reference frame, we can regard the changes (27), (28), (29) as differential, thus getting (from (24))

$$\delta\widetilde{x} = d\mathcal{F}_R^{-1}\widetilde{X}_R + \mathcal{F}_R^{-1}\,d\widetilde{X} \tag{31}$$

from (25) we find

$$d\mathcal{F}_R^{-1} = \begin{bmatrix} 0 & 0 \\ -dA^{-1}Y_R - A^{-1}\,dY & dA^{-1} \end{bmatrix} \tag{32}$$

so that

$$\delta\widetilde{x} = \begin{bmatrix} 0 & 0 \\ -dA^{-1}Y_R - A^{-1}\,dY & dA^{-1} \end{bmatrix} \begin{bmatrix} 1 \\ X_R \end{bmatrix} + \begin{bmatrix} 1 & 0 \\ -A^{-1}Y_R & A^{-1} \end{bmatrix} \begin{bmatrix} 0 \\ dX \end{bmatrix}$$

or

$$\delta\widetilde{x} = \begin{bmatrix} 0 \\ -dA^{-1}Y_R - A^{-1}dY + A^{-1}dX + dA^{-1}X_R \end{bmatrix} \tag{33}$$

having

$$dA^{-1}X_R - dA^{-1}Y_R = dA^{-1}(X_R - Y_R) = dA^{-1}Ax_R$$

and

$$A^{-1}dX = dx$$

we get

$$\delta x = dx - dy + dA^{-1}Ax_R. \tag{34}$$

The only not trivial part of (34) is the infinitesimal matrix

$$S = dA^{-1}A = dA^{\mathrm{T}}A \tag{35}$$

where we have used that differentiation and transposition commute and that A is orthogonal

$$A^{\mathrm{T}}A = I. \tag{36}$$

Differentiation of (36) gives

$$dA^{\mathrm{T}}A + A^{\mathrm{T}}dA = 0$$

or

$$\begin{aligned} S + S^{\mathrm{T}} &= 0 \\ S^{\mathrm{T}} &= -S \end{aligned}$$

i.e. the matrix S is skew-symmetric. If we write

$$S = \begin{bmatrix} 0 & -s_3 & -s_2 \\ s_3 & 0 & -s_1 \\ s_2 & s_1 & 0 \end{bmatrix}$$

our problem is now to express s_1, s_2, s_3 by dx and T. As the column vectors of the matrix $A + dA$ have the directions of the coordinate frame

$$\mathcal{F} = \mathcal{F}_R + d\mathcal{F}$$

the first column vector of $A + dA$ equals the vector $k(\operatorname{grad} W)_Q$ in the global system, where k is a constant $\neq 0$. Therefore this vector in the reference system must equal the first column vector of

$$A^{-1}(A + dA) = I - S$$

or

$$\begin{aligned} k\frac{\partial W}{\partial x_1} &= \frac{1}{\gamma}\frac{\partial U}{\partial x_1} \\ k\frac{\partial W}{\partial x_2} &= -s_3 \\ k\frac{\partial W}{\partial x_3} &= -s_2 \end{aligned} \tag{37}$$

where $\gamma = \|\operatorname{grad} U\|$ is the reference gravity. Using (30) in the local system we can write (37) as

$$\frac{\partial U}{\partial x_1} + dy^{\mathrm{T}}\Big(\operatorname{grad}\frac{\partial U}{\partial x_1}\Big) + \frac{\partial T}{\partial x_1} = \frac{1}{k\gamma}\frac{\partial U}{\partial x_1} \tag{38}$$

$$k\bigg(dy^{\mathrm{T}}\Big(\operatorname{grad}\frac{\partial U}{\partial x_2}\Big) + \frac{\partial T}{\partial x_2}\bigg) = -s_3 \tag{39}$$

$$k\bigg(dy^{\mathrm{T}}\Big(\operatorname{grad}\frac{\partial U}{\partial x_3}\Big) + \frac{\partial T}{\partial x_3}\bigg) = -s_2 \tag{40}$$

from the first of these equations follows, that

$$-k = \frac{1}{\gamma} + 0(dy, T)$$

so that we in the infinitesimal equations (39) and (40) may put $-k = 1/\gamma$ to find

$$s_2 = \frac{1}{\gamma}\bigg(dy^{\mathrm{T}}\Big(\operatorname{grad}\frac{\partial U}{\partial x_3}\Big) + \frac{\partial T}{\partial x_3}\bigg) \tag{41}$$

$$s_3 = \frac{1}{\gamma}\bigg(dy^{\mathrm{T}}\Big(\operatorname{grad}\frac{\partial U}{\partial x_2}\Big) + \frac{\partial T}{\partial x_2}\bigg) \tag{42}$$

a similar argument applied at the second (or third) column vector in the matrix $I - S$ will prove that

$$s_1 = -d\alpha_0$$

and we have

$$\delta x = dx - dy$$

$$-\frac{1}{\gamma}\begin{bmatrix} 0 & -dy^{\mathrm{T}}\big(\operatorname{grad}\frac{\partial U}{\partial x_2}\big)+\frac{\partial T}{\partial x_2} & -dy^{\mathrm{T}}\big(\operatorname{grad}\frac{\partial U}{\partial x_3}\big)+\frac{\partial T}{\partial x_3} \\ dy^{\mathrm{T}}\big(\operatorname{grad}\frac{\partial U}{\partial x_2}\big)+\frac{\partial T}{\partial x_2} & 0 & \gamma d\alpha_0 \\ dy^{\mathrm{T}}\big(\operatorname{grad}\frac{\partial U}{\partial x_3}\big)+\frac{\partial T}{\partial x_3} & -\gamma d\alpha_0 & 0 \end{bmatrix} \times \begin{bmatrix} x_{1R} \\ x_{2R} \\ x_{3R} \end{bmatrix}. \tag{43}$$

To do (b) i.e. to deduce the changes in H and K we differentiate (19) to get

$$dx_1 = \cos K \sin H\, dR - R \sin K \sin H\, dK + R \cos K \cos H\, dH \tag{44}$$

$$dx_2 = \sin K\, dR + R \cos K\, dK \tag{45}$$

$$dx_3 = \cos K \cos H\, dR - R \sin K \cos H\, dK - R \cos K \sin H\, dH \tag{46}$$

and form (44) $\cdot x_{3R}$ − (46) $\cdot x_{1R}$

$$x_{3R}\,dx_1 - x_{1R}\,dx_3 = R^2\cos^2 K\,dH = \left(x_{1R}^2 + x_{3R}^2\right) dH = r^2\,dH$$

to get

$$dH = b^{\mathrm{T}}\,dx \tag{47}$$

where

$$b^{\mathrm{T}} = \begin{bmatrix} \dfrac{x_{3R}}{r} & 0 & -\dfrac{x_{1R}}{r^2} \end{bmatrix}.$$

To get dK we form (44)$\cdot x_{2R}x_{1R}$ + (46) $\cdot x_{2R}x_{3R}$

$$x_{1R}x_{2R}\,dx_1 + x_{2R}x_{3R}\,dx_3 = R^2\cos^2 K\sin K\,dR - R^3\sin^2 K\cos K\,dK$$

and (45) $\cdot r^2$

$$r^2\,dx_2 = R^2\cos^2 K\sin K\,dR + R^3\cos^2 K\cos K\,dK$$

subtraction gives

$$-x_{1R}x_{2R}\,dx_1 + r^2\,dx_2 - x_{2R}x_{3R}\,dx_3 = R^3\cos K\,dK = R^2 r\,dK$$

or

$$dK = c^{\mathrm{T}}\,dx \tag{48}$$

where

$$c^{\mathrm{T}} = \begin{bmatrix} -\dfrac{x_{1R}x_{2R}}{rR^2} & \dfrac{r}{R^2} & -\dfrac{x_{2R}x_{3R}}{rR^2} \end{bmatrix}.$$

The observation equation for an observed horizontal direction now follows, observing that if x_R is changed with δx then K is incremented with δK and as a consequence of the way we have defined the local coordinate system we have

$$K + \delta K + v = 0.$$

As δK can be regarded as a differential quantity, we may apply (48) to get

$$K + v = -c^{\mathrm{T}}\delta x$$

or, using (43)

$$\begin{aligned}\frac{x_{1R}x_{2R}}{rR^2}\,dx_1 - \frac{r}{R^2}\,dx_2 + \frac{x_{2R}x_{3R}}{rR^2}\,dx_3 + \Big(\frac{x_{1R}}{r\gamma}\frac{\partial^2 U}{\partial x_1 \partial x_2} - \frac{x_{1R}x_{2R}}{rR^2}\Big)\,dy_1 \\ + \Big(\frac{x_{1R}}{r\gamma}\frac{\partial^2 U}{\partial x_2^2} + \frac{r}{R^2}\Big)\,dy_2 + \Big(\frac{x_{1R}}{r\gamma}\frac{\partial^2 U}{\partial x_2 \partial x_3} - \frac{x_{2R}x_{3R}}{rR^2}\Big)\,dy_3 \\ + \frac{x_{3R}}{r}\,d\alpha_0 + \frac{x_{1R}}{r\gamma}\frac{\partial T}{\partial x_2} = K + v.\end{aligned} \tag{49}$$

Let H be the complement of an observed zenith distance. Using the same argument as above we can write an observation equation for zenith distances as

$$H_m - H + v = \delta H = b^{\mathrm{T}}\delta x$$

or explicitly

$$\frac{x_{3R}}{r^2}\,dx_1 - \frac{x_{1R}}{r^2}\,dx_3 - \Big(\frac{1}{\gamma}\frac{\partial^2 U}{\partial x_1 \partial x_3} + \frac{x_{3R}x_{2R}}{\gamma r^2}\frac{\partial^2 U}{\partial x_1 \partial x_2} + \frac{x_{3R}}{r^2}\Big)\,dy_1$$
$$-\Big(\frac{1}{\gamma}\frac{\partial^2 U}{\partial x_2 \partial x_3} + \frac{x_{2R}x_{3R}}{\gamma r^2}\frac{\partial^2 U}{\partial x_2^2}\Big)\,dy_2 - \Big(\frac{1}{\gamma}\frac{\partial^2 U}{\partial x_3^2} + \frac{x_{2R}x_{3R}}{\gamma r^2}\frac{\partial^2 U}{\partial x_2 \partial x_3} + \frac{x_{3R}}{r^2}\Big)\,dy_3$$
$$+\frac{x_{1R}x_{2R}}{r^2}d\alpha_0 - \frac{1}{\gamma}\frac{\partial T}{\partial x_3} - \frac{x_{2R}x_{3R}}{\gamma r^2}\frac{\partial T}{\partial x_2} = H_m - H + v. \quad (50)$$

The above observation equations only apply to an object given an absolute position as e.g. a geodetic station on the Earth. Regarding astronomical measurements, the position of a star will be given by its direction cosines or by H, K as defined in equation (18). Now of course we could start at

$$\delta\widetilde{x} = d\mathcal{F}_R^{-1}\widetilde{X}_R + \mathcal{F}_R\,d\widetilde{X} \quad (51)$$

and carry out the calculations along the same line as above (with the obvious modification (26) of $\widetilde{X}_R$), but at this point we will put

$$d\widetilde{X} = 0$$

as

1. we do not believe it to be relevant here to set up observation equations which apply corrections to the positions of the stars, and
2. the measurement carried out to determine the position of the star e.g. a time measurement, will be taken as being exact. The error thereby induced should be treated as a collimation error.

Visualizing (26), (32), (35) and (18) we can write (31) as

$$\begin{bmatrix} d\xi \\ d\eta \\ d\zeta \end{bmatrix} = \begin{bmatrix} 0 & -s_3 & -s_2 \\ s_3 & 0 & -s_1 \\ s_2 & s_1 & 0 \end{bmatrix} \begin{bmatrix} \xi_R \\ \eta_R \\ \zeta_R \end{bmatrix}. \quad (52)$$

Inserting H and K by (18) and carrying out the differentiation and multiplication gives

$$-\sin K \sin H\, dK + \cos K \cos H\, dH = -s_3 \sin K - s_2 \cos K \cos H \quad (53)$$
$$\cos K\, dK = s_3 \cos K \sin H - s_1 \cos K \cos H \quad (54)$$
$$\sin K \cos H\, dK - \cos K \sin H\, dH = s_2 \cos K \sin H + s_1 \sin K \quad (55)$$

$(53) \cdot \cos H - (55) \cdot \sin H$ gives

$$\cos K\, dH = -s_1 \sin H \sin K - s_2 \cos K - s_3 \sin K \cos H$$
$$dH = -(s_1 \sin H + s_3 \cos H) \tan K - s_2$$

or, written out explicitly as an observation equation for an observed passage of the horizontal wire at the fixed zenith distance H_m and the fixed azimuth α_m:

$$\sin H \tan K \, d\alpha_0 - \frac{1}{\gamma}\Big(\frac{\partial^2 U}{\partial x_1 \partial x_2} \cos H \tan K + \frac{\partial^2 U}{\partial x_1 \partial x_3}\Big)\, dy_1$$
$$- \frac{1}{\gamma}\Big(\frac{\partial^2 U}{\partial x_2^2} \cos H \tan K + \frac{\partial^2 U}{\partial x_2 \partial x_3}\Big)\, dy_2 - \frac{1}{\gamma}\Big(\frac{\partial^2 U}{\partial x_2 \partial X_3} \cos H \tan K + \frac{\partial^2 U}{\partial x_3^2}\Big)\, dy_3$$
$$- \frac{1}{\gamma}\frac{\partial T}{\partial x_2} \cos H \tan K - \frac{1}{\gamma}\frac{\partial T}{\partial x_3} = H_m - H + v. \quad (56)$$

K is found by forming (53) $\cdot \sin H+$ (55) $\cdot \cos H$

$$- \sin K \, dK = -s_3 \sin K \sin H + s_1 \sin K \cos H \quad (57)$$

and (54) $\cdot \cos K-$ (57) $\cdot \sin K$

$$dK = s_3 \sin H - s_1 \cos H$$

and the observation equation for the passage of a star through the vertical hair with the azimuth α_m becomes

$$\frac{\sin H}{\gamma}\frac{\partial^2 U}{\partial x_1 \partial x_2} dy_1 + \frac{\sin H}{\gamma}\frac{\partial^2 U}{\partial x_2^2} dy_2 + \frac{\sin H}{\gamma}\frac{\partial^2 U}{\partial x_2 \partial x_3} dy_3 + \frac{\sin H}{\gamma}\frac{\partial T}{\partial x_2}$$
$$+ \cos H \, d\alpha_0 = K + v. \quad (58)$$

We shall once again stress that the equations (56) and (58) are valid no matter what position the observer has on the Earth. The price we have to pay is that we always have to be in possession of corresponding values of α_m and H_m. We do not regard this as a serious draw back, as the "other" measurement by no means has to be exact, say less than 10° which a little consideration will show.

We are now going to leave the local coordinate systems. Suppose that we have carried out a gravity measurement at the station Q. Considering the methods of measurement used today, we can regard the measured gravity g to be an absolute value, i.e.

$$g = \| \operatorname{grad} W \|_Q$$

which can be expanded as

$$g = \| \operatorname{grad} W \|_{Y_R + dY} = \big\| \operatorname{grad} U + dY^{\mathrm{T}} (\operatorname{grad} U) + T \big\|_{Y_R}.$$

If we as reference value use $\gamma = \| \operatorname{grad} U \|_{Y_R}$ we have

$$g - \gamma = \frac{g^2 - \gamma^2}{g + \gamma} = \frac{\sum_i \Big(\frac{\partial U}{\partial x_i} + \sum_j dY_j \frac{\partial^2 U}{\partial x_i \partial x_j} + \frac{\partial T}{\partial x_i}\Big)^2 - \sum_i \Big(\frac{\partial U}{\partial x_i}\Big)^2}{g + \gamma}$$

$$= \frac{2\sum_i \frac{\partial U}{\partial x_i}\left(\sum_j dY_j \frac{\partial^2 U}{\partial x_i \partial x_j} + \frac{\partial T}{\partial x_i}\right) + \text{second order terms}}{2\gamma + \text{first order terms}}.$$

Introducing the 3×3 matrix M consisting of the second order derivatives of U

$$M = \frac{\partial^2 U}{\partial x_i \partial x_j}$$

we get

$$g - \gamma = \frac{1}{\gamma}(\operatorname{grad} U)^{\mathrm{T}}(M\, dY + \operatorname{grad} T) + \text{second order terms}$$

so that the observation equation becomes

$$\frac{1}{\gamma^2}(\operatorname{grad} U)^{\mathrm{T}} M\, dY + (\operatorname{grad} U)^{\mathrm{T}}(\operatorname{grad} T) = \frac{g-\gamma}{\gamma} + v. \tag{59}$$

Spirit levelling between two stations P and Q can be regarded as a measured difference $(W_P - W_Q)$ of the value of the potential at the two points. An expansion gives

$$W_p - W_Q = U_{X_R} + dX^{\mathrm{T}}(\operatorname{grad} U)_{X_R} + T_P - U_{Y_R} - dY^{\mathrm{T}}(\operatorname{grad} U)_{Y_R} - T_Q$$

and the observation equation becomes

$$dX^{\mathrm{T}}(\operatorname{grad} U)_{X_R} - dY^{\mathrm{T}}(\operatorname{grad} U)_{Y_R} + T_P - T_Q = (W_P - W_Q) - (U_{X_R} - U_{Y_R}) + v. \tag{60}$$

The last example given in this section will be the observation equation for an observed distance S_m. It becomes

$$\frac{S_m - R}{R} + v = \frac{X_{1R}}{R^2}\,dX_1 + \frac{X_{2R}}{R^2}\,dX_2 + \frac{X_{3R}}{R^2}\,dX_3 - \frac{Y_{1R}}{R^2}\,dY_1 - \frac{Y_{2R}}{R^2}\,dY_2 - \frac{Y_{3R}}{R^2}\,dY_3. \tag{61}$$

III

The aim of this section is to find an element in the Hilbert space H_B (i.e. a potential) given the values of n linear functionals of the disturbing potential. Visualizing the observation equations, Theorem 12.3 states that the linear functionals l_i used are bounded and have the norm

$$\|l_i\|^2 = l_{i_x}\left(l_{i_y}\left(K(x,y)\right)\right) \tag{62}$$

where the norm used is the usual maximum norm for linear mappings.

For any two of these linear functionals we can define a scalar product by

$$\langle l_i, l_j \rangle = \tfrac{1}{2}\left(\|l_i + l_j\| - \|l_i\| - \|l_j\|\right) \tag{63}$$

thus having

$$\langle l_i, l_j \rangle = l_{i_x}\big(l_{j_y}\big(K(x,y)\big)\big). \tag{64}$$

Suppose that we have carried out n measurements in order to determine the $3m$ coordinates of m different geodetic stations. Let A denote the matrix formed by the n observations with $T =$ the zero-potential, P the weight matrix of the system and r the vector of residuals. The normal equations are

$$A^{\mathrm{T}}PAx = A^{\mathrm{T}}Pr. \tag{65}$$

As the reference coordinates generally not allow the system to be linear (65) must be iterated in the sense that the solution x must be added to the reference coordinates until we have a solution for which $x = 0$.

When we have iterated "to the bottom" a new vector of residuals r' may be calculated using the latest coordinates and we have

$$A^{\mathrm{T}}Pr' = 0.$$

Inserting the disturbing potential T and the matrix L representing the linear functionals we have

$$LT = r'. \tag{66}$$

From this point we have found it convenient to denote the approximating potential in H_B by T, as we do not believe it will cause any confusion. Now, the problem of determining T is a problem of collocation if we require (66) to be satisfied exact. This has in general an infinity of solutions, but if we choose the solution with minimal norm then we have a problem of least-squares collocation and such a problem has generally a unique solution.

Let T be the minimal solution. Any solution T_1 to (66) can be written in the form

$$T_1 = T + T_0$$

where T_0 is a solution to

$$LT_0 = 0 \tag{67}$$

we have

$$\|T_1\| = \langle T + T_0, T + T_0 \rangle = \|T\| + \|T_0\| + 2\langle T, T_0 \rangle \tag{68}$$

if T satisfies not only (66) but also

$$\langle T, T_0 \rangle = 0$$

for every solution T_0 to (67) then it follows from (68) that T is the solution with minimal norm. Let l_i be the linear functional of the i'th row of L. Theorem 12.3 states that it may be represented by $l_{i_x}\big(K(x_j)\big)$ so that we can write (67) as

$$\big\langle l_{i_x}\big(K(x_j)\big), T_0 \big\rangle_H = 0$$

so that T must be orthogonal to the elements

$$l_{i_x}\big(K(x_j)\big) \qquad i = 1, 2, \ldots, n$$

and it follows that T can be expressed as

$$T = \sum_{i=1}^{n} \alpha^i l_{i_x} \left(K(x_j)\right)$$

for a suitable set of constants α^i, $i = 1, 2, \ldots, n$. (66) gives

$$l_{j_y}\left(\sum_{i=1}^{n} \alpha^i l_{i_x} \left(K(x_j)\right)\right) = r'_j$$

or using the condensed notation of (64)

$$\sum_{i=1}^{n} \langle l_i, l_j \rangle \alpha^i = r'_j \tag{69}$$

the matrix $\left[\langle l_i, l_j \rangle\right]$ can be shown to be non-singular for the linear functionals derived in Section II, so $\left[\langle l_i, l_j \rangle\right]^{-1}$ exists and the constants α^i can be found from (69) by

$$\alpha = \left[\langle l_i, l_j \rangle\right]^{-1} r' \tag{70}$$

and the solution of the least-squares collocation problem is

$$T = r'^{\mathrm{T}} \left[\langle l_i, l_j \rangle\right]^{-1} l_x \left(K(x_j)\right) \tag{71}$$

where $l_x\left(K(x_j)\right)$ is the column vector $\left\{l_{i_x}\left(K(x_j)\right)\right\}$. The norm of T becomes

$$\|T\| = r'^{\mathrm{T}} \left[\langle l_i, l_j \rangle\right]^{-1} \left[\langle l_i, l_j \rangle\right] \left[\langle l_i, l_j \rangle\right]^{-1} r' = r'^{\mathrm{T}} \left[\langle l_i, l_j \rangle\right]^{-1} r'. \tag{72}$$

Various reasons speak against a solution as (71). For instance we risk to get an oscillating T and moreover we know that the observation errors are "built in" the vector r'. This leads us to another way of presenting the problem: Find $T \in H_B$ such that the expression

$$\delta = \|T\|_H + (LT - r')^{\mathrm{T}} P (LT - r') \tag{73}$$

is minimum.

Visualizing the previous pages we have that to any vector β the problem

$$LT = \beta$$

has a solution of minimal norm:

$$\|T\| = \beta^{\mathrm{T}} \left[\langle l_i, l_j \rangle\right]^{-1} \beta$$

for such a solution (73) becomes

$$\delta = \beta^{\mathrm{T}} \left[\langle l_i, l_j \rangle\right]^{-1} \beta + (\beta - r')^{\mathrm{T}} P (\beta - r') = \beta^{\mathrm{T}} \left(\left[\langle l_i, l_j \rangle\right]^{-1} + P\right) \beta - 2\beta^{\mathrm{T}} P r' + r'^{\mathrm{T}} P r'$$

this expression attains its minimum for β satisfying

$$\left(\left[\langle l_i, l_j \rangle\right]^{-1} + P\right)\beta = Pr' \tag{74}$$

multiplying (74) by P (the variance-covariance matrix) and using $\beta = \left[\langle l_i, l_j \rangle\right]\alpha$ gives

$$\left(P^{-1} + \left[\langle l_i, l_j \rangle\right]\right)\alpha = r' \tag{75}$$

the matrix $\left(P^{-1} + \left[\langle l_i, l_j \rangle\right]\right)$ is positive definite being the sum of two positive definite matrices and so (75) may be solved for α. Then the solution T is found by

$$T = \sum_{i=1}^{n} \alpha^i l_{i_x}\left(K(x_j)\right) \tag{76}$$

a simple calculation shows that δ is given by

$$\delta = r'^{\mathrm{T}}\left(P^{-1} + \left[\langle l_i, l_j \rangle\right]\right)^{-1} r'. \tag{77}$$

The way to proceed is now to use an iterative method by first calculating the adjusted coordinates ignoring T. Then using these coordinates we find an estimate for T better than zero; then better values for the coordinates etc.

This way of procedure is similar to the classical way of calculating coordinates where the physical information is processed using the knowledge of coordinates and afterwards used to improve these coordinates. The difference is that we in this method have systematized these steps so that the whole calculating process can be carried out in a short time on a computer.

The Concept of Solution in Integrated Geodesy

We have seen how to find the observation equations in integrated geodesy. We saw that the observation equations are a set of linear equations with a constant term, there is one equation corresponding to each observed quantity and it expresses the increment of this quantity as a function of the increments of the coordinates and of the linear functionals on the potential involved in this observation.

It is not at all evident how to define a solution of an adjustment problem in integrated geodesy: even if we had determined the values of the (finitely many) linear functionals on the potential, this potential would not be determined because it depends on infinitely many parameters. We have build up the observation equations in a rather hypothesis-free manner but now we cannot advance further without introducing a hypothesis concerning the variation of the potential.

In classical adjustment we also have to introduce a hypothesis—here concerning the distribution of the increments to the observed values—in order to define a solution. This hypothesis, which is in fact a statistical hypothesis, brings about that we may define the optimal solution of the adjustment problem as that solution of the observation equation which minimizes a certain quadratic form in the increments to the observed values.

In comparing solutions to the observation equation we will intuitively—other things equal—prefer a solution giving a 'smaller' correction to the potential for one which gives a 'larger' correction. But in order to give the words 'smaller' and 'larger' a meaning we must define some form for measure for the potential. In analogy with classical adjustment we will choose a quadratic norm so that we are led to the application of Hilbert space theory.

From a superficial consideration it would seem desirable to choose the norm in such a way that we could work in a Hilbert space $\mathcal{H}$ such that

1. All elements of $\mathcal{H}$ are harmonic functions in the space outside the surface of the Earth.
2. $\mathcal{H}$ contains all the functions which are harmonic outside the surface of the Earth.
3. All linear functionals of forms corresponding to quantities observable in geodesy are continuous functionals on $\mathcal{H}$.

One can however prove that it is impossible to find a Hilbert space which satisfies these three requests. The best we can achieve is one which satisfies the three requests where the second request is replaced by the following one:

2.′ Every function harmonic outside the surface of the Earth may be approximated arbitrarily well (in a sense to be defined more explicitly) by elements of $\mathcal{H}$.

There are many possible choices for a norm which defines spaces satisfying these modified requests and there are many problems related to such a choice. In the next section we shall discuss some of these problems, but here we shall only suppose that we have chosen one such norm $\|\cdot\|$. When such a norm has been chosen we shall prefer a solution which minimizes this norm.

Let us suppose that in a given adjustment problem in integrated geodesy there are involved N linearly independent linear functionals $f_1, f_2, \ldots, f_N$. Suppose moreover that (x_0, ϕ_0, v_0) satisfies the observation equations. x_0 is the vector defining the increments of the coordinates, v_0 one defining the increments of the values of the observations and $\phi_0 \in \mathcal{H}$ is the increment to the reference potential. Then if ϕ is any harmonic function for which

$$f_i \phi = 0, \qquad i = 1, 2, \ldots, N \tag{78}$$

then also $(x_0, \phi_0 + \phi, v_0)$ is a solution. So for the original solution to be optimal we must demand that

$$\|\phi_0\| = \inf_{f_i\phi=0,\ i=1,2,\ldots,N} \|\phi_0 + \phi\|. \tag{79}$$

The set of elements of $\mathcal{H}$ satisfying (78) is a closed linear subspace of $\mathcal{H}$, therefore there exists a closed linear subspace $\mathcal{H}_0$ of $\mathcal{H}$ which is orthogonal to it and $\mathcal{H}_0$ is evidently an N-dimensional Hilbert space. Let us suppose that ϕ_0 already has been chosen such that $\phi_0 \in \mathcal{H}_0$—if it was not so originally we could replace it with its orthogonal projection onto $\mathcal{H}_0$ without effecting the values of the functionals f_i. But then

$$\|\phi_0 + \phi\| = \left(\|\phi_0\|^2 + \|\phi\|^2\right)^{1/2} \geq \|\phi_0\|$$

so that (79) is satisfied for this choice for ϕ_0 and—given the values of $f_i\phi$—only for this choice and we have proved:

> For the triple (x_0, ϕ_0, v_0) satisfying the observation equations to be optimal it is necessary that $\phi_0 \in \mathcal{H}$, the subspace of $\mathcal{H}$ orthogonal to all solutions of the equations (78).

The observation equations may be written as

$$a_i^{\mathrm{T}} x + f_i\phi = C_i + V_i, \qquad i = 1, 2, \ldots, N$$

where a_i are vectors in an n-dimensional vector space (n is the number of unknown coordinates), V_i are the increments to the values of the observations and C_i are constants. Here we have for the sake of simplicity ignored the possibility of having observation equations involving no functionals on the potential or involving such functionals which are linear combinations of the first N ones.

As we may suppose $\phi \in \mathcal{H}_0$ and as the functionals f_i are supposed to be linearly independent we may use $u_i = f_i\phi$ as coordinates in the N-dimensional vector space $\mathcal{H}_0$. $\|\phi\|^2$ must be a quadratic form in the coordinates u_i, say

$$\|\phi\|^2 = \sum_{ij} Q_{ij} u_i u_j.$$

Suppose we have a complete orthonormal system $\{\psi_i\}$, $i = 1, 2, \ldots, N$ in $\mathcal{H}_0$, $\{\psi_i\}$ are then vectors such that $(\psi_i, \psi_j) = \delta_{ij}$ and $(\psi_j, \phi) = 0$ for every $\phi \in \mathcal{H}$ such that $f_i\phi = 0$ for $i, j = 1, 2, \ldots, N$. Then there are constants α_{ij}, $i, j = 1, 2, \ldots, N$ such that $f_i\phi = \sum_{j=1}^{N} \alpha_{ij}(\psi_j, \phi)$ for all $\phi \in \mathcal{H}$. If we write

$$F_i = \sum_{j=1}^{N} \alpha_{ij}\psi_j \tag{80}$$

then $f_i\phi = (F_i, \phi)$, that is F_i is the element in $\mathcal{H}_0$ which represents the linear functional f_i. We can also say that F_i is the element in $\mathcal{H}$ which represents the functional f_i, but as we have seen F_i are actually in $\mathcal{H}_0$.

From (80) follows

$$(F_i, F_j) = \sum_{h,l=1}^{N} \left(\alpha_{ih}\psi_h, \alpha_{jl}\psi_l\right) = \sum_{l=1}^{N} \alpha_{il}\alpha_{jl},$$

or if we write $A = \{\alpha_{ij}\}$

$$(F_i, F_j) = AA^{\mathrm{T}}.$$

We have supposed the functionals f_i to be linearly independent so that A must be non-singular and from

$$u_i = f_i\phi = \sum_{j=1}^{N} \alpha_{ij}(\psi_j, \phi)$$

we deduce that

$$\{(\psi_j, \phi)\} = A^{-1}u,$$

but from the orthonormality and the completeness of the system $\{\psi_i\}$ follows

$$\|\phi\|^2 = \sum_{j=1}^{N}(\psi_j, \phi)^2 = u^{\mathrm{T}}(AA^{\mathrm{T}})^{-1}u,$$

that is $Q = (AA^{\mathrm{T}})^{-1} = \{(F_i, F_j)\}^{-1}$.

Let us rewrite the observation equations in matrix notation

$$Ax + u = c + v \tag{81}$$

these equations differ only from the classical observation equations by the presence of the vector u which similarly to the vector v shall be small. Classically we have $u = 0$ and $v^{\mathrm{T}}Pv = \min$, where P is the weight matrix for the observations.

If the observations were supposed to be exact, then we could put $v = 0$ in (81) and solve (81) for $u^{\mathrm{T}}Qu = \min$, in this case the result would be determined by the normal equations

$$A^{\mathrm{T}}QAx = A^{\mathrm{T}}Qc,$$

whereas we in the classical case would have

$$A^{\mathrm{T}}PAx = A^{\mathrm{T}}Pc.$$

Neither the first nor the second choice are reasonable, instead let us try to solve (81) for

$$(1-\alpha)v^{\mathrm{T}}Pv + \alpha u^{\mathrm{T}}Qu = \min,$$

where $0 \le \alpha \le 1$.

This adjustment problem may easily be reduced to a classical one. For t an N-dimensional vector and I the unit matrix in $\mathbf{R}^N$ we may write (81) as

$$\begin{aligned} Ax + It &= c + v \\ It &= u. \end{aligned}$$

We write P_1 for $(1-\alpha)P$ and Q_1 for αQ and then the normal equations are:

$$\begin{bmatrix} A & I \\ 0 & I \end{bmatrix}^{\mathrm{T}} \begin{bmatrix} P_1 & 0 \\ 0 & Q_1 \end{bmatrix} \begin{bmatrix} A & I \\ 0 & I \end{bmatrix} \begin{bmatrix} x \\ t \end{bmatrix} = \begin{bmatrix} A & I \\ 0 & I \end{bmatrix}^{\mathrm{T}} \begin{bmatrix} P_1 & 0 \\ 0 & Q_1 \end{bmatrix} \begin{bmatrix} c \\ 0 \end{bmatrix}$$

$$\begin{bmatrix} A^{\mathrm{T}}P_1 & 0 \\ P_1 & Q_1 \end{bmatrix} \begin{bmatrix} Ax + t \\ t \end{bmatrix} = \begin{bmatrix} A^{\mathrm{T}}P_1 & 0 \\ P_1 & Q_1 \end{bmatrix} \begin{bmatrix} c \\ 0 \end{bmatrix}$$

$$A^{\mathrm{T}} P_1 A x + A^{\mathrm{T}} P_1 t = A^{\mathrm{T}} P_1 c$$
$$P_1 A x + (P_1 + Q_1) t = P_1 c$$

or by elimination of t:

$$A^{\mathrm{T}} \big(P_1 - P_1 (P_1 + Q_1)^{-1} P_1\big) A x = A^{\mathrm{T}} \big(P_1 - P_1 (P_1 + Q_1)^{-1} P_1\big) c$$

but if both P_1 and Q_1 are non-singular we find:

$$\begin{aligned} P_1 - P_1 (P_1 + Q_1)^{-1} P_1 &= P_1 - P_1 (P_1 + Q_1)^{-1} (P_1 + Q_1 - Q_1) \\ &= P_1 (P_1 + Q_1)^{-1} Q_1 = \big(Q_1^{-1} (P_1 + Q_1) P_1^{-1}\big)^{-1} = \big(P_1^{-1} + Q_1^{-1}\big)^{-1}, \end{aligned}$$

so that the normal equations may be written as

$$A^{\mathrm{T}} \big(\alpha P^{-1} + (1-\alpha) Q^{-1}\big)^{-1} A x = A^{\mathrm{T}} \big(\alpha P^{-1} + (1-\alpha) Q^{-1}\big)^{-1} c. \qquad (82)$$

This result is at the first sight astonishing because it can *almost* be expressed by saying that the normal equations in integrated geodesy differ only from those in geometric geodesy by the use of another weight-matrix.

The matrix P^{-1} in (82) is the variance-covariance matrix of the observations, but what is Q^{-1}? As we have seen above $Q^{-1} = \big\{(F_i, F_j)\big\}$ that is its elements are scalar products of the elements in $\mathcal{H}$ (or $\mathcal{H}_0$) representing the linear functionals $\{f_i\}$. Note that Q^{-1} depends only on the metric and the functionals and not on the orthogonal coordinate system $\{\psi_i\}$.

In the statistical approach to collocation the matrix $\big\{(F_i, F_j)\big\}$ is regarded as a variance-covariance matrix and although we shall not do so here, we see that given the norm the optimal value for the 'weight coefficient' α may be found by statistical considerations.

The Challenge of Integrated Geodesy

No doubt the less satisfactory point of integrated geodesy is that of the choice of the norm: the result is only defined when the norm has been chosen, it depends on the norm, and the choice of the norm is to some extent arbitrary.

The norm in integrated geodesy corresponds to the weight matrix in classical adjustment and we know that if we have sufficiently many observations and these observations are correct then the result is independent of the weight matrix if only the matrix satisfies one formal condition, that of positive definiteness.

The analogue to this result must in integrated geodesy have the form of a theorem of convergence because there must be executed infinitely many observations in order to determine infinitely many unknowns, and we could ask: under which conditions will the solution in integrated geodesy converge to the correct result independently of the choice of the norm when the observations are correct and their number increases without limit. This should give conditions about the class of norms from which the choice may be made and about the nature and distribution of the observations. In order to be meaningful this question must be modified so as to demand not only convergence but also stability i.e. the results must depend continuously on the observation data.

An explicit answer to this question is not known as yet. The problem is in fact a boundary value problem much more general than Molodenskiy's problem. A reasonable answer seems to be that the 'limit conjecture,' that is:

1. if the observations contain information enough to determine the solution,
2. if they are distributed so as to be overall dense on a surface surrounding the surface of the Earth (this condition is necessary for the stability), and
3. if the norm is chosen so as to correspond to what we have called 'a regular Hilbert space of functions harmonic on a Bjerhammar domain' (i.e. the part of the space outside any closed surface inside the Earth)

then the result will converge to the true solution, be stable and independent of the choice of norm, but generally the resulting potential will *not* converge in the chosen norm but in some norm corresponding to a regular Hilbert space of functions harmonic outside the surface on which the observations are dense. This is because the resulting potential need not be harmonic between the Bjerhammar surface and the surface of the Earth.

If this conjecture is correct, as we believe it to be, perhaps with small technical corrections, then the choice of a norm is reduced to a problem of economy: choose the norm such that the result computed from a given set of observations will be the best possible. A complete solution of this problem can scarcely be given by other than statistical means. But a statistical solution seems today to be impossible, and that for two reasons: 1) the statistical approach is based on the supposition of the potential of the Earth as a stochastic process homogeneous with respect to rotations with respect to the centre of the Earth, a supposition which is inconsistent with the deviation of the surface of the Earth from a sphere. Therefore the higher degree-variances, or rather, the asymptotic behaviour of the degree-variances, which determine the behaviour of the potential near the boundary of the Bjerhammar sphere, are statistically meaningless. 2) As [5] has established even for a spherical Earth the determination of the covariance form would be a much larger problem than that of the potential itself. From these objections does in no way follow that the statistical approach is without practical and theoretical value for integrated geodesy on the contrary the lower order degree-variances can be rather well-determined and it is evidently of importance to choose a norm consistent with the well-determined degree-variances.

As a result of this it may be concluded that the 'norm problem' has to be solved by a combination of statistical *and* mathematical and physical methods. The nearest problem is to find out how the higher order degree-variances of the norm effect the result. As yet very little is known about this and in order to get interesting results it is necessary to study thoroughly the theory of regular Hilbert spaces of harmonic functions.

One could ask if this new conception of geodesy really deserves the name of integrated geodesy. Our answer must be: No, not yet! Two forms for information which are of great importance for the potential namely the topography and the geology are almost respectively totally ignored in our approach. There are several possibilities for including information about topography and geology in the computations but probably the most natural way to do it will only present itself when we have a better insight in the connection between norms for harmonic functions and the correspondent mass-distributions.

It is here as it normally is when new ideas turn up that a lot of new problems appear with a challenge for interesting research along new lines.

References

[1] Garrett Birkhoff and Saunders MacLane. *A Survey of Modern Algebra.* New York, 1965.

[2] H. M. Dufour. Toute la géodésie sans ellipsoïde. *Bulletin Géodésique*, 88:125–143, 1968.

[3] Weikko A. Heiskanen and Helmut Moritz. *Physical Geodesy.* W. H. Freeman and Company, San Francisco, 1967.

[4] Torben Krarup. *A Contribution to the Mathematical Foundation of Physical Geodesy.* Meddelelse 44. Geodætisk Institut, København, 1969.

[5] S. L. Lauritzen. *The Probalistic Background of Some Statistical Methods in Physical Geodesy.* Meddelelse 48. Geodætisk Institut, København, 1973.

[6] Herbert Meschkowski. *Hilbertsche Räume mit Kernfunktion*, volume 113 of *Die Grundlehren der mathematischen Wissenschaften in Einzeldarstellungen mit besonderer Berücksichtigung der Anwendungsgebiete.* Springer, Berlin, 1962.

[7] Wolfgang J. Sternberg and Turner L. Smith. *The theory of potential and spherical harmonics*, volume 3 of *Mathematical expositions.* Toronto, Canada, 1946.

[8] C. C. Tscherning. Representation of covariance functions related to the anomalous potential of the earth using reproducing kernels. Technical report, The Danish Geodetic Institute, Internal Report No. 3, Charlottenlund, 1972.

[9] Hermann Weyl. *The Classical Groups.* Princeton University Press, 1946.

13

On Potential Theory

Preface

For years I have missed a book on potential theory in which were considered the needs of a physical geodesist, i.e. in which were collected the tools for least-squares methods for the determination of harmonic functions with given domain of regularity from observed values of linear functionals with support near the boundary and in which was treated—besides the classical boundary value problems—also the oblique derivative problem.

When I had to write my lecture notes for the International Summer School in the Mountains on mathematical methods in physical geodesy this need became acute for me because I could give no reasonable reference to literature for the responsibility for lectures on geodetic aspects of potential theory without having access to a systematic text on the foundation.

Therefore I had during a few months to write down the following pages which I herewith publish despite of their very preliminary character.

I hope to be able to publish the second (last?) volume next year and after a serious overhauling and an addition of instructive examples, which is perhaps the most important, an official edition will perhaps be published.

1. Homogeneous Polynomials in $\mathbf{R}^q$

Let $x \in \mathbf{R}^q$ denote the column vector with coordinates x_i, $i = 1, 2, \ldots, q$. $|x| = (x^{\mathrm{T}}x)^{1/2}$ is the length of the vector x and $x^{\mathrm{T}}y$ is the scalar product of the vectors $x, y \in \mathbf{R}^q$.

It is convenient to use multiindices such that if $x \in \mathbf{R}^q$ and α a multiindex $\alpha = \alpha_1, \alpha_2, \ldots, \alpha_q$, where $\{\alpha_i\}$ is an ordered set of non-negative integers, then

$$x^\alpha = x_1^{\alpha_1} x_2^{\alpha_2} \cdots x_q^{\alpha_q}$$

is a monomial on $\mathbf{R}^q$ of degree $|\alpha| = \alpha_1 + \alpha_2 + \cdots + \alpha_q$.

If we also write

$$\alpha! = \alpha_1!\,\alpha_2!\,\cdots\,\alpha_q!,$$

then we have for $x, y \in \mathbf{R}^q$ and $n > 0$

$$\Big(\sum_{i=1}^{q} x_i\Big)^n = \sum_{|\alpha|=n} \frac{n!}{\alpha!} x^\alpha$$

and

$$(x^{\mathrm{T}} y)^n = \sum_{|\alpha|=n} \frac{n!}{\alpha!} x^\alpha y^\alpha. \tag{1}$$

As every homogeneous polynomial $P(x)$ of degree n in the q variables x_1, x_2, ..., x_q can be written as

$$P(x) = \sum_{|\alpha|=n} a_\alpha x^\alpha \tag{2}$$

for one and only one set of real numbers $\{\, a_\alpha \mid |\alpha| = n \,\}$, this set of polynomials form a linear vector space of finite dimension. The number of dimensions of this vector space is the number M of elements in the base $\{\, x^\alpha \mid |\alpha| = n \,\}$ of the space and by induction with respect to n it is easily proved that

$$M(q,n) = \binom{n+q-1}{q-1} = \frac{(n+q-1)!}{(q-1)!\,n!}. \tag{3}$$

Besides the classical polynomials—polynomials in x—we will introduce polynomials in 'differentiation with respect to x' in the following way: In a polynomial in x such as (2) $\partial/\partial x_i$ is substituted for x_i for each occurrence of x_i for $i = 1, 2, \ldots, q$ and we write

$$P\Big(\frac{\partial}{\partial x}\Big) = \sum_{|\alpha|=n} a_\alpha \frac{\partial^n}{\partial x^\alpha} = \sum_{|\alpha|=n} a_\alpha \frac{\partial^n}{\partial x_1^{\alpha_1} \partial x_2^{\alpha_2} \cdots \partial x_q^{\alpha_q}}. \tag{4}$$

The expression (4) is a homogeneous differential operator of order n. Such an expression will be called a *homogeneous differential operator* of order n. It would be more correct to say a homogeneous differential operator with constant coefficients. If such an operator is applied on a homogeneous polynomial $Q(x) = \sum_{|\alpha|=n} b_\alpha x^\alpha$ of the same degree then the result will be a real number:

$$\begin{aligned} P\Big(\frac{\partial}{\partial x}\Big)Q(x) &= \sum_{|\alpha|=n}\sum_{|\beta|=n} a_\alpha b_\beta \Big(\frac{\partial^{\alpha_1}}{\partial x_1^{\alpha_1}} x_1^{\beta_1}\Big)\Big(\frac{\partial^{\alpha_2}}{\partial x_2^{\alpha_2}} x_2^{\beta_2}\Big)\cdots\Big(\frac{\partial^{\alpha_q}}{\partial x_q^{\alpha_q}} x_q^{\beta_q}\Big) \\ &= \sum_{|\alpha|=n} a_\alpha b_\beta \alpha_1!\,\alpha_2!\,\cdots\alpha_q! = \sum_{|\alpha|=n} a_\alpha b_\alpha \alpha!. \end{aligned} \tag{5}$$

From the symmetry of the result in (5) in a and b follows

$$Q\Big(\frac{\partial}{\partial x}\Big)P(x) = P\Big(\frac{\partial}{\partial x}\Big)Q(x).$$

For P and Q homogeneous polynomials of degree n on $\mathbf{R}^q$ the expression

$$(P,Q)_n = P\Big(\frac{\partial}{\partial x}\Big)Q(x) \tag{6}$$

satisfies the axioms for a scalar product:

$$(P,Q) \text{ is } \quad \text{bilinear in } P \text{ and } Q \quad \text{and} \quad \text{symmetric in } P \text{ and } Q$$

and

$$\|P\|_n^2 = (P,P)_n = \sum_{|\alpha|=n} a_\alpha^2 \alpha! \geq 0,$$

and

$$(P,P)_n > 0 \qquad \text{for } P \neq 0.$$

Therefore we can define the finite dimensional Hilbert space $\mathcal{P}_{qn}$ as the vector space of homogeneous polynomials of degree n in q variables with the scalar product (6).

We see immediately that the set of monomials

$$\left\{\phi_\alpha(x) = \frac{x^\alpha}{\sqrt{\alpha!}} \;\middle|\; |\alpha| = n\right\}$$

constitute a complete set of orthonormal elements.

In accordance with the general theory for orthonormal systems every element $P \in \mathcal{P}_{qn}$ can be expressed as

$$P(x) = \sum_{|\alpha|=n} a_\alpha \phi_\alpha(x) \tag{7}$$

where the coefficients a_α are determined by

$$a_\alpha = (P, \phi_\alpha)_n. \tag{8}$$

The formulae (7) and (8) can be combined in

$$P(x) = \sum_{|\alpha|=n} (P,\phi_\alpha)_n \, \phi_\alpha(x) = \sum_{|\alpha|=n} \frac{1}{\alpha!}\big(P(y), y^\alpha\big)_n x^\alpha = \big(K(x,\cdot), P\big)_n, \tag{9}$$

where

$$K(x,y) = \sum_{|\alpha|=n} \frac{1}{\alpha!} x^\alpha y^\alpha = \frac{1}{n!}\sum_{|\alpha|=n} \frac{n!}{\alpha!} x^\alpha y^\alpha = \frac{1}{n!}\big(x^{\mathrm{T}} y\big)^n \tag{10}$$

(here (1) is used).

The formula

$$P(x) = \big(K(x,\cdot), P\big)_n \qquad \text{for } P \in \mathcal{P}_{qn} \tag{11}$$

expresses the fact that $K(x,y)$ is a reproducing kernel for the Hilbert function space $\mathcal{P}_{qn}$. According to the definition of the scalar product in $\mathcal{P}_{qn}$ it may be read as

$$P(x) = P\Big(\frac{\partial}{\partial y}\Big)\frac{(x^{\mathrm{T}}y)^n}{n!}$$

or as

$$P(x) = \frac{1}{n!}\Big(x^{\mathrm{T}}\frac{\partial}{\partial y}\Big)^n P(y).$$

Every homogeneous polynomial of degree n may be written in the following form

$$P(x) = \sum_{i=1}^{M} a_i (y_i^{\mathrm{T}} x)^n$$

for a_i real numbers and $y_i \in \mathbf{R}^q$ for $i = 1, 2, \ldots, M$. This can be proved directly but it follows also from the following

Theorem 13.1 *If H is a finite dimensional Hilbert space of continuous functions defined on a domain Ω then there exists a set $\{y_i\}$, $i = 1, 2, \ldots, M$, where $y_i \in \Omega$ and M is the number of dimensions of H such that for each $f \in H$ there exists a set $\{a_i\}$ of M real numbers such that*

$$f(x) = \sum_{i=1}^{M} a_i K(x, y_i), \qquad x, y \in \Omega, \tag{12}$$

where $K(y, x)$ is the reproducing kernel

$$K(x, y) = \sum_{i=1}^{M} \phi_i(x)\phi_i(y), \qquad x, y \in \Omega, \tag{13}$$

of H. Here $\{\phi_i\}$, $i = 1, 2, \ldots, M$, is any complete orthonormal set of elements of H.

Proof That $f \in H$ and $\{\phi_i\}$ is a complete orthonormal set of elements of H means that there exists a set of real numbers $\{b_i\}$ such that

$$f(x) = \sum b_i \phi_i(x),$$

on the other hand putting (13) into (12) gives

$$f(x) = \sum_{i,j=1}^{M} a_i \phi_j(x)\phi_j(y_i) = \sum_{i=1}^{M} \phi_i(x)\Big(\sum_{j=1}^{M} a_j \phi_i(y_j)\Big),$$

so that the theorem is proved if we can prove that it is possible to find the set of M points $y_i \in \Omega$ such that the following system of M linear equations with M unknowns $\{a_j\}$ has one solution for every set of M constants $\{b_i\}$:

$$\sum_{j=1}^{M} a_j \phi_i(y_i) = b_i, \qquad i = 1, 2, \ldots, M,$$

i.e. if we can prove that $\{y_j\}$ can be found such that

$$\det_{1\le i,j\le M} \phi_i(y_j) \neq 0. \tag{14}$$

The functions $\phi_i(x)$ are linearly independent—for else they would not be orthonormal—therefore none of them can vanish identically and so there must be a $y_1 \in \Omega$ for which $\phi_1(y_1) \neq 0$. There must also be a y_2 such that

$$\begin{vmatrix} \phi_1(y_1) & \phi_1(y_2) \\ \phi_2(y_1) & \phi_2(y_2) \end{vmatrix} \neq 0,$$

for else we would have for every $y \in \Omega$ that

$$\begin{vmatrix} \phi_1(y_1) & \phi_1(y) \\ \phi_2(y_1) & \phi_2(y) \end{vmatrix} = 0,$$

or

$$a\phi_1(y) + b\phi_2(y) = 0$$

for $a = \phi_2(y_1)$ and $b = -\phi_1(y_1)$ which is different from zero, this is impossible because ϕ_1 and ϕ_2 are linearly independent. In the same way we prove existence of a y_3 such that

$$\begin{vmatrix} \phi_1(y_1) & \phi_1(y_2) & \phi_1(y_3) \\ \phi_2(y_1) & \phi_2(y_2) & \phi_2(y_3) \\ \phi_3(y_1) & \phi_3(y_2) & \phi_3(y_3) \end{vmatrix} \neq 0$$

etc. until (14) is proved.

It would perhaps be convenient to write here a few facts on reproducing kernels for Hilbert spaces.

A reproducing kernel for a Hilbert space H of continuous functions defined on Ω is a function $K(x, y)$ in two variables $x, y \in \Omega$ such that

1. $K(x, y)$ is for any fixed $x \in \Omega$ as a function of y an element $K(x, \cdot)$ of H.
2. $K(x, y)$ is for any fixed $y \in \Omega$ as a function of x an element $K(\cdot, y)$ of H.
3. For every $f \in H$ is

$$f(x) = \big(f, K(\cdot, x)\big),$$

where $(\cdot, \cdot)$ is the scalar product in H.

From 1, 2, and 3 follows

$$K(x, y) = \big(K(\cdot, y), K(\cdot, x)\big)$$

and

$$K(y, x) = \big(K(\cdot, x), K(\cdot, y)\big)$$

i.e.

$$K(x, y) = K(y, x).$$

It is immediately clear that for H finite dimensional

$$K(x,y) = \sum_{i=1}^{N} \phi_i(x)\phi_i(y)$$

is a reproducing kernel where $\{\phi_i\}$, $i = 1, 2, \ldots, N$, is a complete orthonormal system. If $\{\psi_i\}$, $i = 1, 2, \ldots, N$, is another complete orthonormal system in H then

$$K_1(x,y) = \sum_{i=1}^{N} \psi_i(x)\psi_i(y)$$

is also a reproducing kernel but

$$K_1(x,y) = K(x,y),$$

because no Hilbert space has more than one reproducing kernel.

Proof Let both $K(x,y)$ and $K_1(x,y)$ be reproducing kernels for H then for every $f \in H$ and for every $x \in \Omega$

$$\begin{aligned}\left\|K(\cdot,x) - K_1(\cdot,x)\right\|^2 &= \big(K(\cdot,x) - K_1(\cdot,x), K(\cdot,x) - K_1(\cdot,x)\big)\\ &= \big(K(\cdot,x) - K_1(\cdot,x), K(\cdot,x)\big)\\ &\qquad - \big(K(\cdot,x) - K_1(\cdot,x), K_1(\cdot,x)\big)\\ &= K(x,x) - K_1(x,x) - K(x,x) + K_1(x,x) = 0.\end{aligned}$$

If $f(x)$ is any—not necessarily homogeneous—polynomial of degree n in $x \in \mathbf{R}^q$ then it can be written as

$$f(y) = f_0(y) + f_1(y) + \cdots + f_n(y),$$

where $f_i \in \mathcal{P}_{qi}$. By (9) and (10) we have

$$f_i(y) = \frac{1}{i!}\left(y^{\mathrm{T}}\frac{\partial}{\partial z}\right)^i f_i(z) = \frac{1}{i!}\left(\left(y^{\mathrm{T}}\frac{\partial}{\partial z}\right)^i f(z)\right)_{z=0},$$

so that we can write:

$$f(y) = \sum_{i=0}^{n}\left(\frac{1}{i!}\left(y^{\mathrm{T}}\frac{\partial}{\partial z}\right)^i f(z)\right)_{z=0}. \tag{15}$$

This may be generalized as follows: $f(x+y)$ is by fixed x a polynomial of degree n in y and on this function we may apply (15):

$$\begin{aligned}f(x+y) &= \sum_{i=0}^{n}\left(\frac{1}{i!}\left(y^{\mathrm{T}}\frac{\partial}{\partial z}\right)^i f(x+z)\right)_{z=0}\\ &= f(x) + \frac{1}{1!}\left(y^{\mathrm{T}}\frac{\partial}{\partial x}\right)f(x) + \frac{1}{2!}\left(y^{\mathrm{T}}\frac{\partial}{\partial x}\right)^2 f(x) + \cdots + \frac{1}{n!}\left(y^{\mathrm{T}}\frac{\partial}{\partial x}\right)^n f(x)\end{aligned} \tag{16}$$

this is Taylor's formula for polynomials in q variables.

There exists a generalization of this formula for $n+1$ times continuously differentiable functions which need not be polynomials:

Theorem 13.2 (Taylor's formula) *Let $\Omega \subset \mathbf{R}^q$ be an open set, and let $f\colon \Omega \to \mathbf{R}$, $f \in C^{n+1}(\Omega)$. For $x \in \Omega$, let $y \in \mathbf{R}^q$ so that for all $t \in [0,1]$: $x + ty \in \Omega$. Then the following expansion holds:*

$$f(x+y) = \sum_{i=0}^{n} \frac{1}{i!}\Big(y^{\mathrm{T}}\frac{\partial}{\partial x}\Big)^{i} f(x) + \frac{1}{n!}\int_0^1 (1-t)^n \Big(y^{\mathrm{T}}\frac{\partial}{\partial x}\Big)^{n+1} f(x+ty)\,dt. \quad (17)$$

Proof First we notice, that

$$\frac{d}{dt}\Big(y^{\mathrm{T}}\frac{\partial}{\partial x}\Big)^{n} f(x+ty) = \Big(y^{\mathrm{T}}\frac{\partial}{\partial x}\Big)^{n+1} f(x+ty)$$

so that the remainder may be treated according to

$$\begin{aligned}
&\frac{1}{n!}\int_0^1 (1-t)^n \Big(y^{\mathrm{T}}\frac{\partial}{\partial x}\Big)^{n} f(x+ty)\,dt \\
&\quad= \frac{1}{n!}\int_0^1 (1-t)^n d\bigg(\Big(y^{\mathrm{T}}\frac{\partial}{\partial x}\Big)^{n+1} f(x+ty)\bigg) \\
&\quad= \bigg[\frac{1}{n!}(1-t)^n \Big(y^{\mathrm{T}}\frac{\partial}{\partial x}\Big)^{n} f(x+ty)\bigg]_0^1 \\
&\qquad\qquad + \frac{1}{(n-1)!}\int_0^1 (1-t)^{n-1}\Big(y^{\mathrm{T}}\frac{\partial}{\partial x}\Big)^{n} f(x+ty)\,dt \\
&\quad= -\frac{1}{n!}\Big(y^{\mathrm{T}}\frac{\partial}{\partial x}\Big)^{n} f(x) + \frac{1}{(n-1)!}\int_0^1 (1-t)^{n-1} d\bigg(\Big(y^{\mathrm{T}}\frac{\partial}{\partial x}\Big)^{n-1} f(x+ty)\bigg).
\end{aligned}$$

After carrying out integration by parts n times the theorem follows.

For functions $f \in C^{\infty}(\Omega)$ it is sometimes possible to prove that the remainder for $n \to \infty$ converges to 0 and then $f(x+y)$ can be represented by *Taylor's series:*

$$f(x+y) = \sum_{i=0}^{\infty} \frac{1}{i!}\Big(y^{\mathrm{T}}\frac{\partial}{\partial x}\Big)^{i} f(x). \quad (18)$$

2. Harmonic Polynomials

Polynomials $P(x)$ on $\mathbf{R}^q$ for which $\Delta P(x) = 0$ are called *harmonic polynomials.* Here the Laplacian differential operator is defined by

$$\Delta = \frac{\partial^2}{\partial x_1^2} + \frac{\partial^2}{\partial x_2^2} + \cdots + \frac{\partial^2}{\partial x_q^2}.$$

Homogeneous polynomials $P(x)$ of degree n on $\mathbf{R}^q$ for which $\Delta P(x) = 0$ are called *spherical harmonics* of degree n, they constitute a linear subspace $\mathcal{H}_{qn}$ of the Hilbert space $\mathcal{P}_{qn}$ with the same scalar product $(\cdot,\cdot)_n$.

Theorem 13.3 *Every harmonic polynomial $P_n(x)$ of degree n can in one and only one way be written as*

$$P_n = \sum_{i=0}^{n} Q_i, \tag{19}$$

where Q_i is a spherical harmonic of degree i.

Proof It is clear that P_n may be written uniquely as in (19) with Q_i a homogeneous polynomial of degree i, what shall be proved is only that $\Delta Q_i = 0$. But from (19) and that $\Delta P_n = 0$ follows

$$\Delta P_n = \sum_{i=0}^{n} \Delta Q_i = 0. \tag{20}$$

But (20) expresses that the zero polynomial is a sum of homogeneous polynomials $\{\Delta Q_i\}$ of degrees $i-2$ and this is possible only if $\Delta Q_i = 0$ for $i = 2$, $3, \ldots, n$, ΔQ_0 and ΔQ_1 are automatically zero.

On account of this theorem we shall to the present concentrate on spherical harmonics.

We shall now find the subspace of $\mathcal{P}_{qn}$ which is the orthogonal complement to $\mathcal{H}_{qn}$, i.e. the subspace $\mathcal{H}_{qn}^c \subset \mathcal{P}_{qn}$ such that every $P \in \mathcal{P}_{qn}$ may be uniquely decomposed as

$$P = h + h^c, \qquad h \in \mathcal{H}_{qn}, \quad h^c \in \mathcal{H}_{qn}^c$$

such that h is orthogonal to h^c:

$$(h, h^c)_n = 0.$$

Now it is clear that for $Q \in \mathcal{P}_{qn-2}$ the polynomial $|x|^2 Q \in \mathcal{P}_{qn}$. But $|x|^2 Q \in \mathcal{H}_{qn}^c$ for

$$P \in \mathcal{H}_{qn} \Rightarrow \big(|x|^2 Q(x), P(x)\big)_n = Q\Big(\frac{\partial}{\partial x}\Big)\Delta P(x) = 0,$$

i.e. $|x|^2 Q(x)$ is orthogonal to $P \in \mathcal{H}_{qn}$.

If on the other hand $P \in \mathcal{P}_{qn}$ is orthogonal to $|x|^2 Q(x)$ for all $Q \in \mathcal{P}_{qn-2}$ then

$$0 = \big(|x|^2 Q(x), P(x)\big)_n = Q\Big(\frac{\partial}{\partial x}\Big)\Delta P(x) = \big(Q(x), \Delta P(x)\big)_{n-2},$$

i.e.

$$\big(Q(x), \Delta P(x)\big)_{n-2} = 0 \qquad \text{for all } Q \in \mathcal{P}_{qn-2},$$

which is only possible if $\Delta P(x) = 0$, and we have proved:

Theorem 13.4 *The orthogonal complement $\mathcal{H}_{qn}^c$ to the space of spherical harmonics $\mathcal{H}_{qn} \subset \mathcal{P}_{qn}$ is the space of polynomials $|x|^2 Q(x) \subset \mathcal{P}_{qn}$ for all $Q \in \mathcal{P}_{qn}$, i.e. every homogeneous polynomial P of degree n can be expressed uniquely as the sum*

$$P(x) = h(x) + |x|^2 Q(x), \tag{21}$$

where h is a spherical harmonic of degree n and Q is a homogeneous polynomial of degree $n-2$.

Definition 13.1 *$h(x)$ is called the* harmonic projection *of the homogeneous polynomial $P(x)$ if $P(x)$ and $h(x)$ are related as in* (21).

As $\mathcal{P}_{qn}$ is the direct sum of the two mutually orthogonal subspaces $\mathcal{H}_{qn}$ and $|x|^2\mathcal{P}_{qn-2}$ we have

$$\dim \mathcal{P}_{qn} = \dim \mathcal{H}_{qn} + \dim \mathcal{P}_{qn-2}$$

or

$$N_{qn} = \dim \mathcal{H}_{qn} = \binom{n+q-1}{q-1} - \binom{n+q-3}{q-1} = (2n+q-2)\frac{(n+q-3)!}{(q-2)!n!}.$$

We have proved:

Theorem 13.5 *For the number N_{qn} of dimensions of the vector space $\mathcal{H}_{qn}$ we have*

$$N_{qn} = (2n+q-2)\frac{(n+q-3)!}{(q-2)!n!} = \frac{2n+q-2}{q-2}\binom{n+q-3}{q-3};$$

i.e. there exist N_{qn} linearly independent spherical harmonics of degree n.

In particular, for $q = 3$ we have

$$N_{3n} = 2n + 1.$$

If we apply Theorem 13.4 again to $Q \in \mathcal{P}_{qn-2}$ and so on we prove:

Theorem 13.6 *Every homogeneous polynomial of degree n may be expressed by spherical polynomials of degree n and lower degrees by the formula*

$$P(x) = h_n(x) + |x|^2 h_{n-2}(x) + |x|^4 h_{n-4} + |x|^6 h_{n-6} + \cdots, \qquad (22)$$

where $h_n \in \mathcal{H}_{qn}$, $h_{n-2} \in \mathcal{H}_{qn-2}$, $h_{n-4} \in \mathcal{H}_{qn-4}$, ... are uniquely determined.

Part of this result may also be expressed in another way which perhaps gives a justification of the term 'spherical harmonic':

Corollary 13.1 *To every homogeneous polynomial $P(x)$ of degree n there corresponds a harmonic polynomial $Q(x)$ such that $P(x) = Q(x)$ on the unit sphere $|x|^2 = 1$, and*

$$Q(x) = h_n(x) + h_{n-2}(x) + h_{n-4}(x) + h_{n-6}(x) + \cdots$$

where the spherical harmonics h_n, h_{n-2}, ... are the same as above.

Corollary 13.2 *To every polynomial $P(x)$ there corresponds a harmonic polynomial $Q(x)$ of the same degree such that $P(x) = Q(x)$ on the unit sphere.*

The corollary is proved from Corollary 13.1 by writing $P(x)$ as a sum of homogeneous polynomials.

Theorem 13.6 has also another aspect. If $P(\frac{\partial}{\partial x})$ is a homogeneous differential operator of degree n then from (22) follows

$$P\Big(\frac{\partial}{\partial x}\Big) = h_n\Big(\frac{\partial}{\partial x}\Big) + h_{n-2}\Big(\frac{\partial}{\partial x}\Big)\Delta + h_{n-4}\Big(\frac{\partial}{\partial x}\Big)\Delta^2 + \cdots . \tag{23}$$

Here we call $h_n(\frac{\partial}{\partial x})$, $h_{n-2}(\frac{\partial}{\partial x})$, $h_{n-4}(\frac{\partial}{\partial x})$ *harmonic differential operators* of order n, $n-2$, $n-4$, ... If we apply the operator $P(\frac{\partial}{\partial x})$ on a function f on $\mathbf{R}^q$ which satisfies the equation $\Delta f = 0$ then all the right hand terms in (23) except the first one vanish and we have:

Theorem 13.7 *If $\phi(x)$ satisfies Laplace's equation $\Delta\phi = 0$ then the result of the application of the homogeneous operator $P(\frac{\partial}{\partial x})$ on ϕ depends only on the harmonic projection $h(x)$ of $P(x)$ and of ϕ:*

$$P\Big(\frac{\partial}{\partial x}\Big)\phi = h\Big(\frac{\partial}{\partial x}\Big)\phi. \tag{24}$$

Proof Follows directly from (21).

The proofs we have given of the Theorems 13.4–13.7 are not constructive proofs, so we have no method for computing the harmonic projection directly. Although it is not difficult from Theorem 13.6 to find the following formula for this projection

$$h(x) = P(x) - \frac{|x|^2\Delta P(x)}{2(q+2n-4)} + \frac{|x|^4\Delta^2 P(x)}{2\cdot 4(q+2n-4)(q+2n-6)}\cdots . \tag{25}$$

I have chosen to go an indirect way and first prove the classical *Hobson's theorem*:

Theorem 13.8 *For $P \in \mathcal{P}_{qn}$ we have*

$$\begin{aligned} P\Big(\frac{\partial}{\partial x}\Big)\frac{1}{|x|^{q-2}} &= (-1)^n(q-2)q\cdots(q+2n-4)\frac{1}{|x|^{q+2n-2}} \\ &\quad\times\left(1 - \frac{|x|^2\Delta}{2(q+2n-4)} + \frac{|x|^4\Delta^2}{2\cdot 4(2+2n-4)(q+2n-6)}\cdots\right)P(x). \end{aligned} \tag{26}$$

Proof Direct computation gives for $P(x) = x_i^n$, $i = 1, 2, \ldots, q$:

$$\begin{aligned} \frac{\partial^n}{\partial x_i^n}\frac{1}{|x|^{q-2}} &= (-1)^n(q-2)q\cdots(q+2n-4)\frac{1}{|x|^{q+2n-2}} \\ &\quad\times\left(x_i^n - \frac{n(n-1)|x|^2x_i^{n-2}}{2(q+2n-4)} + \frac{n(n-1)(n-2)(n-3)|x|^4x_i^{n-4}}{2\cdot 4(2+2n-4)(q+2n-6)}\cdots\right). \end{aligned}$$

But $n(n-1)x_i^{n-2} = \frac{\partial^2}{\partial x_i^2} x_i^n = \Delta x_i^n$; $n(n-1)(n-2)(n-3)x_i^{n-4} = \frac{\partial^4}{\partial x_i^4} x_i^n = \Delta^2 x_i^n$, etc., so that (26) is valid for these special cases. Then (26) must also be valid for $P(x) = (\alpha^{\mathrm{T}} x)^n$ for α any unit vector in $\mathbf{R}^q$ because the differential operator Δ is invariant with respect to rotations, and multiplication with a constant shows that then it is valid also for $P(x) = (y^{\mathrm{T}} x)^n$ where y is any vector in $\mathbf{R}^q$. From the linearity of the equation (26) follows then that (26) is valid for

$$P(x) = \sum_{i=1}^{M} a_i \left(y_i^{\mathrm{T}} x\right)^n, \quad a_i \text{ real numbers}, \ y_i \in \mathbf{R}^q, \ i = 1, 2, \ldots, M. \tag{27}$$

But as we have proved in Section 1 that any homogeneous polynomial of degree n may be written in the form (27), (26) is proved for the general case.

If we take for $P\left(\frac{\partial}{\partial x}\right)$ any harmonic differential operator $h\left(\frac{\partial}{\partial x}\right)$ of order n we find

$$h\left(\frac{\partial}{\partial x}\right) \frac{1}{|x|^{q-2}} = (-1)(q-2) \cdots (q+2n-4) \frac{h(x)}{|x|^{q+2n-2}}. \tag{28}$$

The function $|x|^{2-q}$ is harmonic for $x \neq 0$ i.e. $\Delta \frac{1}{|x|^{q-2}} = 0$ for $x \neq 0$:

$$\frac{\partial}{\partial x_i} \frac{1}{|x|^{q-2}} = -\frac{q-2}{2} \frac{1}{|x|^q} 2x_i = -(q-2) \frac{x_i}{|x|^q}$$

$$\frac{\partial^2}{\partial x_i^2} \frac{1}{|x|^{q-2}} = -(q-2) \frac{1}{|x|^q} + (q-2)q \frac{x_i^2}{|x|^{q+2}}$$

$$\Delta \frac{1}{|x|^{q-2}} = \sum_{i=1}^{q} \frac{\partial^2}{\partial x_i^2} \frac{1}{|x|^{q-2}} = -(q-2)q \frac{1}{|x|^q} + (q-2)q \frac{|x|^2}{|x|^{q+2}} = 0,$$

therefore we can apply Theorem 13.7 on our $P(x)$ and $\phi = \frac{1}{|x|^{q-2}}$ to the result that if $h(x)$ is the harmonic projection of $P(x)$ then

$$P\left(\frac{\partial}{\partial x}\right) \frac{1}{|x|^{q-2}} = (-1)^n (q-2)q \cdots (q+2n-4) \frac{h(x)}{|x|^{q+2n-2}},$$

from which we again deduce the formula (25).

The operator

$$\Pi : \mathcal{P}_{qn} \to \mathcal{H}_{qn} : P(x) \mapsto h(x) = P(x) - \frac{|x|^2 \Delta P(x)}{2(q+2n-2)} + \frac{|x|^4 \Delta^2 P(x)}{2 \cdot 4(q+2n-4)(q+2n-6)} \cdots \tag{29}$$

which projects the Hilbert space of homogeneous polynomial of degree n orthogonally onto its subspace consisting of spherical harmonics of degree n is an ortogonal projector and it may be a good exercise for the reader to prove directly that it satisfies the formal conditions for being so:

$$\Pi^2 = \Pi \quad \text{and} \quad (\Pi P, Q)_n = (P, \Pi Q)_n \qquad \text{for all } P, Q \in \mathcal{P}_{qn},$$

and also that

$$\Pi\left(|x|^2 Q(x)\right) = 0 \qquad \text{for all } Q \in \mathcal{P}_{qn-2},$$
$$\Pi P(x) = P(x) \qquad \text{for all } P \in \mathcal{H}_{qn}.$$

Now we want to find the reproducing kernel $H_n(x,y)$ of $\mathcal{H}_{qn}$. For $\{\phi_i\}$, $i = 1$, 2, ..., N, any complete orthonormal system in $\mathcal{H}_{qn}$

$$H_n(x,y) = \sum_{i=1}^{N} \phi_i(x)\phi_i(y).$$

Let $\{\psi_j\}$, $j = 1, 2, \ldots, M-N$, be any complete orthonormal system for $\mathcal{H}^{c}_{qn}$ then

$$\{\phi_i\} \cup \{\psi_j\} \qquad i = 1, 2, \ldots, N, \quad j = 1, 2, \ldots, M-N$$

must be a complete orthonormal system in $\mathcal{P}_{qn}$ such that

$$\frac{1}{n!}\left(x^{\mathrm{T}} y\right)^n = \sum_{i=1}^{N} \phi_i(x)\phi_i(y) + \sum_{j=1}^{M-N} \psi_j(x)\psi_j(y)$$

because $\frac{1}{n!}(x^{\mathrm{T}}y)^n$ is the reproducing kernel for $\mathcal{P}_{qn}$.

From the properties of the projector Π proved above follows

$$\Pi\phi_i = \phi_i \qquad i = 1, 2, \ldots, N$$
$$\Pi\psi_j = 0 \qquad j = 1, 2, \ldots, M-N$$

so that by Π_x denote Π as operating on x we have

$$\Pi_x\left(\frac{1}{n!}\left(x^{\mathrm{T}}y\right)^n\right) = \sum_{i=1}^{N} \phi_i(x)\phi_i(y) = H_n(x,y),$$

or by (29)

$$H_n(x,y) = \frac{1}{n!}\left(\left(x^{\mathrm{T}}y\right)^n - \frac{n(n-1)|x|^2|y|^2}{2(q+2n-4)}(x^{\mathrm{T}}y)^{n-2}\right.$$
$$\left. + \frac{n(n-1)(n-2)(n-3)|x|^4|y|^4}{2\cdot 4(q+2n-4)(q+2n-6)}(x^{\mathrm{T}}y)^{n-4}\cdots\right) \tag{30}$$

here we have used

$$\frac{\partial}{\partial x_i}\left(x^{\mathrm{T}}y\right)^n = n\left(x^{\mathrm{T}}y\right)^{n-1} y_i$$
$$\frac{\partial^2}{\partial x_i^2}\left(x^{\mathrm{T}}y\right)^n = n(n-1)\left(x^{\mathrm{T}}y\right)^{n-2} y_i^2$$
$$\Delta\left(x^{\mathrm{T}}y\right)^n = n(n-1)|y|^2\left(x^{\mathrm{T}}y\right)^{n-2}.$$

If we by θ denote the angle between the two vectors x and y in $\mathbf{R}^q$ then

$$\left(x^{\mathrm{T}} y\right)^m = |x|^m |y|^m \cos^m \theta,$$

so that we may write (30) as

$$H_n(x,y) = \frac{|x|^n |y|^n}{n!} \left(\cos^n \theta - \frac{n(n-1)}{2(q+2n-4)} \cos^{n-2}\theta \right.$$
$$\left. + \frac{n(n-1)(n-2)(n-3)}{2\cdot 4(q+2n-4)(q+2n-6)} \cos^{n-4}\theta - \cdots \right). \qquad (31)$$

As we see $H_n(x,y)$ is a product of $|x|^n$, $|y|^n$, and a polynomial of degree n in $\cos\theta$. This polynomial is even if n is even and odd if n is odd. It is customary to rewrite (31) in the form

$$H_n(x,y) = |x|^n |y|^n \frac{\binom{n+q-3}{q-3}}{(q-2)q(q+2)\cdots(q+2n-4)} P_n(\cos\theta) \qquad (32)$$

where

$$P_n(t) = \frac{(q-2)q(q+2)\cdots(q+2n-4)}{(q-2)(q-1)q\cdots(q+n-3)}$$
$$\times \left(t^n - \frac{n(n-1)}{2(q+2n-4)} t^{n-2} + \frac{n(n-1)(n-2)(n-3)}{2\cdot 4(q+2n-4)(q+2n-6)} t^{n-4} - \cdots \right)$$

is called the *Legendre polynomial* of degree n (in $\mathbf{R}^q$) and where the constant, as we shall see later, is chosen so as to make $P_n(1) = 1$.

The formula

$$\phi_1(x)\phi_1(y) + \phi_2(x)\phi_2(y) + \cdots + \phi_N(x)\phi_N(y)$$
$$= |x|^n |y|^n \frac{\binom{n+q-3}{q-3}}{(q-2)q(q+2)\cdots(q+2n-4)} P_n(\cos\theta) \qquad (33)$$

is called the *addition theorem* for the spherical harmonics and is of great importance in the theory and the applications of spherical harmonics. It is valid for $\{\phi_i\}$, $i = 1, 2, \ldots, N$, *any* complete orthonormal system in $\mathcal{H}_{qn}$. We have proved:

Theorem 13.9 *If $f(x)$ is a function such that there exists a neighbourhood U of $x_0 \in \mathbf{R}^q$ such that for $x \in U$: $\Delta f(x) = 0$, then again*

$$\frac{1}{n!}\left(y^{\mathrm{T}} \frac{\partial}{\partial x}\right)^n f(x) = H_n\left(y, \frac{\partial}{\partial y}\right) f(x), \qquad x \in U.$$

We shall now give another, and more conventional, expression for the scalar product in $\mathcal{H}_{qn}$, which has relation to the interplay between the spherical harmonics and that we will call the *outer spherical harmonics*.

We have already seen that if we in the Hobson formula (26) for $P(x)$ put a spherical harmonic $h(x)$ of degree n we find

$$h\Big(\frac{\partial}{\partial x}\Big)\frac{1}{|x|^{q-2}} = (-1)^n(q-2)q(q+2)\cdots(q+2n-4)\frac{h(x)}{|x|^{q+2n-2}}. \tag{34}$$

The function

$$\widetilde{h}(x) = \frac{h(x)}{|x|^{q+2n-2}} \tag{35}$$

is obviously not a polynomial, but it is an algebraic function which is positively homogeneous of degree $-(n+q-2)$.

A function $f(x)$, $x \in \mathbf{R}^q \setminus \{0\}$, is said to be positively homogeneous of degree d if for all $t > 0$ we have

$$f(tx) = t^d f(x).$$

Evidently a homogeneous polynomial of degree d is positively homogeneous of degree d. We shall use Euler's lemma on positively homogeneous functions:

If $f(x)$, $x \in \mathbf{R}^q \setminus \{0\}$, is positively homogeneous of degree d then

$$\sum_{i=1}^{q} x_i \frac{\partial f}{\partial x_i} = d\ f. \tag{36}$$

Proof

$$\Big(\frac{\partial f(tx)}{\partial t}\Big)_{t=1} = \sum_{i=1}^{q} x_i \frac{\partial f}{\partial x_i}, \tag{37}$$

and

$$\frac{\partial f(tx)}{\partial t} = \frac{\partial}{\partial t}(t^d f(x)) = dt^{d-1} f(x). \tag{38}$$

By putting $t = 1$ in (38), (36) follows from (37) and (38).

$\Delta\widetilde{h}(x) = 0$ for $x \neq 0$ i.e. overall where $\widetilde{h}$ is defined. This follows from the fact that all derivatives—also of higher orders—with respect to the Cartesian coordinates of harmonic functions are harmonic, but we shall prove it directly by using Euler's lemma:

$$\frac{\partial}{\partial x_i}\widetilde{h}(x) = \frac{1}{|x|^{q+2n-2}}\frac{\partial h}{\partial x_i} - (q+2n-2)\frac{x_i}{|x|^{q+2n}}h(x)$$

$$\begin{aligned}\frac{\partial^2}{\partial x_i^2}\widetilde{h}(x) = {} & \frac{1}{|x|^{q+2n-2}}\frac{\partial^2 h}{\partial x_i^2} - 2(q+2n-2)\frac{1}{|x|^{q+2n}}x_i\frac{\partial h}{\partial x_i} \\ & - (q+2n-2)\frac{h(x)}{|x|^{q+2n}} + (q+2n-2)(q+2n)\frac{x_i^2 h(x)}{|x|^{q+2n+2}}\end{aligned}$$

$$\begin{aligned}\Delta\widetilde{h}(x) &= \frac{1}{|x|^{q+2n-2}}\Delta h(x) - 2(q+2n-2)\frac{n}{|x|^{q+2n}}h(x)\\ &\quad - (q+2n-2)q\frac{h(x)}{|x|^{q+2n}} + (q+2n-2)(q+2n)\frac{h(x)}{|x|^{q+2n}}\\ &= \frac{\Delta h(x)}{|x|^{q+2n-2}} + (q+2n-2)(-2n-q+q+2n)\frac{h(x)}{|x|^{q+2n}} = 0\end{aligned}$$

because $\sum x_i \frac{\partial h}{\partial x_i} = nh$ and $\Delta h = 0$.

The functions $\widetilde{h}(x)$ defined by (35) will be called *outer spherical harmonics* of degree n.

Let x and y be two different points of $\mathbf{R}^q$, then from (34) evidently follows

$$\begin{aligned}h\Big(\frac{\partial}{\partial x}\Big)\frac{1}{|x-y|^{q-2}} &= (-1)^n(q-2)q(q+2)\cdots(q+2n-4)\frac{h(x-y)}{|x-y|^{q+2n-2}}\\ &= (q-2)q(q+2)\cdots(q+2n-4)\frac{h(y-x)}{|x-y|^{q+2n-2}}, \qquad (39)\end{aligned}$$

because the polynomial h is odd for n odd and even for n even. By putting $x = 0$ (39) becomes

$$\left(h\Big(\frac{\partial}{\partial x}\Big)\frac{1}{|y-x|^{q-2}}\right)_{x=0} = (q-2)q(q+2)\cdots(q+2n-4)\widetilde{h}(y). \qquad (40)$$

From Corollary 13.5 follows for any $x \in \mathbf{R}^q$, any $k \in \mathcal{H}_{qm}$ and for any surface ω with $x \in \omega$ and where 0 is in the interior of ω:

$$k(x) = \frac{1}{\omega_q}\int\limits_{\omega}\left(k(y)\frac{\partial}{\partial n_y}\rho_x(y) - \rho_x(y)\frac{\partial}{\partial n_y}k(y)\right)d\omega_y \qquad (41)$$

where

$$\rho_x(y) = \frac{1}{(q-2)|x-y|^{q-2}}.$$

For $x \neq y$, $\rho_x(y)$ is arbitrarily often differentiable, therefore we may in differentiating (41) reverse the order of differentiation and integration so that if h is any spherical harmonic of order n we have

$$\left(h\Big(\frac{\partial}{\partial x}\Big)k(x)\right)_{x=0} = \frac{(q-2)q(q+2)\cdots(q+2n-4)}{(q-2)\omega_q}\int\limits_{\omega}\Big(k(y)\frac{\partial}{\partial n}\widetilde{h}(y) - \widetilde{h}(y)\frac{\partial}{\partial n}k(y)\Big)d\omega \qquad (42)$$

where we have used (41); here, ω denotes any smooth surface surrounding 0.

Let us consider (42) first for $m = n$. Then $h, k \in \mathcal{H}_{qn}$ and

$$\left(h\Big(\frac{\partial}{\partial x}\Big)k(x)\right)_{x=0} = h\Big(\frac{\partial}{\partial x}\Big)k(x) = (h,k)_n$$

and (42) reads:

$$(h,k)_n = \frac{(q-2)q(q+2)\cdots(q+2n-4)}{(q-2)\omega_q} \int_\omega \left(k\frac{\partial \widetilde{h}}{\partial n} - \widetilde{h}\frac{\partial k}{\partial n} \right) d\omega$$

that is, if we for $h, k \in \mathcal{H}_{qn}$ define

$$(h,k) = \frac{1}{\omega_q} \int_\omega \left(k\frac{\partial \widetilde{h}}{\partial n} - \widetilde{h}\frac{\partial k}{\partial n} \right) d\omega$$

where the closed smooth surface ω surrounds 0 then

$$(h,k) = \frac{q-2}{(q-2)q(q+2)\cdots(q+2n-4)}(h,k)_n$$

so that $(h,k) = (k,h)$ is a scalar product for spherical harmonics of degree n, in fact the two scalar products $(\cdot,\cdot)$ and $(\cdot,\cdot)_n$ differ only from one another by a positive constant factor. We will call the Hilbert space of spherical harmonics of degree n with the scalar product $(\cdot,\cdot)$ $\overline{\mathcal{H}}_{qn}$, it has evidently the same elements as $\mathcal{H}_{qn}$ and when two elements are orthogonal in one of these spaces the same two polynomials are also orthogonal in the other space. If $\{\phi_i\}$ is a complete orthonormal system in $\mathcal{H}_{qn}$ then $\{\sqrt{(q-2)q(q+2)\cdots(q+2n-4)/(q-2)}\,\phi_i\}$ is a complete orthonormal system in $\overline{\mathcal{H}}_{qn}$, and the reproducing kernel $L_n(x,y)$ of $\overline{\mathcal{H}}_{qn}$ must be

$$L_n(x,y) = \frac{(q-2)q(q+2)\cdots(q+2n-4)}{q-2} H_n(x,y)$$

where $H_n(x,y)$ is the reproducing kernel for $\mathcal{H}_{qn}$. From (32) follows then:

$$L_n(x,y) = |x|^n|y|^n \frac{1}{q-2}\binom{n+q-3}{q-3} P_n(\cos\theta), \tag{43}$$

and the formula expressing the addition theorem for spherical harmonics (33) now becomes

$$\sum_{i=1}^{N} \phi_i(x)\phi_i(y) = |x|^n|y|^n \frac{1}{q-2}\binom{n+q-3}{q-3} P_n(\cos\theta), \tag{44}$$

where $\{\phi_i\}$ now is a complete orthonormal system in $\overline{\mathcal{H}}_{qn}$.

For $m < n$ the expression

$$h\Big(\frac{\partial}{\partial x}\Big)k(x) \tag{45}$$

where $h \in \mathcal{H}_{qn}$ and $k \in \mathcal{H}_{qm}$ is identically 0, for $m > 0$ (45) is a homogeneous polynomial of degree $m-n$ and we must have for $m \neq n$:

$$\left(h\Big(\frac{\partial}{\partial x}\Big)k(x) \right)_{x=0} = 0$$

or, by (42)

$$\frac{1}{\omega_q}\int\limits_{\omega}\left(k\frac{\partial\widetilde{h}}{\partial n}-\widetilde{h}\frac{\partial h}{\partial n}\right)d\omega=0 \tag{46}$$

for h and k spherical harmonics of different degrees.

We can extend the operation $\widetilde{\ }$ defined in (35) for spherical harmonics to harmonic polynomials e.g. $l(x)$ by expressing $l(x)$ as a sum of spherical harmonics h_i, $i = 0, 2, \ldots, n$, where n is the degree of l and the degree of h_i is i:

$$l=\sum_{i=0}^{n}h_i$$

and defining

$$\widetilde{l}=\sum_{i=0}^{n}\widetilde{h}_i=\frac{1}{|x|^{q-2}}\sum_{i=0}^{n}\frac{h_i(x)}{|x|^{2i}}.$$

This definition is evidently unique, and using it we may define

$$(l_1,l_2)=\frac{1}{\omega_q}\int\limits_{\omega}\left(l_1\frac{\partial\widetilde{l}_2}{\partial n}-\widetilde{l}_2\frac{\partial l_1}{\partial n}\right)d\omega \tag{47}$$

for all pairs of harmonic polynomials l_1, l_2. It has many of the properties of the scalar product but the space of all harmonic polynomials is *not* a Hilbert space, it is only a pre-Hilbert space. With respect to the scalar product $(\cdot,\cdot)$ two spherical harmonics of different orders are always orthogonal.

The scalar product $(\cdot,\cdot)$ is that which we shall use normally as the standard scalar product in the following although it is not the scalar product for spherical harmonics commonly used in the geodetic literature.

Not only for historical reasons we must also say a little about the *spherical surface harmonics* and the commonly used normalization.

If we in (35) put $|x| = 1$ we see that on the unit sphere $|x| = 1$ the values of $h(x)$ and $\widetilde{h}(x)$ are the same and we will call

$$h\Big(\frac{x}{|x|}\Big)=\frac{1}{|x|^n}h(x)\qquad\text{for } x\neq 0\quad\text{and}\quad h\in\overline{\mathcal{H}}_{qn}$$

a spherical surface harmonic of degree 1, it is a function defined on the unit sphere, and the outer spherical harmonics as well as the spherical harmonics may be expressed by means of them and $|x|$ as follows:

$$h(x)=|x|^n h\Big(\frac{x}{|x|}\Big)\qquad\text{and}\qquad\widetilde{h}(x)=\frac{h\big(\frac{x}{|x|}\big)}{|x|^{q+n-2}}.$$

For points at the unit sphere we have for $r = |x|$ and $\partial/\partial r$ denoting partial differentiation with respect to r with constant of $x/|x|$:

$$\frac{\partial h(x)}{\partial r}=\frac{\partial}{\partial r}\left(r^n h\Big(\frac{x}{r}\Big)\right)=nr^{n-1}h\Big(\frac{x}{r}\Big)=\frac{n}{|x|}h(x)$$

$$\frac{\partial \widetilde{h}(x)}{\partial r} = -\frac{q+n-2}{|x|}\widetilde{h}(x)$$

so that if we in (47) for ω choose the unit sphere we find for $h \in \overline{\mathcal{H}}_{qn}$ and $k \in \overline{\mathcal{H}}_{qm}$:

$$(h,k) = -\frac{1}{\omega_q} \int\limits_{S(0,1)} \left[h\Big(\frac{x}{|x|}\Big)\Big(-(q+m-2)k\Big(\frac{x}{|x|}\Big)\Big) - k\Big(\frac{x}{|x|}\Big) nh\Big(\frac{x}{|x|}\Big) \right] d\omega_q$$

$$= (m+n+q-2)\frac{1}{\omega_q} \int\limits_{S(0,1)} h\Big(\frac{x}{|x|}\Big) k\Big(\frac{x}{|x|}\Big)\, d\omega_q. \tag{48}$$

$S(a,r)$ denotes the sphere $|x-a| = r$ with centre a and radius r. Here the expression

$$\frac{1}{\omega_q} \int\limits_{S(0,1)} h\Big(\frac{x}{|x|}\Big) k\Big(\frac{x}{|x|}\Big)\, d\omega_q,$$

the mean value over the unit sphere of the product of the values of two spherical surface harmonics, is the 'traditional' scalar product of the two harmonics used to define the 'fully normalized spherical harmonics.' It is not without regret that I try to introduce into geodesy another normalization convention which in my eyes has important advantages relative to the old one and which—as far as I understand—is that preferred e.g. in geophysics.

For both of these normalization conventions the scalar product of spherical harmonics of different degrees is zero, and the 'geometry' in the single $\mathcal{H}_{qn}$ spaces are essentially the same for the two normalizations.

We can however use the equation (48) to deduce an important result.

Let us in (44) put $y = x$ and therefore also $\theta = 0$:

$$\sum_{i=1}^{N} \left(\phi_i\Big(\frac{x}{|x|}\Big) \right)^2 = \frac{1}{q-2} \binom{n+q-3}{q-3} P_n(1), \tag{49}$$

where $\{\phi_i\}$ is a complete orthonormal system in $\overline{\mathcal{H}}_{qn}$, and where we have divided with $|x|^{2n}$ on both sides.

Next we will on both sides in (49) take the mean value over $S(0,1)$. The right hand is unchanged being constant. On the left hand we use (48) and the fact that $(\phi_i, \phi_i) = 1$:

$$\frac{N}{2n+q-2} = \frac{1}{q-2} \binom{n+q-3}{q-3} P_n(1)$$

but as we have found (Theorem 13.5) that $N = \frac{2n+q-2}{q-2}\binom{n+q-3}{q-3}$,

$$P_n(1) = 1$$

results as I have promised above.

From this result and (49) we find

$$\sum_{i=1}^{N}\left(\phi_i(x)\right)^2 = \frac{1}{q-2}\binom{n+q-3}{q-3}|x|^{2n}$$

from which follows

$$|\phi_i(x)| \le |x|^n\sqrt{\frac{1}{q-2}\binom{n+q-3}{q-3}}, \qquad \text{for} \quad i = 1, 2, \ldots, N \tag{50}$$

where the expression under the root sign is a polynomial in n of degree $q-3$, for $q=3$ the square root is 1, evidently (50) is important for the discussion of convergence for series in spherical harmonics. The corresponding formula for outer spherical harmonics is:

$$|\widetilde{\phi}_i(x)| \le \frac{1}{|x|^{n+q-1}}\sqrt{\frac{1}{q-2}\binom{n+q-3}{q-3}}, \qquad \text{for} \quad i = 1, 2, \ldots, N. \tag{51}$$

We can prove more than $P_n(1)=1$ namely that $|P_n(\cos\theta)| \le 1$: $L_n(x,y)$ is the reproducing kernel for $\overline{\mathcal{H}}_{qn}$, therefore

$$\big|L_n(x,y)\big| = \big|\big(L_n(x,\cdot), L_n(y,\cdot)\big)\big| \le \|L_n(x,\cdot)\|\|L_n(y,\cdot)\| = \big(L_n(x,x)L_n(y,y)\big)^{1/2}$$

or by (43)

$$\left||x|^n|y|^n\frac{1}{q-2}\binom{n+q-3}{q-3}P_n(\cos\theta)\right| \le \left||x|^n|y|^n\frac{1}{q-2}\binom{n+q-3}{q-3}P_n(1)\right|,$$

i.e. $P_n(\cos\theta) \le 1$ q.e.d.

As I mentioned above it is the expression $(\cdot,\cdot)$ that we shall use here as the standard scalar product but I have to introduce many other scalar products. In the few last pages of this section two new scalar products shall be defined.

The first one $(\cdot,\cdot)_{B_0}$ is defined as

$$(h,k)_{B_0} = \frac{1}{\omega_q}\int\limits_{B(0,1)} h(x)k(x)\,dx, \qquad h \in \overline{\mathcal{H}}_{qn}, \quad k \in \overline{\mathcal{H}}_{qm},$$

i.e. the $L_2\big(B(0,1)\big)$ scalar product where $B(0,1) = \big\{x \in \mathbf{R}^q \,\big|\, |x| < 1\big\}$. We want to express it by $(\cdot,\cdot)$.

The computation is straightforward:

$$(h,k)_{B_0} = \frac{1}{\omega_q}\int\limits_{B(0,1)} h(x)k(x)\,dx = \frac{1}{\omega_q}\int_0^1 r^{q-1}\,dr\int\limits_{S(0,1)} r^n h\Big(\frac{x}{r}\Big)r^m k\Big(\frac{x}{r}\Big)\,d\omega_q$$

$$= \frac{1}{\omega_q}\int_0^1 r^{m+n+q-1}\,dr\int\limits_{S(0,1)} h\Big(\frac{x}{r}\Big)k\Big(\frac{x}{r}\Big)\,d\omega_q$$

$$= \frac{1}{m+n+q}\frac{1}{\omega_q}\int\limits_{S(0,1)} h\Big(\frac{x}{r}\Big)k\Big(\frac{x}{r}\Big)\,d\omega_q$$

$$= \frac{1}{(m+n+q)(m+n+q-2)}(h,k),$$

the last step follows from (48). We restate the result:

$$(h,k)_{B_0} = \frac{1}{(m+n+q)(m+n+q-2)}(h,k) \qquad \text{for } h\in\overline{\mathcal{H}}_{qn},\ k\in\overline{\mathcal{H}}_{qm}. \tag{52}$$

Again the scalar product of spherical harmonics of different degrees is 0, and within $\overline{\mathcal{H}}_{qn}$ the new scalar product is a constant multiple of the standard scalar product, the constant being $1/\big((2n+q)(2n+q-2)\big)$.

The other scalar product $(h,k)_{B_1}$ is defined by

$$(h,k)_{B_1} = \frac{1}{\omega_q}\int\limits_{B(0,1)}\sum_{i=1}^{q}\frac{\partial h}{\partial x_i}\frac{\partial k}{\partial x_i}\,dx = -\frac{1}{\omega_q}\int\limits_{S(0,1)} h\frac{\partial k}{\partial n}\,d\omega_q$$

$$= \frac{1}{\omega_q}\int\limits_{S(0,1)}|r|^n h\Big(\frac{x}{r}\Big)\frac{\partial r^m}{\partial r}k\Big(\frac{x}{r}\Big)\,d\omega_q = \frac{m}{\omega_q}\int\limits_{S(0,1)} h\Big(\frac{x}{r}\Big)k\Big(\frac{x}{r}\Big)\,d\omega_q$$

$$= \frac{m}{n+m+q-2}(h,k), \tag{53}$$

where we have used Green's formula.

On the other hand:

$$(h,k)_{B_1} = \frac{1}{\omega_q}\int\limits_{B(0,1)}\sum_{i=1}^{q}\frac{\partial h}{\partial x_i}\frac{\partial k}{\partial x_i}\,dx = \sum_{i=1}^{q}\frac{1}{\omega_q}\int\limits_{B(0,1)}\frac{\partial h}{\partial x_i}\frac{\partial k}{\partial x_i}\,dx = \sum_{i=1}^{q}\Big(\frac{\partial h}{\partial x_i},\frac{\partial k}{\partial x_i}\Big)_{B_0}.$$

For $h\in\overline{\mathcal{H}}_{qn}$ we thus find (note that $\frac{\partial h}{\partial x_i}\in\mathcal{H}_{qn-1}$)

$$\sum_{i=1}^{q}\Big\|\frac{\partial h}{\partial x_i}\Big\|^2 = (2n+q-2)(2n+q-4)\sum_{i=1}^{q}\Big\|\frac{\partial h}{\partial x_i}\Big\|^2_{B_0}$$

$$= (2n+q-2)(2n+q-4)\|h\|^2_{B_1} = n(2n+q-4)\|h\|^2. \tag{54}$$

From this follows at last:

$$\Big\|\frac{\partial h}{\partial x_i}\Big\| \le \sqrt{n(2n+q-4)}\,\|h\|, \qquad i=1,\,2,\,\ldots,\,q.$$

3. Series in Spherical Harmonics

In this section, let $\{\phi_{ni}\}$, $i=1,\,2,\,\ldots,\,N_{qn}$ for every fixed $n=0,\,1,\,2,\,\ldots,$ denote a complete orthonormal system in $\overline{\mathcal{H}}_{qn}$.

Definition 13.2 *The* radius of convergence R *for the series*

$$\sum_{n=0}^{\infty}\sum_{i=1}^{N_{qn}} a_{ni}\phi_{ni}(x), \qquad a_{ni} \in \mathbf{R}, \tag{55}$$

is defined by

$$R = \sup\left\{ r \;\middle|\; r > 0;\ \sum_{n=0}^{\infty}\sum_{i=1}^{N_{qn}} \left(r^n a_{ni}\right)^2 < \infty \right\}$$

and the sphere of convergence *for the series* (55) *is* $S(0, R)$.

We shall prove

Theorem 13.10 *The series in spherical harmonics* (55) *converges absolutely in the interior of its sphere of convergence and on and inside every sphere* $S(0,r)$ *where* $r <$ *the radius of convergence is uniform.*

Proof The series (55) may be written as

$$\sum_{n=0}^{\infty} A_n \Phi_n(x), \tag{56}$$

where

$$A_n = \left(\sum_{i=1}^{N_{qn}} a_{ni}^2\right)^{1/2}$$

and

$$\Phi_n(x) = \frac{1}{A_n}\sum_{i=1}^{N_{qn}} a_{ni}\phi_{ni}(x),$$

where obviously $\Phi_n \in \overline{\mathcal{H}}_{qn}$ for $n = 0, 1, \ldots$.

Then

$$R = \sup\left\{ r \;\middle|\; r > 0;\ \sum_{n=0}^{\infty} (A_n r^n)^2 < \infty \right\}.$$

Let $r < R$ and choose r_1 such that $r < r_1 < R$. Then the series $\sum_{n=0}^{\infty}(A_n r_1^n)^2$ converges and therefore we must have $\lim_{n\to\infty} A_n r_1^n = 0$ and then the series

$$\sum_{n=0}^{\infty} A_n r^n = \sum_{n=0}^{\infty} A_n r_1^n \left(\frac{r}{r_1}\right)^n \tag{57}$$

must converge. (As $|r/r_1| < 1$ it converges because $\{A_n r_1^n\}$ is bounded.) This convergence is not spoiled by multiplying every term by a polynomial in n and therefore also the series

$$\frac{1}{\sqrt{q-2}}\sum_{n=0}^{\infty} A_n \sqrt{\binom{n+q-3}{q-3}}\, r^n \tag{58}$$

converges.

But from (50) follows

$$\left|\Phi_n(x)\right| \le |x|^n \sqrt{\frac{1}{q-2}\binom{n+q-3}{q-3}}$$

so that (56) and hence (55) are dominated by the converging series with non-negative coefficients (58), and the theorem is proved.

Definition 13.3 *If a function $f(x)$ is defined for $x \in \Omega$, where Ω is any open set in $\mathbf{R}^q$ and*

$$\int_\Omega \left(f(x)\right)^2 dx < \infty,$$

then f is said to be square integrable *in Ω and we write*

$$f \in L^2(\Omega).$$

Corollary 13.3 (of Theorem 13.10) *With the notations of the theorem*

$$\sum_{n=0}^{\infty}\sum_{i=1}^{N_{qn}} a_{ni}\phi_{ni} \in L^2\left(B(0,r)\right),$$

where $B(0,r)$ is the interior of the sphere $S(0,r)$, $r < R$.

Proof Follows from the uniform convergence of the series in $\overline{B}(0,r)$.

Theorem 13.11 *For $|y|<|x|$ the function $\rho_y(x)=\frac{1}{q-2}\frac{1}{|x-y|^{q-2}}$ may be represented by the following convergent series*

$$\begin{aligned}\rho_y(x) = \rho_0(x)\Bigg[1 &+ \binom{q-2}{q-3}\frac{|y|}{|x|}P_1(\cos\theta) + \binom{q-1}{q-3}\frac{|y|^2}{|x|^2}P_2(\cos\theta)\\ &+ \binom{q}{q-3}\frac{|y|^3}{|x|^3}P_3(\cos\theta) + \cdots\\ &+ \binom{n+q-3}{q-3}\frac{|y|^n}{|x|^n}P_n(\cos\theta) + \cdots\Bigg]\end{aligned} \tag{59}$$

where θ is the angle between the vectors x and y.

Proof By Taylor's formula (17)

$$\begin{aligned}\frac{1}{|x-y|^{q-2}} &= \frac{1}{|x|^{q-2}} - \frac{1}{1!}\left(y^{\mathrm{T}}\frac{\partial}{\partial x}\right)\frac{1}{|x|^{q-2}} + \frac{1}{2!}\left(y^{\mathrm{T}}\frac{\partial}{\partial x}\right)^2\frac{1}{|x|^{q-2}} + \cdots\\ &+ (-1)^n\frac{1}{n!}\left(y^{\mathrm{T}}\frac{\partial}{\partial x}\right)^n\frac{1}{|x|^{q-2}}\\ &+ (-1)^{n+1}\frac{1}{n!}\int_0^1(1-s)^n\left(y^{\mathrm{T}}\frac{\partial}{\partial x}\right)^{n+1}\frac{1}{|x-sy|^{q-2}}\,ds\end{aligned} \tag{60}$$

$\frac{1}{|x-sy|^{q-2}}$ satisfies Laplace's equation $\Delta\frac{1}{|x-sy|^{q-2}} = 0$ for all $x \in \mathbf{R}^q$ such that $x - sy \neq 0$, therefore we may apply Theorem 13.9 for $m = 0, 1, 2, \ldots$:

$$\frac{1}{m!}\left(y^{\mathrm{T}}\frac{\partial}{\partial x}\right)^m \frac{1}{|x-sy|^{q-2}} = H_m\left(y, \frac{\partial}{\partial x}\right)^m \frac{1}{|x-sy|^{q-2}} \qquad \text{for } x \neq sy. \quad (61)$$

$H_m(y, \frac{\partial}{\partial x})$ is for all y a harmonic differential operator of order m, therefore by Hobson's theorem (Theorem 13.8):

$$H_m\left(y, \frac{\partial}{\partial x}\right)\frac{1}{|x|^{q-2}} = (-1)^m(q-2)q(q+2)\cdots(q+2m-4)\frac{H_m(x,y)}{|x|^{q+2m-2}}, \quad (62)$$

but by (32), $H_m(x,y) = |x|^m|y|^m\frac{\binom{m+q-3}{q-3}}{(q-2)q(q+2)\cdots(q+2m-4)}P_m(\cos\theta)$; therefore (62) becomes

$$H_m\left(y, \frac{\partial}{\partial x}\right)\frac{1}{|x|^{q-2}} = (-1)^m\binom{m+q-3}{q-3}\frac{1}{|x|^{q-2}}\frac{|y|^m}{|x|^m}P_m(\cos\theta),$$

so that if we can prove that the remainder in (60)

$$R_n = (-1)^{n+1}\frac{1}{n!}\int_0^1 (1-s)^n\left(y^{\mathrm{T}}\frac{\partial}{\partial x}\right)^{n+1}\frac{1}{|x-sy|^{q-2}}\,ds \to 0$$

for $n \to \infty$ then the theorem is proved.

In order to find an estimate for the remainder we apply (62) for $m = n+1$ at $x - sy$ instead of x and find

$$\begin{aligned}&\frac{1}{(n+1)!}\left(y^{\mathrm{T}}\frac{\partial}{\partial x}\right)^{n+1}\frac{1}{|x-sy|^{q-2}}\\ &\quad= (-1)^{n+1}(q-2)q(q+2)\cdots(q+2m-2)\frac{H_{n+1}(x-sy,y)}{|x-sy|^{q+2n}}\\ &\quad= (-1)^{n+1}\binom{n+q-2}{q-3}\frac{|x-sy|^{n+1}|y|^{n+1}}{|x-sy|^{q+2n}}P_{n+1}(\cos\theta')\end{aligned}$$

where θ' is the angle between y and $x - sy$. From $|P_{n+1}(\cos\theta)| \leq 1$ follows

$$\begin{aligned}\left|\frac{1}{n!}\left(y^{\mathrm{T}}\frac{\partial}{\partial x}\right)^{n+1}\frac{1}{|x-sy|^{q-2}}\right| &\leq (n+1)\binom{n+q-2}{q-3}\frac{|y|^{n+1}}{|x-sy|^{q+n-1}}\\ &\leq (n+1)\binom{n+q-2}{q-3}\frac{|y|^{n+1}}{(|x|-s|y|)^{q+n-1}}\end{aligned}$$

because $|x - sy| \geq |x| - s|y|$ so that

$$\begin{aligned}|R_n| &\leq \frac{n+1}{(|x|-|y|)^{q-2}}\binom{n+q-2}{q-3}\int_0^1(1-s)^n\frac{|y|^{n+1}}{(|x|-s|y|)^{n+1}}\,ds\\ &= \frac{n+1}{(|x|-|y|)^{q-2}}\binom{n+q-2}{q-3}\frac{k^{n+1}}{1-k}\int_0^1\left(\frac{1-s}{1-ks}\right)^n ds\end{aligned}$$

where $k = |y|/|x| < 1$.

The integrand $\left(\frac{1-s}{1-ks}\right)^n < 1$ so that

$$|R_n| \le \frac{n+1}{(|x|-|y|)^{q-2}}\binom{n+q-2}{q-3}\frac{k^{n+1}}{1-k} \to 0 \qquad \text{for } n \to \infty,$$

and the theorem follows.

Corollary 13.4 *Under the assumptions of the theorem we can develop $\rho_y(x)$ for $|y| < |x|$ in a series in spherical harmonics in y and outer spherical harmonics in x as follows:*

$$\rho_y(x) = \sum_{n=0}^{\infty}\sum_{i=1}^{N_{qn}} \widetilde{\phi}_{ni}(x)\phi_{ni}(y).$$

Proof

$$\frac{1}{q-2}\frac{1}{|x|^{q-2}}\binom{n+q-3}{q-3}\frac{|y|^n}{|x|^n}P_n(\cos\theta)$$

$$= \frac{1}{|x|^{q+2n-2}}|x|^n|y|^n\frac{1}{q-2}\binom{n+q-3}{q-3}P_n(\cos\theta) = \frac{1}{|x|^{q+2n-2}}\sum_{i=1}^{N_{qn}}\phi_{ni}(x)\phi_{ni}(y)$$

by (44) and

$$\frac{\phi_{ni}(x)}{|x|^{q+2n-2}} = \widetilde{\phi}_{ni}(x)$$

by (35) and the corollary follows.

4. Harmonic Functions on Bounded Sets

With point sets we shall here and overall in the following mean sets of points in $\mathbf{R}^q$, where $q \ge 3$.

Definition 13.4 *A function $f(x)$ defined on any point set is said to be* harmonic *in a point x_0 if there exists a neighbourhood of x_0 where f is two times differentiable and satisfies Laplace's equation*

$$\Delta f = 0.$$

A function f is said to be harmonic in a bounded point set if it is harmonic in all points of this point set. The set of harmonic functions in the bounded set S is denoted by $H(S)$.

Remark 13.1 From these definitions follows that if $A \subset \mathbf{R}^q$ is any closed bounded point set then from f harmonic in A follows that there exists a bounded open set Ω containing A in its interior such that f is harmonic in Ω.

Example 13.1

1. For every bounded set $S \subset \mathbf{R}^q$, every harmonic polynomial is harmonic in S.
2. For $y \notin S$, S any bounded set in $\mathbf{R}^q$, the function

$$\rho_y(x) = \frac{1}{(q-2)|x-y|^{q-2}} \tag{63}$$

is harmonic in S as a function of x.

Since linear combinations of harmonic functions are harmonic, $H(S)$ is obviously a linear vector space.

Now we suppose that the bounded set Ω as a boundary which is a two times continuously differentiable hypersurface ω, then Green's formula reads

$$\int_{\omega} \left(u\frac{\partial v}{\partial n} - v\frac{\partial u}{\partial n} \right) d\omega = \int_{\Omega} (u\Delta v - v\Delta u)\, d\Omega, \tag{64}$$

where $u, v \in C^2(\overline{\Omega})$ i.e. u and v are two times continuously differentiable on Ω and its boundary and where $\partial/\partial n$ denotes differentiation along the outgoing normal.

We shall here and in the following use the symbol ω_q to denote the surface area of the unit sphere $S(0,1)$ in $\mathbf{R}^q$. ω_3 is 4π, for general q is

$$\omega_q = \frac{2\pi^{q/2}}{\Gamma(\frac{q}{2})}.$$

Theorem 13.12 *Let Ω be a bounded open set of $\mathbf{R}^q$ with two times differentiable boundary ω and let $u \in C^2(\overline{\Omega})$ then*

$$u(y) = \frac{1}{\omega_q}\left[\int_{\omega} \left(\rho_y \frac{\partial u}{\partial n} - u\frac{\partial \rho_y}{\partial n} \right) d\omega - \int_{\Omega} \rho_y \Delta u\, d\Omega \right], \qquad \textit{for } y \in \Omega. \tag{65}$$

Corollary 13.5 *If the function ϕ is harmonic in $\overline{\Omega}$ it follows immediately from Theorem 13.12 that*

$$\phi(y) = \frac{1}{\omega_q} \int_{\omega} \left(\rho_y \frac{\partial \phi}{\partial n} - \phi\frac{\partial \rho_y}{\partial n} \right) d\omega \qquad \textit{for } y \in \Omega. \tag{66}$$

Proof We apply (64) to the open set $\Omega_\epsilon = \{ x \in \Omega \mid |x-y| > \epsilon \}$, where $0 < \epsilon <$ the distance from y to the complement of Ω and $v = \rho_y$, it is clear that $\rho_y \in C^2(\overline{\Omega}_\epsilon)$:

$$\int_{\omega} \left(u\frac{\partial \rho_y}{\partial n} - \rho_y \frac{\partial u}{\partial n} \right) d\omega + \int_{S(y,\epsilon)} \left(u\frac{\partial \rho_y}{\partial n} - \rho_y \frac{\partial u}{\partial n} \right) d\omega = \int_{\Omega_\epsilon} (u\Delta\rho_y - \rho_y \Delta u)\, d\Omega. \tag{67}$$

In the second integral we introduce spherical coordinates with origin in y and write $d\omega_q$ for the surface element of the unit sphere in $\mathbf{R}^q$ and r for $|x-y|$:

$$\int\limits_{S(y,\epsilon)} \left(u\frac{\partial \rho_y}{\partial n} - \rho_y \frac{\partial u}{\partial n}\right) d\omega = \frac{1}{q-2} \int\limits_{S(y,\epsilon)} \left(u\left(-\frac{\partial}{\partial r}\frac{1}{r^{q-2}}\right)_{r=\epsilon} - \frac{1}{\epsilon^{q-2}}\frac{\partial u}{\partial n}\right) \epsilon^{q-1}\, d\omega_q$$

$$= \int\limits_{S(y,\epsilon)} u\frac{1}{\epsilon^{q-1}}\epsilon^{q-1}\, d\omega_q - \frac{1}{q-2}\int\limits_{S(y,\epsilon)} \epsilon \frac{\partial u}{\partial n}\, d\omega_q$$

$$= \omega_q \frac{1}{\omega_q \epsilon^{q-1}} \int\limits_{S(y,\epsilon)} u\epsilon^{q-1}\, d\omega_q - \frac{\omega_q}{q-2}\frac{1}{\omega_q\epsilon^{q-1}}\int\limits_{S(y,\epsilon)} \frac{\partial u}{\partial n}\epsilon^{q-1}\, d\omega_q$$

$$= \omega_q \mathcal{M}_{S(y,\epsilon)}(u) - \frac{\omega_q \epsilon}{q-2}\mathcal{M}_{S(y,\epsilon)}\left(\frac{\partial u}{\partial n}\right)$$

where $\mathcal{M}_{S(y,\epsilon)}(f)$ denotes the mean value of the function f over the surface of the sphere $S(y,\epsilon)$. If f is continuous we have $\lim_{\epsilon\to 0}\mathcal{M}_{S(y,\epsilon)}(f) = f(y)$ and so we have:

$$\lim_{\epsilon\to 0} \int\limits_{S(y,\epsilon)} \left(u\frac{\partial \rho_y}{\partial n} - \rho_y\frac{\partial u}{\partial n}\right) d\omega = \omega_q u(y).$$

In the last integral in (67) $\Delta \rho_y = 0$ so that

$$\int\limits_{\Omega_\epsilon} (u\Delta\rho_y - \rho_y \Delta u)\, d\Omega = -\int\limits_{\Omega_\epsilon} \rho_y \Delta u\, d\Omega$$

from all this results

$$\omega_q u(y) = \int\limits_{\omega} \left(u\frac{\partial \rho_y}{\partial n} - \rho_y\frac{\partial u}{\partial n}\right) d\omega - \lim_{\epsilon\to 0}\int\limits_{\Omega_\epsilon} \rho_y \Delta u\, d\Omega.$$

Now the theorem follows if

$$\lim_{\epsilon\to 0}\int\limits_{\Omega_\epsilon} \rho_y \Delta u\, d\Omega = \int\limits_{\Omega} \rho_y\Delta u\, d\Omega.$$

To show this it suffices to prove that ρ_y is integrable over Ω since Δu is bounded on Ω. By transforming to spherical coordinates relative to the pole y

$$\int\limits_{B(y,\epsilon)} \rho_y\, d\Omega \le \omega_q \int_0^\epsilon \frac{1}{q-2}\frac{1}{r^{q-2}} r^{q-1}\, dr$$

where $B(y,\epsilon) = \{\, x \mid |x-y| \le \epsilon \,\}$.

Corollary 13.6 *If ϕ is a harmonic polynomial then (66) is valid for any $y \in \mathbf{R}^q$ if ω is any smooth surface with y in its interior.*

This follows from the fact that a harmonic polynomial is harmonic in any bounded point set.

Theorem 13.13 (The mean value theorem) *If $u \in H(\Omega)$ then its value in the point $x \in \Omega$ is the arithmetic mean of its values on each sphere $S(x,r)$ so that $0 < r <$ the distance from x to $\complement\Omega$. [$\complement\Omega$ is the complement of S i.e. $\complement\Omega = \{ y \in \mathbf{R}^q \mid y \notin \Omega \}$.]*

Proof (64) gives for $u \in H(\Omega)$ and $v = 1$

$$\int\limits_{\omega} \frac{\partial u}{\partial n}\, d\omega = 0$$

for each closed smooth surface in the interior of Ω. Therefore (63) gives for $u \in H(\Omega)$ and ω the sphere $S(x,r)$:

$$u(x) = \frac{1}{q-2}\frac{1}{\omega_q}\left[\int\limits_{S(x,r)} r^{2-q}\frac{\partial u}{\partial r}\, d\omega - \int\limits_{S(x,r)} (2-q)r^{1-q} u\, d\omega\right] = \frac{1}{r^{q-1}\omega_q}\int\limits_{S(x,r)} u\, d\omega$$

q.e.d.

Theorem 13.14 *Every $u \in H(\Omega)$ is in $C^\infty(\Omega)$ and every derivative of $u \in H(\Omega)$ with respect to Cartesian coordinates is harmonic in Ω. In particular, every harmonic derivative of u is harmonic in Ω.*

Proof For $x \in \Omega$ choose a sphere $S(x,r)$ as in Theorem 13.13. Then for every y so that $|y - x| < r$ we have

$$u(y) = \frac{1}{\omega_q}\int\limits_{S(x,r)} \left(\rho_y(z)\frac{\partial u(z)}{\partial n} - u(z)\frac{\partial \rho_y(z)}{\partial n}\right) d\omega_z. \tag{68}$$

For $|y - x| > r/2$, $\rho_y(x)$ and all its derivatives are bounded and continuous, therefore we may differentiate $u(y)$ by differentiating under the integral sign in (68) and $u \in C^\infty(\Omega)$ follows from x being an arbitrary point of Ω. That $D^\alpha u$ is harmonic follows from

$$\Delta D^\alpha u = D^\alpha \Delta u = 0.$$

We shall now use a function $\phi \in C^\infty(\mathbf{R}^q)$ which depends only on $|x|$, which is zero for $|x| > 1$, and for which $\int_{\mathbf{R}^q} \phi(x)\, dx = 1$. The existence of such a function is not obvious but a classical example is $\phi(x) = f(|x|^2 - 1)$, where

$$f(t) = Ke^{1/t} \quad \text{for } t < 0 \qquad \text{and} \qquad f(t) = 0 \quad \text{for } t \geq 0,$$

where the constant K is determined such that $\int_{\mathbf{R}^q} \phi(x)\,dx = 1$. If u and r satisfy the conditions of Theorem 13.13 it follows from the same theorem that

$$u(x) = \frac{1}{r^q} \int\limits_{\mathbf{R}^q} u(x-y)\phi\Big(\frac{y}{r}\Big)\,dy = \frac{1}{r^q} \int\limits_{\mathbf{R}^q} u(y)\phi\Big(\frac{x-y}{r}\Big)\,dy.$$

Here we may differentiate under the integral sign and we find for any multi-index α:

$$\Big(\frac{\partial}{\partial x}\Big)^{\alpha} u(x) = \frac{1}{r^q} \int\limits_{\mathbf{R}^q} u(y)\Big(\frac{\partial}{\partial x}\Big)^{\alpha} \phi\Big(\frac{x-y}{r}\Big)\,dy.$$

From Schwartz' inequality we then find

$$\left|\Big(\frac{\partial}{\partial x}\Big)^{\alpha} u(x)\right| \le \left[r^{-2q} \int\limits_{\mathbf{R}^q} \bigg(\Big(\frac{\partial}{\partial x}\Big)^{\alpha} \phi\Big(\frac{x-y}{r}\Big)\bigg)^2 dy \int\limits_{B(x,r)} |u(y)|^2\,dy \right]^{1/2}$$

where $B(x,r)$ is $\{\, y \in \mathbf{R}^q \mid |y-x| < r \,\}$, and from this follows the existence of constants A_α such that

$$\left|\Big(\frac{\partial}{\partial x}\Big)^{\alpha} u(x)\right| \le A_\alpha r^{-q/2-|\alpha|} \left[\int\limits_{B(x,r)} |u(y)|^2\,dy \right]^{1/2}. \tag{69}$$

Theorem 13.15 *Let Ω be a bounded open subset of $\mathbf{R}^q$ and $u \in H(\Omega)$. For every closed $K \subset \Omega$ and every open neighbourhood $\Omega' \subset \Omega$ of K there exist constants C_α such that*

$$\sup_{z\in K} \big|D_\alpha u(z)\big| \le C_\alpha \left[\int\limits_{\Omega'} \big|u(x)\big|^2\,dx \right]^{1/2}. \tag{70}$$

Proof This follows from (69) and the fact that every $z \in K$ is the centre of a sphere which has positive radius and is contained in Ω' and as K is compact these spheres may be chosen so that these radii have a lower bound.

Corollary 13.7 *If $u_n \in H(\Omega)$, $n = 1, 2, \ldots,$ and u_n converges to u in $L_2(\Omega')$ for all $\Omega' \Subset \Omega$ for $n \to \infty$ then $u \in H(\Omega)$. ($\Omega' \Subset \Omega$ means that there exists a compact set C such that $\Omega' \subset C \subset \Omega$.)*

Proof From (70) follows that the second derivatives of u_n converge uniformly on compact subsets of Ω. Hence $u \in C^2(\Omega)$ and $\Delta u = \lim_{n\to\infty} u_n = 0$ and so $u \in H(\Omega)$.

Corollary 13.8 (I. Harnack) *If $u_n \in H(\Omega)$ converges uniformly on compact subsets of Ω then $\lim_{n\to\infty} u_n = u \in H(\Omega)$ and all derivatives of u_n converge to the corresponding derivatives of u.*

Proof We have for any $\Omega' \Subset \Omega$ that $\overline{\Omega}' \subset \Omega$ and $\overline{\Omega}'$ is compact. In

$$\left[\int_{\Omega} |\phi(x)|^2 \, dx \right]^{1/2} \leq K \sup_{x \in \Omega'} |\phi(x)|$$

where K^2 is the volume of Ω'. Put $\phi = u - u_n$ to show that uniform convergence implies convergence in $L_2(\Omega')$. The rest follows from Corollary 13.8.

Theorem 13.16 *The series*

$$u(x) = \sum_{n=0}^{\infty} \sum_{i=1}^{N_{qn}} a_{ni} \phi_{ni}(x), \qquad a_{ni} \in \mathbf{R}, \tag{71}$$

where $\{\phi_{ni}\}$, $i = 1, 2, \ldots, N$ for every $n = 0, 1, 2, \ldots,$ is a complete orthonormal system in $\overline{\mathcal{H}}_{qn}$ converges to a harmonic function in the interior of its sphere of convergence and it is term-wise differentiable there.

Proof Follows immediately from Theorem 13.3 and from Theorem 13.15.

Theorem 13.17 *If $u \in H(\Omega)$ where $\Omega = \{ x \in \mathbf{R}^q \mid |x| < r \}$, we have for $x \in \Omega$*

$$u(x) = (q-2) \sum_{n=0}^{\infty} \sum_{i=1}^{N_{qn}} \frac{\phi_{ni}(x)}{(q-2)q(q+2)\cdots(q+2n-4)} \left(\phi_{ni}\left(\frac{\partial}{\partial y}\right) u(y) \right)_{y=0} \tag{72}$$

where $\{\phi_{ni}\}$ is as in Theorem 13.16. The series (72) converges uniformly on compact subsets of Ω.

Proof (66) gives

$$u(x) = -\frac{1}{\omega_q} \int_{|y|=r_2} \left(\rho_x \frac{\partial u}{\partial r} - u \frac{\partial \rho_x}{\partial r} \right) d\omega, \qquad \text{for } |x| \leq r_1 < r_2 < r.$$

But from Theorem 13.11 follows that the series

$$\rho_x(y) = \sum_{n=0}^{\infty} \sum_{i=1}^{N} \phi_{ni}(\phi) \widetilde{\phi}_{ni}(y), \qquad |x| \leq r_1, |y| = r_1$$

converges uniformly and absolutely so that we may integrate term by term:

$$u(x) = -\sum_{n=0}^{\infty} \sum_{i=1}^{N} \phi_{ni}(x) \langle \widetilde{\phi}_{ni}, u \rangle \tag{73}$$

where we have used the notation

$$\langle \widetilde{\phi}_{ni}, u \rangle = \frac{1}{\omega_q} \int_{|y|=r_2} \left(\widetilde{\phi}_{ni} \frac{\partial u}{\partial r} - u \frac{\partial}{\partial r} \widetilde{\phi}_{ni} \right) d\omega. \tag{74}$$

Harmonic differentiation of (73) and putting $x=0$ gives for every $m=0$, 1, ... and $j=1, 2, \ldots, N_{qm}$:

$$\left(\phi_{mj}\left(\frac{\partial}{\partial x}\right)u(x)\right)_{x=0} = -\left(\phi_{mj}\left(\frac{\partial}{\partial x}\right)\phi_{mj}(x)\right)_{x=0}\langle\widetilde{\phi}_{mj}, u\rangle, \tag{75}$$

because $\left(\phi_{mj}\left(\frac{\partial}{\partial x}\right)\phi_{mj}(x)\right)_{x=0} = 0$ for $m, j \neq n, i$.

Now $\{\phi_{ni}\}$ are normed with respect to the standard scalar product $\langle\cdot,\cdot\rangle$ and we have

$$\phi_{mj}\left(\frac{\partial}{\partial x}\right)\phi_{mj}(x) = \langle\phi_{mj}, \phi_{mj}\rangle_m = \frac{(q-2)q(q+2)\cdots(q+2n-4)}{q-2}\langle\phi_{mj}, \phi_{mj}\rangle$$

and we can write (75) as

$$\left(\phi_{mj}\left(\frac{\partial}{\partial x}\right)u(x)\right)_{x=0} = \frac{(q-2)q(q+2)\cdots(q+2n-4)}{q-2}\langle\widetilde{\phi}_{mj}, u\rangle$$

or

$$-\langle\widetilde{\phi}_{mj}, u\rangle = \frac{q-2}{(q-2)q(q+2)\cdots(q+2n-4)}\left(\phi_{mj}\left(\frac{\partial}{\partial x}\right)u(x)\right)_{x=0}$$

so that we may write (73) as

$$u(x) = (q-2)\sum_{n=0}^{\infty}\sum_{i=1}^{N_{qn}}\frac{\left(\phi_{ni}\left(\frac{\partial}{\partial y}\right)u(y)\right)_{y=0}}{(q-2)q(q+2)\cdots(q+2n-4)}\phi_{ni}(x)$$

so that the theorem is proved.

Corollary 13.9 *If $u \in H\big(B(y,r)\big)$ where $B(y,r) = \big\{\, x \in \mathbf{R}^q \mid |x-y| < r \,\big\}$, we have for $x \in B(y,r)$:*

$$u(x) = (q-2)\sum_{n=0}^{\infty}\sum_{i=1}^{N_{qn}}\frac{\phi_{ni}(x-y)}{(q-2)q(q+2)\cdots(q+2n-4)}\left(\phi_{ni}\left(\frac{\partial}{\partial y}\right)u(y)\right)_{y=x}$$

where $\{\phi_{ni}\}$ is as in Theorem 13.16. This series converges uniformly on compact subsets of $B(y,r)$.

This is the analogue to Taylor's formula.

Corollary 13.10 (The uniqueness of *harmonic continuation*) *If $u \in H(\Omega)$ and there is a point $x \in \Omega$ where*

$$\phi_{ni}\left(\frac{\partial}{\partial x}\right)u(x) = 0 \qquad \textit{for all } i = 1, 2, \ldots, N_{qn} \textit{ and all } n = 0, 1, \ldots \tag{76}$$

it follows that $u = 0$ in Ω if Ω is connected.

Proof The set of all $x \in \Omega$ for which (76) is satisfied is closed and by Corollary 13.10 it is also open. Since it is non-empty by assumption it must be equal to Ω. (Remember that an open set Ω is said to be connected if there does not exist a pair Ω_1 and Ω_2 of open sets such that $\Omega = \Omega_1 \cup \Omega_2$ and $\Omega_1 \cap \Omega_2 = \emptyset$.)

Corollary 13.11 *If $u_1, u_2 \in H(\Omega)$ and Ω is connected then $u_1 = u_2$ if every harmonic derivative of u_1 at x is equal to the corresponding harmonic derivative of u_2 at x.*

Proof From Corollary 13.11 follows that $u_1 - u_2 = 0$.

Theorem 13.18 *Ω is as everywhere in this section open and bounded. Let $u \in C(\overline{\Omega})$ and $u \in H(\Omega)$. Then the maximum and the minimum of u in $\overline{\Omega}$ is attained on the boundary of $\overline{\Omega}$.*

Proof If the maximum (minimum) is attained in an interior point x then from Theorem 13.13 follows that the same value is attained at *all* points of every sphere contained in Ω and having its centre at x i.e. u must be constant in a ball with centre at x and therefore it must be constant in the component of Ω which contains x according to Corollary 13.10 and so it will attain its maximum (minimum) at the boundary. (A component of an open set is an open connected subset $\Omega_1 \subset \Omega$ such that $\Omega \setminus \Omega_1$ is open.)

5. Harmonic Functions on Complements to Bounded Sets

In geodesy we are particularly interested in harmonic functions in the space outside the Earth. This is not a bounded set but its complement is: The surface of the Earth and the interior of the Earth. We shall now define harmonic functions on complements to bounded sets such that the essential results from Section 4 remain valid for those functions and such that they should be relevant for the study of the potential of the Earth.

First we shall prove

Theorem 13.19 *The following two properties for functions f defined in a subset of $\mathbf{R}^q$ are equivalent:*

1. *there exists a neighbourhood of x where f is two times differentiable and where $\Delta x = 0$*
2. *there exists an $r > 0$ such that for $|y - x| < r$ $f(y)$ can be expressed by a convergent series in $\{\phi_{ni}(y - x)\}$.*

Here $x \in \mathbf{R}^q$ and $\{\phi_{ni}\}$ is a complete orthonormal system in $\overline{\mathcal{H}}_{qn}$.

Proof That (1) $\Rightarrow$ (2) follows directly from Theorem 13.17. That (2) $\Rightarrow$ (1) follows from Theorem 13.16.

From this theorem follows that we could have used (2) to define harmonic functions. It is therefore natural to define

Definition 13.5 *A function f defined in a subset of* $\mathbf{R}^q$ *is said to be* harmonic in ∞ *if there exists an $r > 0$ such that for $|x| > r$ $f(x)$ can be expressed by a convergent series in $\{\widetilde{\phi}_{ni}(x)\}$, where $\{\phi_{ni}\}$ is as in Theorem 13.19.*

A function f defined in S, the complement to a bounded point set in $\mathbf{R}^q$, *is said to be harmonic in S if it is harmonic in every point of S and in ∞.*

In order to convince ourselves that these definitions make sense we have to get some information about series in $\{\widetilde{\phi}_{ni}(x)\}$. The tool we shall use here is the *Rayleigh transformation.*

Definition 13.6 *If*

$$u(x) = \sum_{n=0}^{\infty}\sum_{i=1}^{N} a_{ni}\phi_{ni}(x), \qquad \phi_{ni} \in \overline{\mathcal{H}}_{qn}, \tag{77}$$

is a series in orthonormalized spherical harmonics its Rayleigh transform $\mathcal{R}(u)(x)$ is defined as

$$\mathcal{R}(u)(x) = \sum_{n=0}^{\infty}\sum_{i=1}^{N} a_{ni}\widetilde{\phi}_{ni}(x). \tag{78}$$

(77) *will also be called the Rayleigh transform of* (78).

For $h(x)$ a spherical polynomial of degree n we have defined

$$\widetilde{h}(x) = \frac{h(x)}{|x|^{2n+q-2}},$$

which—as $h(x)$ is homogeneous of degree n—may also be written as

$$\widetilde{h}(x) = \frac{1}{|x|^{q-2}} h\Big(\frac{x}{|x|^2}\Big); \tag{79}$$

so that we have for u given by (77):

$$\mathcal{R}(u)(x) = \frac{1}{|x|^{q-2}} u\Big(\frac{x}{|x|^2}\Big).$$

It is easy to see that from (79) follows

$$h(x) = \frac{1}{|x|^{q-2}} \widetilde{h}\Big(\frac{x}{|x|^2}\Big),$$

therefore it was reasonable to define that u was the Rayleigh transform of $\mathcal{R}(u)$. Note, in order to have $\mathcal{R}^2 = I$ we must define

$$\frac{x}{|x|^2} = \begin{cases} \infty & \text{for } x = 0, \\ 0 & \text{for } x = \infty. \end{cases}$$

An immediate consequence of the above is:

Theorem 13.20 *If r is the radius of convergence of the series u in (77) then the series $\mathcal{R}(u)$ in (78) converges absolutely and uniformly for $|x| \geq 1/R$ for any $0 < R < r$, and the series represents the harmonic function $\frac{1}{|x|^{q-2}} u\left(\frac{x}{|x|^2}\right)$ for $|x| > 1/r$. The series (78) may be differentiated term by term.*

The Rayleigh transformation can be generalized to harmonic functions not necessarily defined in neighbourhoods of 0 or ∞ by the formula $\mathcal{R}\big(u(x)\big) = \frac{1}{|x|^{q-2}} u\left(\frac{x}{|x|^2}\right)$. By straightforward computation it can be proved that the Rayleigh transformed of a function harmonic in one domain Ω is a harmonic function in another domain

$$\Omega' = \left\{ x \in \mathbf{R}^q \;\middle|\; \frac{x}{|x|^2} \in \Omega \right\},$$

in fact

$$\Delta_x \left(\frac{1}{|x|^{q-2}} u(y) \right) = \frac{\Delta_y u_y}{|x|^{q+2}},$$

where $y = x/|x|^2$.

Theorem 13.18 (the maximum-minimum theorem) is not directly valid for unbounded domains, but we shall prove

Theorem 13.21 *Let Ω be an open set in $\mathbf{R}^q$ bounded or the complement to a bounded set and let $u \in C(\overline{\Omega})$ and $u \in H(\Omega)$. Then the maximum of $|u|$ is attained on the boundary of $\overline{\Omega}$.*

Proof If Ω is bounded this follows from Theorem 13.18. If Ω is the complement to a bounded set A choose R such that $A \subset B(0, R)$. Then apply Theorem 13.18 on the bounded set $\Omega \cap B(0, r)$ for $r > R$. As u is harmonic at ∞ we have $u(x) \sim O(|x|^{2-q})$ so that also the $\max_{|x|>r} |u(x)| \sim O(|r|^{2-q})$, and the theorem follows.

Closed sets which are complements to bounded sets are not compact, therefore bounded functions on such sets need not be quadratic integrable and we cannot apply the way of reasoning as in Theorem 13.15. But instead of the integrals of type $\int_{\Omega'} u^2 \, dx$ we can use integrals as $\int_{\Omega'} w(x) u^2 \, dx$ where the weight function $w(x) \geq 0$ for all $x \in \Omega'$ and where there exists a positive constant c and a neighbourhood of the boundary of Ω' such that $w(x) > c$ for all points in that neighbourhood. If we use such weight functions the assumption of compactness of sets may often be replaced by that of compactness of the boundaries of the sets. This will often be used tacitly in the following.

6. Harmonic Functionals and the Green Transform

We shall in the following regard only functions which are harmonic on bounded sets or on complements to bounded sets and in the last case harmonic in ∞.

Then if such a function ϕ is harmonic in an open set Ω then for every closed set $A \subset \Omega$

$$\|\phi\|_A = \left[\int_A \frac{\phi^2}{1+|x|^{q+1}} \, dx \right]^{1/2} \tag{80}$$

is finite and if $\{\phi_t\}$, $t > 0$, is a family of functions such that $\phi_t \to \phi_0$ uniformly on A for $t \to 0$ then $\|\phi_t - \phi_0\|_A \to 0$, i.e. ϕ_t converges also with respect to the norm $\|\cdot\|_A$. The notation $\|\phi\|_A$ defined by (80) will be used several times in the following.

Definition 13.7 *Let A be a closed subset of $\mathbf{R}^q$. If the linear functional f is defined for all harmonic functions on A and there is a constant $k \geq 0$ such that*

$$|f\phi| \leq k\|\phi\|_A, \tag{81}$$

then f is said to be a harmonic functional *with* support *A.*

We recall that to say that a function ϕ is harmonic in a closed set A means that there exists an open set $\Omega \supset A$ such that ϕ is harmonic on Ω.

If f is a harmonic functional

$$A = \operatorname{supp} f$$

means that f has support A. Notice that the support of f is not uniquely determined.

Example 13.2 Let ev_{x_0} be the linear functional defined by

$$ev_{x_0}\phi = \phi(x_0),$$

this functional is defined for all functions harmonic at x_0. From Theorem 13.15 follows that any closed ball with centre at x_0 is a support for ev_{x_0}

For y any point in the complement to the closed set A the function $\rho_y(x) = \frac{1}{q-2}\frac{1}{|x-y|^{q-2}}$ is harmonic in A so that the following definition has a meaning:

Definition 13.8 *Let f be a harmonic functional. The function*

$$\widetilde{f}(y) = f\rho_y(x) \qquad y \in \complement \operatorname{supp} f,$$

is called the Green transform *of f.*

Example 13.3 Let us find the Green transform of ev_{x_0}.

$$\widetilde{ev}_{x_0}(y) = ev_{x_0}\rho_y(x) = \rho_y(x_0) = \frac{1}{q-2}\frac{1}{|y-x_0|^{q-2}}.$$

We observe that not only is $\widetilde{ev}_{x_0}(y)$ defined for $y \neq x_0$, but it is a harmonic function for y in the complement to x_0.

Theorem 13.22 *Let f be a harmonic functional with support A. Then*

(a) $\widetilde{f} \in H(\complement A)$, *and*

(b)

$$\frac{\partial^{|\alpha|}}{\partial y^{\alpha}} f(y) = f_x\left(\frac{\partial^{|\alpha|}}{\partial y^{\alpha}} \rho_y(x)\right), \tag{82}$$

('differentiation under the functional sign' is permitted).

For the proof we shall need two lemmas.

Lemma 13.1 *Let $\{\phi_t\}$, $-c < t < c$, be a family of functions harmonic on A which converge for $t \to 0$ uniformly to ϕ which then must be harmonic on A also, and let f be a harmonic functional with support A then $f\phi_t \to f\phi$ for $t \to 0$.*

Proof From the uniform convergence of $\{\phi_t\}$ follows that $\|\phi_t - \phi\|_A$ for $t \to 0$ and so

$$|f\phi - f\phi_t| = |f(\phi - \phi_t)| \leq k\|\phi_t - \phi\|_A \to 0,$$

and this proves the lemma.

Lemma 13.2 *Let f be a function $f(x, y)$ of two variables $x \in A$ and $y \in \complement A$, where A is compact and let $f \in C^2(A \times \complement A)$, then for any fixed $y_0 \in \complement A$ there exists a $c > 0$ such that for $|h| \leq c$ and $i = 1, 2, \ldots, q$*

$$\frac{f(x, y_0 + e_i h) - f(x, y_0)}{h} \to \left(\frac{\partial f(x, y)}{\partial y_i}\right)_{y_0} \qquad \textit{uniformly in } x \textit{ on } A.$$

Here e_i denotes unit vector $\{0, 0, \ldots, 1, \ldots, 0\}^{\mathrm{T}} \in \mathbf{R}^q$ with 1 in the ith coordinate.

Proof As $y_0 \in \complement A$ there exists a c such that the closed ball $B(y_0, c)$ has no point in common with A and then for $|h| \leq c$ $f(x, y_0 + e_i h)$ is defined for all $x \in A$. From the mean value theorem of differential calculus follows

$$f(x, y_0 + e_i h) = f(x, y_0) + h\left(\frac{\partial f(x, y))}{\partial y_i}\right)_{y_0} + \frac{h^2}{2}\left(\frac{\partial^2 f(x, y)}{\partial y_i^2}\right)_{y+\theta e_i h},$$

where $0 \leq \theta \leq 1$, or

$$\frac{f(x, y_0 + e_i h) - f(x, y_0)}{h} = \left(\frac{\partial f(x, y)}{\partial y_i}\right)_{y_0} + \frac{h^2}{2}\left(\frac{\partial^2 f(x, y)}{\partial y_i^2}\right)_{y+\theta e_i h}, \tag{83}$$

it should be remembered that θ depends on h and x; but the second derivative shall in all cases be taken for $x \in A$ and $-c \leq h \leq c$ which is a compact set in $A \times \complement A$. We have supposed that all second derivatives of f are continuous on $A \times \complement A$ and therefore they are uniformly bounded on $A \times \{\, y_0 + e_i h \mid -c \leq h \leq c \,\}$ and therefore the second term on the right side of (83) converges uniformly to 0 for $h \to 0$ and so the lemma follows.

Proof of Theorem 13.22. Evidently (a) follows from (b), therefore we can concentrate on (b).

For $|\alpha| = 0$ (82) is trivial, therefore if we can prove that if (82) is satisfied for $|\alpha| = n$ then it is also satisfied for $|\alpha| = n+1$ then we have proved (b).

Therefore suppose that for all multiindices α such that $|\alpha| = n$

$$\frac{\partial^{|\alpha|}}{\partial y^\alpha} f_x \rho_y(x) = f_x\left(\frac{\partial^{|\alpha|}}{\partial y^\alpha}\rho_y(x)\right), \tag{84}$$

we shall keep $y \in \complement A$ fixed and choose $c > 0$ as in the proof of Lemma 13.2. For $|h| \leq c$ we deduce first from (84) then from the linearity of f:

$$\begin{aligned}\frac{\frac{\partial^{|\alpha|}}{\partial y^\alpha} f_x \rho_{y+e_i h}(x) - \frac{\partial^{|\alpha|}}{\partial y^\alpha} f_x \rho_y(x)}{h} &= \frac{f_x\left(\frac{\partial^{|\alpha|}}{\partial y^\alpha}\rho_{y+e_i h}(x)\right) - f_x\left(\frac{\partial^{|\alpha|}}{\partial y^\alpha}\rho_y(x)\right)}{h} \\ &= f_x\left(\frac{\frac{\partial^{|\alpha|}}{\partial y^\alpha}\rho_{y+e_i h}(x) - \frac{\partial^{|\alpha|}}{\partial y^\alpha}\rho_y(x)}{h}\right),\end{aligned} \tag{85}$$

we observe that the arguments for f are always harmonic with respect to x on A so that the values of f are really defined.

Now

$$\frac{\partial^{|\alpha|}}{\partial y^\alpha}\rho_y(x) = \frac{1}{q-2}\frac{\partial^{|\alpha|}}{\partial y^\alpha}\frac{1}{|x-y|^{q-2}}$$

is for $x \neq y$ in C^∞ as a function of $x - y$ and therefore also in C^∞ as function of x and y on $A \times \complement A$ so that we may apply Lemma 13.2 to the result that

$$\frac{\frac{\partial^{|\alpha|}}{\partial y^\alpha}\rho_{y+e_i h}(x) - \frac{\partial^{|\alpha|}}{\partial y^\alpha}\rho_y(x)}{h} \to \frac{\partial^{|\alpha|+1}}{\partial y_i \partial y^\alpha}\rho_y(x) \tag{86}$$

uniformly in x on A for $h \to 0$.

At last we apply Lemma 13.1 which in connection with (86) proves that everything in (84) converges to $f_x\left(\frac{\partial^{|\alpha|+1}}{\partial y_i \partial y^\alpha}\rho_y(x)\right)$ but this is equivalent to the result that

$$\frac{\partial^{|\alpha|+1}}{\partial y_i \partial y^\alpha}\left(f_x \rho_y(x)\right) = f_x\left(\frac{\partial^{|\alpha|+1}}{\partial y_i \partial y^\alpha}\rho_y(x)\right)$$

q.e.d.

We have in the Green transform defined a mapping of the harmonic functionals with support in A into $H(\complement A)$.

On the other hand we shall see that every $\psi \in H(\complement A)$ defines a linear functional defined for functions harmonic on A. This functional need not however be supported by A.

If the closed set A is connected let ω be a smooth closed surface with A in its interior, if A consists of several connected closed subsets A_i let ω consist of one smooth closed ω_i surface for every A_i such that every A_i is in the interior of ω_i and in no ω_j for $j \neq i$. Then for any ϕ which is harmonic on A it is possible to find an ω such that all points on ω are in the open set where ϕ is harmonic. For any $\psi \in H(\complement A)$ define

$$\langle \psi, \phi \rangle = \frac{1}{\omega_q} \int\limits_{\omega} \left(\psi \frac{\partial \phi}{\partial n} - \phi \frac{\partial \psi}{\partial n} \right) d\omega. \tag{87}$$

Although the 'surface' ω depends on ϕ and is not uniquely defined this definition is meaningful because it gives the same results for different permissible ω, which can be proved by applying Green's formula on ψ and ϕ on the 'volume' between two permissible surfaces ω_1 and ω_2. That (87) given $\psi \in H(\complement A)$ defines a linear functional is evident, from Theorem 13.15 follows that for any closed set B such that there exists an open set Ω with $A \subset \Omega \subset B$, $\langle \psi, \cdot \rangle$ defines a harmonic functional with support B.

We shall prove

Theorem 13.23 $\widetilde{\langle \psi, \cdot \rangle} = \psi$.

Proof Let $y \in \complement A$ and choose ω such that y is in the exterior of ω then

$$\widetilde{\langle \psi, \cdot \rangle}(y) = \langle \psi, \rho_y \rangle = \frac{1}{\omega_q} \int\limits_{\omega} \left(\psi \frac{\partial \rho_y}{\partial n} - \rho_y \frac{\partial \psi}{\partial n} \right) d\omega = \psi(y)$$

where we have used Corollary 13.6.

The facts expressed in Theorem 13.23 and in the following Theorem 13.24 have evidently contributed to the choice of the name 'Green transform.'

Theorem 13.24 *Let f be a harmonic functional with support A then*

$$\langle \widetilde{f}, \cdot \rangle = f.$$

Proof Let ϕ be harmonic in A, then from Theorem 13.22 and from

$$\phi(y) = \frac{1}{\omega_q} \int\limits_{\omega} \left(\rho_y \frac{\partial \phi}{\partial n} - \phi \frac{\partial \rho_y}{\partial n} \right) d\omega \qquad y \in \Omega$$

follows

$$f\phi = \frac{1}{\omega_q} \int\limits_{\omega} \left(f\phi \frac{\partial \phi}{\partial n} - \phi \frac{\partial \widetilde{f}}{\partial n} \right) d\omega = \langle \widetilde{f}, \phi \rangle$$

q.e.d.

Example 13.4 For the continuous linear functional $f = h\left(\frac{\partial}{\partial x}\right)_{x=0}$ where $h(x)$ is a spherical harmonic we find

$$\widetilde{f}(y) = \left(h\left(\frac{\partial}{\partial x}\right) \rho_y(x) \right)_{x=0} = \frac{(q-2)q(q+2)\cdots(q+2n-4)}{q \quad 2} \widetilde{h}(y).$$

We shall prove that $B(0,R)$ for $R>0$ is a support for f: For $\phi \in H(\Omega)$ where $\Omega \supseteq B(0,R)$ we have

$$\int_{B(0,R)} \phi(x)h(x)\,dx = \int_0^R \left[r^n \int_{S(0,r)} \phi(x) h\Big(\frac{x}{|x|}\Big)\,d\omega \right] dr$$

and for $r<R$

$$f\phi = \langle \widetilde{f}, \phi\rangle = \frac{1}{\omega_q}\frac{(q-2)q(q+2)\cdots(q+2n-4)}{q-2}\int_{S(0,r)} \Big(\widetilde{h}(x)\frac{\partial\phi}{\partial n} - \phi\frac{\partial\widetilde{h}}{\partial n}\Big)\,d\omega.$$

Now

$$\begin{aligned}
\int_{S(0,r)} \Big(\widetilde{h}(x)\frac{\partial\phi}{\partial n} - \phi\frac{\partial\widetilde{h}}{\partial n}\Big)\,d\omega &= \int_{S(0,r)} \Big(\frac{h(x)}{|x|^{2n+q-2}}\frac{\partial\phi}{\partial n} + (n+q-2)\phi\frac{h(\frac{x}{|x|})}{|x|^{n+q-1}}\Big)\,d\omega \\
&= \frac{1}{r^{2n+q-2}}\int_{S(0,r)} h(x)\frac{\partial\phi}{\partial n}\,d\omega + \frac{n+q-2}{r^{n+q-1}}\int_{S(0,r)} \phi(x)h\Big(\frac{x}{|x|}\Big)\,d\omega \\
&= \frac{1}{r^{2n+q-2}}\int_{S(0,r)} \phi(x)\frac{\partial h}{\partial n}\,d\omega + \frac{n+q-2}{r^{n+q-1}}\int_{S(0,r)} \phi(x)h\Big(\frac{x}{|x|}\Big)\,d\omega \\
&= \frac{2n+q-2}{r^{n+q-1}}\int_{S(0,r)} \phi(x)h\Big(\frac{x}{|x|}\Big)\,d\omega
\end{aligned}$$

substituting this result in the first equation we find

$$\int_{B(0,R)} \phi(x)h(x)\,dx = \frac{\omega_q(q-2)f\phi}{(2n+q-2)(q-2)q\cdots(q+2n-4)}\int_0^R r^{2n+q-1}\,dr$$

and using Schwartz' inequality we find

$$\left[\int_{B(0,R)} \phi^2\,dx\right]^{1/2}\left[\int_{B(0,R)} h^2\,dx\right]^{1/2} \geq \frac{\omega_q(q-2)R^{2n+q}}{(2n+q)(2n+q-2)(q-2)q\cdots(q+2n-4)}|f\phi|.$$

Now

$$\begin{aligned}
\left[\int_{B(0,R)} h^2(x)\,dx\right]^{1/2} &= \left[\int_0^R \left[r^{2n+q-1} + \int_{S(0,1)} h^2\Big(\frac{x}{|x|}\Big)\,d\omega\right] dr\right]^{1/2} \\
&= R^{n+q/2}\sqrt{\omega_q}\frac{1}{\sqrt{2n+q-1}}\|h\|\frac{1}{\sqrt{2n+q}}
\end{aligned}$$

where $\|\cdot\|$ is the standard norm for spherical harmonics so we get

$$|f\phi| \leq \frac{\sqrt{(2n+q-2)(2n+q)}(q-2)q(q+2)\cdots(q+2n-4)}{\sqrt{\omega_q}(q-2)} \times R^{-n-q/2}\|h\| \left[\int_{B(0,R)} \phi^2\, dx\right]^{1/2}$$

and the statement follows, observing that

$$\frac{1}{\sqrt{1+R^{q+1}}} \left[\int_{B(0,R)} \phi^2\, dx\right]^{1/2} \leq \|\phi\|_{B(0,R)} \leq \left[\int_{B(0,R)} \phi^2\, dx\right]^{1/2}.$$

A method similar to this may be used to find constants corresponding to the A_α in (7) explicitly for harmonic differentiation.

7. Regular Hilbert Spaces of Harmonic Functions — Duality 1

In geodesy we are mainly interested in Hilbert spaces of harmonic functions because this is the natural field for the solution of least-squares problems in connection with the gravity field of the Earth. It seems to be expedient to delimit the set of such spaces to a smaller set which is well-behaved and which seems to be sufficiently general for applications, and in lack of originality I have given it the name of regular Hilbert spaces of harmonic functions.

In this section Ω shall denote an open subset of $\mathbf{R}^q$ which is bounded or the complement to a bounded set. Normally we shall assume both Ω and $\complement\Omega$ to be connected in order to make the proofs less verbose.

Definition 13.9 *A* regular Hilbert space of harmonic functions *on Ω is a Hilbert space H the elements of which are harmonic functions on Ω such that:*

1. *H contains all functions which are harmonic on any open set with $\overline{\Omega}$ in its interior, and for every closed $A \supset \Omega$ the restriction*

$$R\colon H(A) \to H\colon \phi \mapsto \phi|\Omega \qquad \phi \in H(A)$$

   is continuous,
2. *for every closed $A \subset \Omega$ the restriction*

$$R\colon H \to H(A)\colon \phi \mapsto \phi|A \qquad \phi \in H$$

   is continuous.

Remark 13.2 The definition has a meaning because the restriction in question will always exist.

Theorem 13.25 *A quadratic norm $\|\cdot\|$ is the norm for a regular Hilbert space of harmonic functions on Ω if and only if:*

1. *$\|\phi\|$ is defined for all $\phi \in H(\overline{\Omega})$, and for every closed $A \supset \Omega$ there exists a constant k such that*

$$\|\phi\| \leq k\|\phi\|_A \qquad \textit{for all } \phi \in H(A), \tag{88}$$

2. *for every closed $A \subset \Omega$ there exists a constant k such that*

$$\|\phi\|_A \leq k\|\phi\| \qquad \textit{for all } \phi \in H(\Omega). \tag{89}$$

Proof The 'only if' part is evident from the definition. The 'if' part is proved as follows: From (1) follows that the elements of $H(\overline{\Omega})$ with the norm $\|\cdot\|$ constitute a pre-Hilbert space H', from (2) follows that a Cauchy sequence in H' has a limit which is element of $H(A)$ for any closed $A \subset \Omega$, i.e. the completion H and H' consists of harmonic functions on Ω. The continuity of the restrictions follows directly from the existence of the two constants.

Remark 13.3 The regular Hilbert space which we have constructed in this way from the norm $\|\cdot\|$ will be called the minimal Hilbert space H_{min} corresponding to the norm $\|\cdot\|$. We could also have constructed a Hilbert space by defining H as the elements ϕ of $H(\Omega)$ for which $\|\phi\| < \infty$. We shall call this Hilbert space H_{max}, the maximal Hilbert space corresponding to the norm $\|\cdot\|$. That it is in fact a Hilbert space and not merely a pre-Hilbert space follows again from (2) in the theorem. It is clear that $H_{\text{min}} \subset H_{\text{max}}$ and also that H_{min} is a closed subspace of H_{max}. We shall in the following when the contrary is not directly expressed by a normal Hilbert space with norm $\|\cdot\|$ mean the corresponding minimal space.

We shall now investigate the dual space to H, i.e. the space of continuous linear functionals on H, and the realization of this dual space as a Hilbert space H' the elements of which are the Green transforms of these continuous linear functionals.

We know that a linear functional f defined on elements of $H(\Omega)$ is a continuous linear functional on H if there exists a constant k such that

$$|f\phi| \leq k\|\phi\| \qquad \text{for all } \phi \in H. \tag{90}$$

Theorem 13.26 *Every harmonic functional with support A in Ω is continuous on H.*

Proof That A is a support of f means that there exists a constant k_1

$$|f\phi| \leq k_1\|\phi\|_A.$$

By (89) there exists a constant k_2 such that

$$\|\phi\|_A \leq k_2\|\phi\|,$$

such that (90) with $k = k_1k_2$ results, and we have:

Corollary 13.12 *If f is a harmonic functional with support in Ω then $\widetilde{f} \in H(\complement\Omega)$, and all elements ψ of $H(\complement\Omega)$ are Green transforms of continuous harmonic functionals on H: $f\phi = \langle\psi, \phi\rangle$ for $\phi \in H$.*

The last part follows by noticing that any closed $A \subset \Omega$ such that A contains the surface ω over which the integral in $\langle\cdot,\cdot\rangle$ is taken is a support for the harmonic functional $\langle\psi,\cdot\rangle$.

Theorem 13.27 *Every closed $A \supset \Omega$ is support for every continuous harmonic functional on H.*

Proof Let A be as in the theorem and let R be the restrictions $R\phi = \phi|\Omega$ for $\phi \in H(A)$. For f any continuous linear functional on H f operates also on $H(A)$ by

$$f\phi = f(\phi|\Omega), \qquad \phi \in H(A),$$

and by (90) and (88):

$$|f\phi| \le k_1 \big\|\phi|\Omega\big\| \le k_1 k_2 \|\phi\|_A.$$

Corollary 13.13 *The Green transforms $\widetilde{f}$ of all continuous harmonic functionals f on H are elements of $H(\complement\Omega)$ and the definition of the pairing $\langle\cdot,\cdot\rangle$ can be extended by continuity so that the original harmonic functional f can be found from its Green transform $\widetilde{f}$ by*

$$f = \langle\widetilde{f},\cdot\rangle.$$

Proof The first part is evident. As we have defined H as the minimal harmonic Hilbert space with the norm $\|\cdot\|$, to every $\phi \in H$ there corresponds a sequence $\{\phi_i\}$, $i = 1, 2, \ldots$, such that each ϕ_i is harmonic in an open neighbourhood of $\overline{\Omega}$ and $\lim_{i\to\infty}\phi_i = \phi$. For each ϕ_i there exists a smooth closed surface ω_i with Ω in its interior such that ω_i is in the domain of harmonicity of ϕ_i and such that $\langle\widetilde{f}, \phi_i\rangle$ can be computed by means of ω_i. It is evident that the sequence $\{\langle\widetilde{f}, \phi_i\rangle\}$ is a Cauchy sequence. From the continuity of f the result follows.

Through the scalar product $(\cdot,\cdot)$ in H every element $\phi \in H$ defines a continuous linear functional $(\phi,\cdot)$ on H. The Green transform of this functional

$$\widetilde{\phi} = (\phi, \rho_x)$$

shall be called the *Green transform of ϕ with respect to H*. It is evident that $\widetilde{\phi+\psi} = \widetilde{\phi} + \widetilde{\psi}$ for $\phi, \psi \in H$ and that $\widetilde{a\phi} = a\widetilde{\phi}$ for a a real number and $\phi \in H$, that is the Green transform of the elements of H constitutes a linear vector space. We can define a Hilbert space structure on this vector space by defining

$$(\widetilde{\phi}, \widetilde{\psi})^{\sim} = (\phi, \psi) \qquad \text{for } \phi, \psi \in H$$

where $(\cdot,\cdot)^{\sim}$ is the scalar product in $\widetilde{H}$. It is evident that

$$\widetilde{\lim_{i\to\infty} \phi_i} = \lim_{i\to\infty} \widetilde{\phi}_i$$

if one of the two limits exists. This Hilbert space will be called $\widetilde{H}$, the Green transform of H.

We shall prove later on that $\widetilde{H}$ is a regular Hilbert space of harmonic functions on $\complement\Omega$ and that H is the Green transform of $\widetilde{H}$, i.e.

$$\widetilde{\widetilde{H}} = H.$$

Now let $\{\phi_i\}$, $i = 1, 2, \ldots$, be a complete orthonormal system in H, then $\{\widetilde{\phi}_i\}$ is evidently an orthonormal system in $\widetilde{H}$ and from the definition of $\widetilde{H}$ follows that it is complete.

For the sake of symmetry we shall write $\rho(x,y)$ instead of $\rho_x(y)$ as before i.e.

$$\rho(x,y) = \frac{1}{q-2}\frac{1}{|x-y|^{q-2}}.$$

For $y \in \complement\overline{\Omega}$, $\rho(x,y)$ is as a function of x an element of H and so ρ may be expressed by the complete orthonormal system $\{\phi_i\}$ as

$$\rho(x,y) = \sum_{i=1}^{\infty} a_i(y)\phi_i(x), \qquad x \in \Omega,\ y \in \complement\overline{\Omega},$$

where

$$a_i(y) = \big(\phi_i(x), \rho(x,y)\big)_x = \widetilde{\phi}_i(y),$$

i.e.

$$\rho(x,y) = \sum_{i=1}^{\infty} \phi_i(x)\widetilde{\phi}_i(y), \qquad x \in \Omega,\ y \in \complement\overline{\Omega}. \tag{91}$$

In addition to the orthogonality relations

$$(\phi_i, \phi_j) = \delta_{ij}, \qquad (\widetilde{\phi}_i, \widetilde{\phi}_j)^{\sim} = \delta_{ij}$$

we have also, as the reader should try to convince himself,

$$\langle \widetilde{\phi}_i, \phi_j \rangle = \delta_{ij}.$$

It is also evident that

$$\big|\langle \widetilde{\psi}_1, \psi_2 \rangle\big| \le \|\widetilde{\psi}_1\|\|\psi_2\|$$

for $\psi_1, \psi_2 \in H$, i.e. the pairing $\langle\cdot,\cdot\rangle$ is jointly continuous in its arguments.

We know from the general theory for Hilbert spaces that $(\phi,\cdot)$ is not merely an example on continuous linear functionals but that to every continuous linear functional f on a Hilbert space H there corresponds an element ϕ, of that space so that

$$f\psi = (\phi, \psi) \qquad \text{for all } \psi \in H,$$

and that $\|\phi\|$ equals the smallest number k such that

$$|f\phi| \leq k\psi \qquad \text{for all } \psi \in H,$$

(Riesz' theorem). As the Green transformation is one to one and an isometry we see that $\widetilde{H}$ is exactly the space of the Green transforms of continuous harmonic functionals on H. Therefore we will call $\widetilde{H}$ the dual to H although it is only one of several different realizations for the dual H' to H, the space of continuous linear functionals on H.

One could ask why we have chosen $\widetilde{H}$ and not the realization given by Riesz of the dual as H itself. The answer must be that $\widetilde{H}$ as the dual is more geometrically lucid and more invariant: The element of H representing a linear functional depends on the norm of H but the Green transform of a harmonic functional is independent of the norm.

8. The Fundamental Kernels — Duality 2

We shall for simplicity call a function $H(x, y)$ for $x, y \in \mathbf{R}^q$ *a (harmonic) kernel* if for x fixed $H(x, y)$ is harmonic as a function of y: $\Delta_y H(x, y) = 0$ and for y fixed $H(x, y)$ is harmonic as a function of x: $\Delta_x H(x, y) = 0$ or as we also shall write it $\Delta H(x, \cdot) = \Delta H(\cdot, y) = 0$.

The most important example is $\rho(x, y)$, which is a harmonic kernel for $x \neq y$.

In connection with a minimal regular Hilbert space H of harmonic functions on an open set Ω and its dual $\widetilde{H}$ there exist three kernels which are of great importance.

The first of them is again $\rho(x, y)$ which for fixed $x \in \widetilde{\Omega} = \complement\overline{\Omega}$ as a function of y is an element of H and for fixed $y \in \Omega$ as a function of x is an element of $\widetilde{H}$ and vice versa.

As we have seen $\rho(x, y)$ has the following reproducing properties

$$\phi(y) = \langle \rho(\cdot, y), \phi \rangle \qquad \text{for every } \phi \in H \text{ and } y \in \Omega$$
$$\psi(y) = \langle \rho(x, \cdot), \psi \rangle \qquad \text{for every } \psi \in \widetilde{H} \text{ and } x \in \widetilde{\Omega}$$

which are special cases of

$$f\phi = \langle \widetilde{f}, \phi \rangle$$

where f is any continuous linear functional on H and

$$\widetilde{f} = f\rho(\cdot, y)$$

is the Green transform of f.

The second fundamental kernel is defined by

$$K(x, y) = \big(\rho(\cdot, x), \rho(\cdot, y)\big)^{\sim} \qquad \text{for } x, y \in \Omega \tag{92}$$

$K(x, y)$ may for x fixed be looked upon as the Green transform of $\rho(\cdot, x)$ with respect to $\widetilde{H}$ and from this follows that $K(x, y)$ is a kernel.

We have seen (91) that $\rho(x, y)$ may be written as

$$\rho(x, y) = \sum_{i=1}^{\infty} \widetilde{\phi}_i(x)\phi_i(y), \qquad x \in \widetilde{\Omega},\ y \in \Omega,$$

where $\{\phi_i\}$ is any complete orthonormal system in H. From this we find

$$K(x, y) = \Bigl(\sum_{i=1}^{\infty} \widetilde{\phi}_i\phi_i(x), \sum_{i=1}^{\infty} \widetilde{\phi}_i\phi_i(y)\Bigr)^{\sim} = \sum_{i=1}^{\infty} \phi_i(x)\phi_i(y), \tag{93}$$

because $\{\widetilde{\phi}_i\}$ is an orthonormal system in H.

Also $K(x, y)$ has a reproducing property:

$$\psi(x) = \bigl(K(x, \cdot), \psi\bigr). \tag{94}$$

This may be proved from the representation of ψ by the orthonormal system $\{\phi_i\}$:

$$\psi = \sum_{i=1}^{\infty} a_i\phi_i, \qquad a_i = (\phi_i, \psi)$$

combined with the representation (93) of $K(x, y)$. It is the property (94) which has given $K(x, y)$ the name the reproducing kernel for the Hilbert space H.

By remembering that

$$\widetilde{ev}_x = \rho(\cdot, x)$$

we can write (92) as

$$K(x, y) = (\widetilde{ev}_x, \widetilde{ev}_y)^{\sim} \qquad x, y \in \Omega.$$

This formula may be generalized. Let f and g be two continuous harmonic functionals on H then

$$f_x\bigl(g_y K(x, y)\bigr) = (\widetilde{f}, \widetilde{g})^{\sim} \tag{95}$$

where f_x etc. means the functional f working on its argument as a function of x. (95) follows directly from the definition of $K(x, y)$ and expresses the important fact that the scalar product of two functionals on H or more correctly the scalar product of the Green transforms of these functionals can be found simply by letting the functionals operate on $K(x, y)$ through each one of the arguments x and y.

As the simplest application of the reproducing kernel on maximum problems let us find for a given point $x \in \Omega$

$$\max_{\|\phi\|=1} |\phi(x)| \qquad \text{for } \phi \in H.$$

We have

$$|\phi(x)| \le \|\widetilde{ev}_x\|^{\sim} \|\phi\| = \|\widetilde{ev}_x\|^{\sim} = \bigl((\widetilde{ev}_x, \widetilde{ev}_x)^{\sim}\bigr)^{1/2} = \bigl(K(x, x)\bigr)^{1/2},$$

this can naturally also be found by using the reproducing property of $K(x, y)$ directly.

Now let $A \subset \Omega$ be closed. Then $K(x,x)$ must have a maximum value K_A on A and we find for any $\phi \in H$:

$$\|\phi\|_A^2 = \int\limits_A \frac{\phi(x)\phi(x)}{1+|x|^{q+1}}\,dx \le K_A \int\limits_A \frac{\|\phi\|^2}{1+|x|^{q+1}}\,dx, \tag{96}$$

which proves that the condition (2) in the definition of regular Hilbert spaces of harmonic functions is necessary for the existence of a reproducing kernel for H.

The third fundamental kernel is $\widetilde{K}(x,y)$ for $x, y \in \widetilde{\Omega}$, the reproducing kernel for $\widetilde{H}$ defined by

$$\widetilde{K}(x,y) = \big(\rho(\cdot,x), \rho(\cdot,y)\big) \tag{97}$$

or

$$\widetilde{K}(x,y) = \widetilde{\rho(\cdot,x)}.$$

It may be expressed by $\{\widetilde{\phi}_i\}$ as

$$\widetilde{K}(x,y) = \sum_{i=1}^{\infty} \widetilde{\phi}_i(x)\widetilde{\phi}_i(y), \qquad x, y \in \widetilde{\Omega}. \tag{98}$$

Its reproducing property is expressed by

$$\widetilde{\psi}(y) = \big(\widetilde{K}(\cdot,y), \widetilde{\psi}\big)^{\sim} \qquad y \in \widetilde{\Omega} \tag{99}$$

and its properties with respect to $\widetilde{H}$ are analogue to the properties of $K(x,y)$ with respect to H.

By a formula analogue to (96) we can prove that also $\widetilde{H}$ satisfies point (2) of the definition of regular Hilbert spaces etc.

The reader should be able to find several proofs for the formula

$$\rho(x,y) = \big\langle \widetilde{K}(x,\cdot), K(y,\cdot)\big\rangle \tag{100}$$

connecting the three fundamental kernels.

From (100) or from $\rho(x,y) = \sum_{i=1}^{\infty} \widetilde{\phi}_i(x)\phi_i(y)$ and (93) and (93) one proves

$$\big|\rho(x,y)\big|^2 \le \widetilde{K}(x,x)K(y,y), \qquad x \in \widetilde{\Omega},\ y \in \Omega,$$

from which follows that for x on the closure of both Ω and $\widetilde{\Omega}$ can $\widetilde{K}(x,x)$ and $K(x,x)$ not both have a finite limit.

There exist very many simple formulas around the fundamental kernels and I shall not take from the reader the joy of discovery, only one more formula has to be mentioned here:

$$(\psi,\psi)^{\sim} = \big\langle \langle \psi(x), K(x,y)\rangle_x, \psi(y)\big\rangle_y, \qquad \text{for } \psi, \psi \in \widetilde{H}. \tag{101}$$

Let $A \supset \widetilde{\Omega}$ be a closed set and ϕ be harmonic in an open set O such that $A \supset O \Supset \widetilde{\Omega}$. Then the pairings in (101) may be computed over a surface ω in O outside $\widetilde{\Omega}$. The kernel $K(x,y)$ and its derivatives are bounded on ω because in fact $\omega \subset \Omega$, therefore there exists a constant c such that

$$\|\phi\|^{\sim} = \big\langle \langle \phi(x), K(x,y) \rangle_x, \phi(y) \big\rangle^{1/2} \leq c \|\phi\|_A$$

and we have proved that $\widetilde{H}$ satisfies also point (1) in the definition of a regular Hilbert space of harmonic functions on $\widetilde{\Omega}$ and finished the proof of the fact that $\widetilde{H}$ is a regular Hilbert space of harmonic functions, and we are now able to formulate the main theorem on duality:

Theorem 13.28 *Let H be a minimal regular Hilbert space of harmonic functions on an open set Ω. Then the dual to H, the space $\widetilde{H}$ consisting of the Green transforms of the continuous linear functionals on H, is a minimal regular Hilbert space of harmonic functions on the open set $\widetilde{\Omega} = \complement\Omega$, and H is the dual to $\widetilde{H}$. The Hilbert spaces H and $\widetilde{H}$ have reproducing kernels K and $\widetilde{K}$ respectively which are connected with ρ by the formulas:*

$$K(x,y) = \big(\rho(\cdot,x), \rho(\cdot,y)\big)^{\sim} \qquad x,y \in \Omega$$

$$\widetilde{K}(x,y) = \big(\rho(\cdot,x), \rho(\cdot,y)\big) \qquad x,y \in \widetilde{\Omega}$$

$$\rho(x,y) = \langle K(x,\cdot), K(y,\cdot) \rangle \qquad x \in \widetilde{\Omega},\ y \in \Omega.$$

9. Runge's Theorem — Duality 3

In this section we shall apply the theory of duality between regular Hilbert spaces of harmonic functions to the proof of Runge's theorem, i.e. to the problem of approximation of functions harmonic in one domain Ω_2 by functions harmonic in a greater domain $\Omega_1 \Supset \Omega_2$. But first we must study a generalization to Hilbert space of transposed matrices, the dual operators.

Suppose now that we have two open sets Ω_1 and Ω_2 and so two minimal regular Hilbert spaces of harmonic functions H_1 and H_2 and their dual spaces $\widetilde{H}_1$ and $\widetilde{H}_2$. We denote the pairing between H_1 and $\widetilde{H}_1$ by $\langle \cdot,\cdot \rangle_1$ and the pairing between H_2 and $\widetilde{H}_2$ by $\langle \cdot,\cdot \rangle_2$.

Let there be given a continuous linear operator

$$A \colon H_1 \to H_2,$$

then we define A', *the dual operator* to A, as the operator

$$A' \colon \widetilde{H}_2 \to \widetilde{H}_1,$$

defined by

$$\langle A'\widetilde{\psi}, \phi \rangle_1 = \langle \widetilde{\psi}, A\phi \rangle_2 \qquad \text{for all } \phi \in H_1 \text{ and } \widetilde{\psi} \in \widetilde{H}_2. \tag{102}$$

It is clear that A' is defined by (102), for the right term of (102) is defined for all ϕ and $\widetilde{\psi}$, so that for any fixed $\widetilde{\psi}$ the left term is a continuous linear functional f on H, so that $A'\widetilde{\psi} = \widetilde{f}$, and so A' is a linear operator. From (102) follows also that A and A' have the same bound i.e. A' is continuous because A is so.

Now suppose that $\Omega_2 \Subset \Omega_1$ from which follows that $H_1 \subset H_2$ i.e. every element in H_1 is also an element in H_2, or more correctly if $\phi \in H_1$ then its restriction $R\phi$ to Ω_2 is in H_2.

But from $\Omega_2 \Subset \Omega_1$ follows $\complement\overline{\Omega}_1 \Subset \complement\Omega_2$ and for every $\widetilde{\psi} \in \widetilde{H}_2$ its restriction $R'\widetilde{\psi}$ is an element of $\widetilde{H}_1$. We have here two operators

$$R\colon H_1 \to H_2 \qquad \text{and} \qquad R'\colon \widetilde{H}_2 \to \widetilde{H}_1,$$

which are dual to one another because

$$\langle R'\widetilde{\psi}, \phi\rangle_1 = \langle\widetilde{\psi}, R\phi\rangle_2 \qquad \text{for all } \widetilde{\psi} \in \widetilde{H}_2 \text{ and all } \phi \in H_1 \tag{103}$$

expresses only that $\langle\widetilde{\psi}, \phi\rangle$ is computed over two different smooth closed surfaces which separate the point sets where the functions ϕ and $\widetilde{\psi}$ are not defined (or not harmonic).

Now R is continuous. This is proved by using a closed set A such that $\Omega_2 \subset A \subset \Omega_1$. From the continuity conditions in the definition of regular Hilbert spaces of harmonic functions follows that the restrictions

$$R_1\colon H_1 \to H(A), \qquad \phi \mapsto \phi|A,$$

and

$$R_2\colon H(A) \to H_2, \qquad \psi \mapsto \psi|\Omega_2$$

are continuous and so is

$$R = R_2 R_1.$$

From the continuity of R follows that R', its dual, is continuous. This result is not surprising as it follows from the fact that H_1 and H_2 are regular Hilbert spaces of harmonic functions.

In finite dimensional vector spaces dual operators are expressed by transposed matrices. Let

$$A\colon \mathbf{R}^n \to \mathbf{R}^m$$

be a linear mapping defined by a matrix A, then the transposed matrix A^{T} defines a mapping

$$A^{\mathrm{T}}\colon \mathbf{R}^m \to \mathbf{R}^n.$$

A is a matrix with m rows and n columns while A^{T} has n rows and m columns. If the mapping A of $\mathbf{R}^n$ onto its image in $\mathbf{R}^m$ is *one to one* then the mapping A^{T} of $\mathbf{R}^m$ is *onto* $\mathbf{R}^n$, this follows directly from the fact that the mapping defined by a matrix is one to one if the columns of A are linearly independent and the mapping is onto if the rows are linearly independent.

In infinite dimensional space we cannot from the mapping R' one to one infer that R is onto as we shall see soon from an example but we can conclude something. Let us look at (103) again and suppose that R' is one to one, which is equivalent to

$$R'\widetilde{\psi} = 0 \quad \Rightarrow \quad \widetilde{\psi} = 0$$

or

$$\langle R'\widetilde{\psi}, \phi\rangle_1 = 0 \qquad \text{for all } \phi \in H_1 \;\Rightarrow\; \widetilde{\psi} = 0,$$

but by (103) this is equivalent to saying

$$\langle \widetilde{\psi}, R\phi\rangle_2 = 0 \qquad \text{for all } \phi \in H_1 \;\Rightarrow\; \widetilde{\psi} = 0, \tag{104}$$

or a continuous linear functional on $\widetilde{H}_2$ which vanishes on the image of R $RH_1 \subset H_2$ is identically zero.

We shall now use the following

Lemma 13.3 *Let F be a Hilbert space, E a closed linear subspace of F and a an element of F not in E, then there exists a continuous, linear functional f on F such that $fa = 1$ and $fb = 0$ for all $b \in E$.*

The proof of this lemma will be given later.

It is clear that the image of R is linear, therefore its closure is a closed linear subspace of H_2, let us call it I and suppose that there is an element $\mu \in H_2$ which is not in I. Then by the lemma, there would be a continuous linear functional on H_2 such that $f\mu = 1$ and $f\nu = 0$ for all $\nu \in I$. But (104) expressed exactly that this is impossible, i.e. $I = H_2$ or every element of H_2 is the limit of a sequence of elements of the form $\{R\phi_i\}$, $\phi_i \in H_1$, $i = 1, 2, \ldots$, or again said with plain words every element of H_2 may be approximated by elements of the form $R\phi$, $\phi \in H_1$.

But what does that mean for our special operators R and R'?

R is the restriction to Ω_2 of harmonic functions on Ω_1, that is if we can prove that R' is one to one then it follows from the considerations above that all harmonic functions on Ω_2 may be approximated by those which are harmonic in Ω_1.

If Ω_1 and Ω_2 are open point sets of $\mathbf{R}^q$ with $\Omega_2 \Subset \Omega_1$, H any minimal regular Hilbert space of harmonic functions on Ω_2, and all harmonic functions on Ω_2 may be approximated in the topology of H by such harmonic functions on Ω_1 which are restrictions (to Ω_2) of harmonic functions on Ω_1, then Ω_2 is said to be a (harmonic) *Runge domain* of Ω_1. So what we have proved so far is that if R', the restriction to $\widetilde{\Omega}_1$ of harmonic functions on $\widetilde{\Omega}_2$, is one to one, that is if two functions on $\widetilde{\Omega}_2$ have the same restriction to $\widetilde{\Omega}_1$ then they are identical. But we have seen in Corollary 13.11 that if $\widetilde{\Omega}_2$ is connected then the harmonic continuation from $\widetilde{\Omega}_1$ to $\widetilde{\Omega}_2$ is unique and so R' is one to one, and we have proved

Theorem 13.29 *Let Ω_1 and Ω_2 be open point sets in $\mathbf{R}^q$ and $\Omega_2 \Subset \Omega_1$, then Ω_2 is a Runge domain of Ω_1 if $\widetilde{\Omega}_2 = \complement\overline{\Omega}_2$ is connected.*

Remark 13.4 From the definition of regular Hilbert spaces of harmonic functions and the results of Section 4 follows that convergence in the topology of H implies not only uniform convergence of the harmonic functions themselves on compact subsets of Ω_2 but also that of any harmonic derivative of harmonic functions and of any harmonic functional with compact support in Ω_2.

As a classical example of the application of Runge's theorem let Ω_1 be the part of $\mathbf{R}^3$ outside the surface of the Earth and Ω_2 the part of the same space outside any Bjerhammar sphere, i.e. a sphere totally inside the Earth. Here again $\Omega_1 \Subset \Omega_2$ and $\complement\Omega_1$ are connected, it is the surface and the interior of the Earth and we have the 'geodetic' Runge theorem: for every function ϕ harmonic outside the Earth, for every closed surface surrounding the Earth, for every Bjerhammar sphere, and for every $\epsilon > 0$ there exists a function ϕ_0 which is harmonic outside the Bjerhammar sphere and such that $|\phi - \phi_0| < \epsilon$ overall on and outside the given surface.

As a counter example let us take a variant of the foregoing one. Let Ω_2 be as before but let now Ω_1 be the part of space outside the Earth and the Moon, then $\complement\Omega_1$ is the Earth and the Moon and they are not connected, therefore we may not apply the theorem here. In fact Ω_2 is here *not* a Runge domain of Ω_1.

It is now clear what I mentioned above that the dual to a one to one operator is not necessarily onto—not every function harmonic outside the Earth may be continued to a harmonic function outside a given Bjerhammar sphere—we can only say that the image is overall dense.

Now we come to the proof of the lemma.

As a is supposed not to be in the closed linear subspace $E \subset F$, a may be written as

$$a = a_0 + b_0, \qquad \text{where } b_0 \in E \quad \text{and} \quad (a_0, b) = 0 \ \text{ for all } b \in E$$

and where $\|a_0\| > 0$ (because $a \notin E$).

Every $x \in F$ may now be written as

$$x = x_0 + x_1, \qquad \text{where } x_0 = \frac{(x, a_0)}{\|a_0\|^2} a_0 \quad \text{and} \quad (x_1, a_0) = 0$$

i.e.

$$x_1 = x - \frac{(x, a_0)}{\|a_0\|^2} a_0.$$

Now define f by

$$fx = \frac{(x, a_0)}{\|a_0\|^2} \qquad \text{for } x \in F.$$

Then we find:

$$fa = \frac{(a_0 + b_0, a_0)}{\|a_0\|^2} = 1$$

and

$$fb = \frac{(b, a_0)}{\|a_0\|^2} = 0 \qquad \text{for all } b \in E$$

as it should be.

14

La Formule de Stokes Est-Elle Correcte?

Commentaires sur le papier de W. Baranov

— by T. Krarup, Geodætisk Institut, Charlottenlund, Denmark

Most readers of this journal will certainly after a rereading of e.g. the relevant pages of *Heiskanen and Moritz: Physical Geodesy* answer the question: "Is Stokes' formula correct" in the affirmative. Nevertheless it may be instructive to see that "the new Stokes' formula" is certainly not correct.

More specifically we shall prove that the T furnished by this formula is not harmonic outside the Earth.

W. Baranov's new Stokes' formula—(17) and (18) in his paper—may be written as

$$S(M,P) = \frac{2}{|M-P|} - 3aW(M,P),$$

where

$$W(M,P) = \frac{1}{a}\left(\frac{r-a}{L^2} + \frac{a\mu}{L^2}\ln\frac{r+L-a\mu}{a(1-\mu)}\right) = \int_a^\infty \frac{d\rho}{\rho^2\sqrt{\rho^2 - 2R\rho\mu + R^2}}$$

that is $W(M,P)$ is the value at P of the potential generated by a mass distribution with the density $1/\rho^2$ on the radius vector through the point P for $a < \rho < \infty$, where ρ is the distance of the mass point in question from the centre of the sphere.

Therefore

$$T(M) = \frac{1}{4\pi}\iint\limits_{|P|=a} g(P)S(M,P)\,d\sigma \tag{1}$$

is the potential generated by

1. the single layer mass distribution $\frac{g(P)}{2\pi}$ and
2. the space mass distribution $-\frac{3a^3}{4\pi\rho^4}g(P)$, $\rho > a$, outside the sphere. Here ρ is again the distance of the mass point from the centre and P is the point where the radius vector through the mass point meets the sphere.

And therefore

$$\Delta T = \frac{3a^2}{L^4} g(P_M),$$

where P_M is the point on the sphere "under" M, that is outside the sphere T is only harmonic in points "over" those points of the sphere where $g = 0$.

On the other hand P as determined by (1) is harmonic *inside* the sphere—in fact "the new Stokes' formula" is the solution of "Stokes' problem for the interior sphere" which is without geodetical interest.

Received: 25.5.1976

— by B. Zondek, Dahlgren Naval Laboratory, Dahlgren, Va., U.S.A.

A recent paper (Baranov 1975) casts doubt on the validity of the classic Stokes formula of physical geodesy. We wish to point out a mathematical error early in Baranov's paper, which invalidates his conclusions. We use Baranov's own notation.

A result of Baranov's equations (5) and (6) is

$$\iint g(P) \left(L\frac{dS}{dL} + 2S + \frac{L^2 - a^2}{r^3} \right) d\sigma = 0. \tag{2}$$

This equation is satisfied by the gravity anomaly $g(P)$ and the Stokes kernel S. Baranov concludes from it that

$$L\frac{dS}{dL} + 2S + \frac{L^2 - a^2}{r^3} = 0 \tag{3}$$

because the integral vanishes "quelle que soit la fonction $g(P)$" i.e. whatever function $g(P)$. This conclusion is erroneous because the gravity anomaly is arbitrary function. In fact, as is well known, its surface harmonic expansion lacks a dipole term (Heiskanen & Moritz, 1967), p. 89.

Received: 5.5.1976

Reponse de W. Baranov

Dans sa lettre du 7 mai 1975, le Dr. T. Krarup m'a écrit ceci:

"When working in mathematics one often need the comforting fact, that it is accepted that it is human to make errors. But this comfort is rather feeble."

L'erreur que j'ai commise se trouve dans la phrase suivante (page 29 du Bulletin Géodésique n° 115):

"Ce noyau remplit la condition (7), on vérifie qu'il est harmonique et nous verrons tout à l'heure que la condition aux limites (3) est également respectée ...."

Or, s'il est vrai que $\Delta S = 0$, la fonction S que j'ai définie n'est pas harmonique pour autant, car elle admet une ligne singulière pour $\mu = 1$. Dr. Krarup a déjà signalé ce défaut d'harmonicité dans sa lettre circulaire ("Last letter to the S.S.G. n° 4.11") et je suis d'accord avec son analyse.

En ce qui concerne les conclusions de B. Zondek, il est impossible de les accepter. En effet, il semble que B. Zondek fait une erreur logique en affirmant qu'une fonction particulière, telle que l'anomalie Δg, n'appartient pas à l'ensemble de fonctions quelconques $g(P)$.

On est d'accord que l'anomalie Δg théorique ne doit pas comprendre l'harmonique d'ordre un. Cependant, les géophysiciens qui font les mesures et les réductions ne se préoccupent pas de cette question. Il est donc légitime d'examiner ce qui se passe dans le cas où la fonction $g(P)$ est quelconque.

Received: 16.6.1976

References

[1] W. Baranov. La formule de stokes est-elle correcte? *Bulletin Géodésique*, 115:27–34, 1975.

[2] Weikko A. Heiskanen and Helmut Moritz. *Physical Geodesy*. W. H. Freeman and Company, San Francisco, 1967.

15

Some Remarks About Collocation

Abstract

It is here proposed to apply the statistical theory not to the gravity field itself but to its cause, the mass distribution inside the Earth. Through an example, the paper presents part of mathematical tools necessary for the realization of this idea. In connection with the same example, it also is demonstrated how an error estimation not using statistical arguments can be found.

Introduction

> "Die Geodäsie ist im allgemeinen ein glänzendes Beispiel dafür, was man mit der Mathematik in den Anwendungen machen kann und wie man es machen soll. Man bekommt selbstverständlich alles nur approximativ bestimmt, zugleich aber hat man überall da, wo die Untersuchung als zu Ende geführt gilt, das Maß der Annäherung festgestellt." *Felix Klein, 1902*

It is a curious thing for one who has followed collocation from its very beginning to attend the discussion at this summer school. I see that my complaint that "(collocation) gives an answer to even the most foolish question if it is only asked in a formally correct way" has proved to be too pessimistic, in fact I have been glad to see that the struggle for reasonable output has forced the users of collocation to formulate their problems in an adequate way, but I cannot evade the impression that collocation is in a state of crisis. Let it be said at once: I believe that this state of crisis has to do with the really amazing fact that we talk so much about statistics and so little about physics.

Surely the statistical approach to collocation is very important, it gives us at least the illusion that we can compute meaningful mean square errors, which "fix the measure for the approximation."

In this paper I have tried to find out what can be done in collocation without statistics especially in order to find the measure for the approximation. In this investigation I have been forced to use a (very simple) physical argument, a more advanced physical argumentation could perhaps give a better result.

The main part of the paper is Section 2 where the approximation error is estimated. If the error estimates calculated from the statistical approach are too good to be true, my estimate is too true to be good, but first I hope that the method may be improved and second I find the interplay between potential theory and collocation interesting in itself and it can perhaps open new ways and reveal new problems. Some ideas in this direction are shortly discussed in Section 4 and I hope that the imagination of the reader will be stimulated and work further with many ideas.

Section 3 only treats a technical detail and may be skipped on first reading.

In Section 1 the theory of least-squares collocation is presented so far as is necessary for the following. The principal attitude of mine which has been determining for the presentation is the following: If we have made, say, n observations related to the anomalous potential T then we can (in infinitely many ways) write the possible candidates to T as a sum:

$$T = T_n + T'$$

where T_n is n-dimensional and T' infinite dimensional, and where T_n is uniquely defined by linear operations from the observations (and the way we have defined the decomposition) and we know nothing about T'. If given a norm in the space of which T must be a member then it is clear that if the spaces of which T and T' are members are orthogonal then the norm of T is minimum if and only if we put $T' = 0$.

This is the geometry of least-squares collocation and I find it important to understand that before statistics is introduced. Statistics says that if certain hypotheses are satisfied then we can find a norm such that T' is minimized in mean and we can say a little more about its statistics.

But there are other possible reasons for the choice of a norm:

1. Numerical reasons, e.g. in order to have a better condition number for the normal equations.
2. For a local use of collocation it may seem unreasonable to use a homogeneous norm, i.e. a norm which forces a minimizing of equal weight all over the surface of the Earth. Why not try to focus on the region in which we are interested? This could be done by a conformal mapping of the whole space in itself and of the Bjerhammar sphere in itself by using the product of two Kelvin-transformations such that the region in question is mapped into one which occupies a much greater part of the image of the surface of the Earth. In the image space we could then use collocation with homogeneous norm. This would be a very obvious idea but one has sometimes the impression that the prevailing statistical approach to collocation in some way confines the initiative and the imagination.

It is possible however to define the fundamental decomposition of T into T_n and T' without any norm simply by defining T_n as a linear combination of n given harmonic functions and T' as any harmonic function for which the n observations would give the value zero. I find it reasonable to call collocation also methods built upon such a decomposition. Then the "single layer"-method becomes a form for collocation; by the way it is possible even to prove that the "single layer"-method (at least when it is slightly idealized) minimizes a certain norm, so that it really is at least an approximation to least-squares collocation. Also Bjerhammar's "reflexive prediction" is—as far as I have understood it—a form of collocation. I wonder if even this method should minimize some norm.

In this paper I have treated only exact collocation, the essential points may be generalized to smoothing collocation in obvious ways.

1. Definition of Exact Collocation and of Least-Squares Collocation —The Main Theorem for Least-Squares Collocation

Let H be a Hilbert-space, let i be 1, 2, ..., n, let $\{l_i\}$ be real numbers and let $\{L_i\}$ be n linearly independent functionals defined on H (in this paper functional will always mean bounded linear functional).

Definition 15.1 Exact collocation (EC) *is to find a solution $\phi \in H$ to the equations*

$$L_i\phi = l_i \qquad i = 1, 2, \ldots, n \tag{1}$$

or in vector form

$$\mathcal{L}\phi = \ell \tag{2}$$

as a linear combination

$$\phi = \sum_1^n a_j\psi_j = \boldsymbol{a}^{\mathrm{T}}\boldsymbol{\psi} \tag{3}$$

where $\{\psi_j\}$ are n given linearly independent elements of H.

By substituting (3) into (1) we find the equations

$$\sum_1^n L_i\psi_j a_j = l_j, \qquad i = 1, 2, \ldots, n$$

i.e. the collocation problem has a unique solution if and only if the matrix $\{L_i\psi_i\}$ is regular. Our assumption about the linear independence of $\{\psi_j\}$ is necessary but not sufficient.

Definition 15.2 Exact least-squares collocation (ELSC) *is to find among all the $\phi \in H$ for which* (2) *is satisfied one for which the norm $\|\phi\|$ is as small as possible.*

That ELSC is a form for EC will follow from Theorem 15.1, but first we must introduce a few notations.

We shall suppose that all elements of H are functions (in geodetic practice even harmonic functions) in an open point set Ω and that H has a reproducing kernel $K(x,y)$; $x \in \Omega, y \in \Omega$. It is well known that $K(\cdot,y) \in H$ and $K(x,\cdot) \in H$ and that $K(x,y) = K(y,x)$. If $F \in H'$ (i.e. for F a functional on H) and $\phi \in H$ we shall sometimes instead of

$$F\phi$$

write

$$\phi(F).$$

This convention makes the notations $K(F,\cdot), K(\cdot,G)$ and $K(F,G)$ for $F,G \in H'$ clear.

Accordingly $K(L_i, L_j)$ becomes an $n \times m$ matrix and we will denote it as $K(\mathcal{L},\mathcal{L})$. Moreover $K(\cdot,\mathcal{L})$ is a row vector the entries of which are elements of H and

$$K(\mathcal{L},\cdot)^{\mathrm{T}} = K(\cdot,\mathcal{L}).$$

Theorem 15.1 *To every $\psi \in H$ there exists a unique element $\phi \in H_{\mathcal{L}} \subset H$ where*

$$H_{\mathcal{L}} = \{\, K(\cdot,\mathcal{L})\xi \mid \xi \in R^n \,\}$$

is an n-dimensional subspace of H, such that

1. $\mathcal{L}\phi = \mathcal{L}\psi$
2. $\|\phi\| \le \|\psi\|$

with $\|\phi\| = \|\psi\|$ if and only if $\psi \in H_{\mathcal{L}}$.

Corollary 15.1 *Under the given assumptions the exact least-squares problem has a unique solution.*

Proof of the corollary. As $\{L_i\}$ are linearly independent there exists a $\psi \in H$ such that that

$$\mathcal{L}\psi = \ell.$$

From this ψ Theorem 15.1 implies the existence of a unique $\phi \in H$ such that

1. $\mathcal{L}\phi = \ell$
2. $\|\phi\| <$ the norm of any other element of H satisfying 1).

Proof of the theorem. First we prove that 1) has a unique solution $\phi \in H_{\mathcal{L}}$:

Putting

$$\phi = K(\cdot,\mathcal{L})\xi, \qquad \xi \in \mathbf{R}^n,$$

1) is equivalent with

$$K(\mathcal{L},\mathcal{L})\xi = \mathcal{L}\psi. \tag{4}$$

These linear equations in $\{\xi_j\}$ have the unique solution

$$\xi = \big(K(\mathcal{L},\mathcal{L})\big)^{-1}\mathcal{L}\psi$$

because $K(\mathcal{L},\mathcal{L})$ is regular, therefore

$$\phi = K(\cdot,\mathcal{L})\big(K(\mathcal{L},\mathcal{L})\big)^{-1}\mathcal{L}\psi \tag{5}$$

is the unique solution of 1) in $H_{\mathcal{L}}$.

Next we prove that if ϕ' is any element of H, but not in $H_{\mathcal{L}}$ and such that $\mathcal{L}\phi' = \mathcal{L}\psi$ then $\|\phi'\| > \|\phi\|$.

ϕ' can be written as $\phi' = \phi + \phi_0$ where $\phi \in H_{\mathcal{L}}$ is given by (5) and $\mathcal{L}\phi_0 = 0$. We have

$$\langle \phi_0, \phi\rangle = \big\langle \phi_0, K(\cdot,\mathcal{L})\xi\big\rangle = \phi_0\mathcal{L}^{\mathrm{T}}\xi = \xi\mathcal{L}\phi_0 = 0,$$

i.e. ϕ_0 is orthogonal to ϕ (and evidently to every element of $H_{\mathcal{L}}$), therefore

$$\|\phi'\|^2 = \|\phi\|^2 + \|\phi_0\|^2 > \|\phi\|^2$$

or

$$\|\phi'\| > \|\phi\|.$$

Corollary 15.2 *The norm of ϕ (the solution of the ELSC problem) is determined by*

$$\|\phi\|^2 = \ell^{\mathrm{T}}\big(K(\mathcal{L},\mathcal{L})\big)^{-1}\ell.$$

Proof From (5) we find

$$\begin{aligned}\|\phi\|^2 = \langle\phi,\phi\rangle &= \big\langle (K\cdot,\mathcal{L})\big(K(\mathcal{L},\mathcal{L})\big)^{-1}\mathcal{L}\psi, (K\cdot,\mathcal{L})\big(K(\mathcal{L},\mathcal{L})\big)^{-1}\mathcal{L}\psi\big\rangle\\ &= (\mathcal{L}\psi)^{\mathrm{T}}\big(K(\mathcal{L},\mathcal{L})\big)^{-1}\mathcal{L}\psi = \ell^{\mathrm{T}}\big(K(\mathcal{L},\mathcal{L})\big)^{-1}\ell.\end{aligned}$$

Our result so far can partly be expressed by saying that the element $\phi \in H$ given by (5) is in a certain sense (namely with respect to the norm $\|\cdot\|$ in H) the best approximation to ψ to be found from our limited knowledge about ψ (namely the values of the functionals L_i of ψ). We will in this paper use the symbol $\widetilde{\psi}$ for this approximation:

$$\widetilde{\psi} = K(\cdot,\mathcal{L})\big(K(\mathcal{L},\mathcal{L})\big)^{-1}\mathcal{L}\psi. \tag{6}$$

Applying the functional F on $\widetilde{\psi}$ gives

$$F\widetilde{\psi} = K(F,\mathcal{L})\big(K(\mathcal{L},\mathcal{L})\big)^{-1}\mathcal{L}\psi$$

which gives the value of F applied on $\widetilde{\psi}$ as a linear combination of the values of the functionals L_i applied on ψ, if we define

$$\widetilde{F}\psi = F\widetilde{\psi}, \tag{7}$$

then $\widetilde{F}$ is a linear functional which may be written as

$$\widetilde{F} = K(F, \mathcal{L})\big(K(\mathcal{L}, \mathcal{L})\big)^{-1}\mathcal{L}, \tag{8}$$

and we could call $\widetilde{F}$ the ELSC approximation to F. We will however give another definition of "the best approximation" $\widetilde{F}$ to F which makes evident some minimum property of $F - \widetilde{F}$:

Definition 15.3 *The exact least-squares collocation approximation to any $F \in H'$ is that linear combination of the given functionals $\{L_i\}$, $i = 1, 2, \ldots, n$, for which $\|F - \widetilde{F}\|'$ is as small as possible. Here $\|\cdot\|'$ is the norm in H', the dual space to H.*

We could also express the definition by saying that F is the orthogonal projection of F into $H'_{\mathcal{L}}$, the n-dimensional subspace of H' spanned by the n functionals $\{L_i\}$. So we want to find $\xi \in \mathbf{R}^n$ such that

$$F - \widetilde{F} = (F - \xi^{\mathrm{T}}\mathcal{L}) \perp \eta^{\mathrm{T}}\mathcal{L} \qquad \text{for all } \eta \in \mathbf{R}^n,$$

or

$$\langle \eta^{\mathrm{T}}\mathcal{L}, F - \xi^{\mathrm{T}}\mathcal{L}\rangle' = 0 \qquad \text{for all } \eta \in \mathbf{R}^n. \tag{9}$$

But we know that for any $G_1, G_2 \in H'$

$$\langle G_1, G_2\rangle' = K(G_1, G_2),$$

therefore we can write (9) as

$$\eta^{\mathrm{T}}\big(K(\mathcal{L}, F) - K(\mathcal{L}, \xi\mathcal{L})\big) = 0 \qquad \text{for all } \eta \in \mathbf{R}^n,$$

that is

$$K(\mathcal{L}, \mathcal{L})\xi = K(\mathcal{L}, F)$$
$$\xi = \big(K(\mathcal{L}, \mathcal{L})\big)^{-1}K(\mathcal{L}, F)$$

or

$$\xi^{\mathrm{T}} = K(F, \mathcal{L})\big(K(\mathcal{L}, \mathcal{L})\big)^{-1}$$

such that

$$\widetilde{F} = \xi^{\mathrm{T}}\mathcal{L} = K(F, \mathcal{L})\big(K(\mathcal{L}, \mathcal{L})\big)^{-1}\mathcal{L}$$

which is the same as (8) and we have proved that Definition 15.3 gives the same F as the Definition (7).

We can express this result in

Theorem 15.2 *The approximation problem given by Definition 15.3 has a unique solution, the minimum is determined by*

$$\left(\|F - \widetilde{F}\|'\right)^2 = K(F, \mathcal{L})\left(K(\mathcal{L}, \mathcal{L})\right)^{-1} K(\mathcal{L}, F) \tag{10}$$

and the solutions for any pair of functionals $F, G \in H'$ *are* consistent, *i.e. there exists an element* $\widetilde{\phi} \in H$ *such that*

$$\widetilde{F}\phi = F\widetilde{\phi} \qquad \text{and} \qquad \widetilde{G}\phi = G\widetilde{\phi}.$$

This $\widetilde{\phi}$ *depends only on the linear functionals* $\{L_i\}$ *and their values* $\{l_i\}$ *and is the solution of the approximation problem given by Definition 15.2.*

(10) is proved by a trivial calculation.

(If the kernel K may be identified with a covariance form, then Theorem 15.2 is the mathematical machinery to be applied in order to arrive at the Gauss-Markov lemma and then it is the mean-square error which is minimized.)

2. An Upper Bound for the Approximation Error of a Functional in Least-Squares Collocation

From the definition of the norm in H', the dual space to H, follows that

$$\left|(F - \widetilde{F})\phi\right| \leq \|F - \widetilde{F}\|' \|\phi\| \qquad \text{for } \phi \in H.$$

Using (10) we find

$$\left((F - \widetilde{F})\phi\right)^2 \leq K(F, \mathcal{L})\left(K(\mathcal{L}, \mathcal{L})\right)^{-1} K(\mathcal{L}, F)\|\phi\|^2. \tag{11}$$

This formula can be improved:

$$(F - \widetilde{F})\phi = (F - \widetilde{F})(\phi - \widetilde{\phi})$$

because

$$(F - \widetilde{F})\widetilde{\phi} = 0.$$

Therefore

$$\left((F - \widetilde{F})\phi\right)^2 \leq K(F, \mathcal{L})\left(K(\mathcal{L}, \mathcal{L})\right)^{-1} K(\mathcal{L}, F)\|\phi - \widetilde{\phi}\|^2$$

or

$$\left((F - \widetilde{F})\phi\right)^2 \leq K(F, \mathcal{L})\left(K(\mathcal{L}, \mathcal{L})\right)^{-1} K(\mathcal{L}, F)\left(\|\phi\|^2 - \ell^{\mathrm{T}}\left(K(\mathcal{L}, \mathcal{L})\right)^{-1}\ell\right). \tag{12}$$

If we try to apply this inequality on errors for the approximation of functionals on the anomalous gravity field T of the Earth we are faced with the unpleasant fact that we cannot find an upper bound for $\|T\|$ in a Hilbert space of functions harmonic outside a Bjerhammar sphere and we normally use such Hilbert spaces in geodetic collocation.

As a first step on the way to a practical estimate of the approximation error for geodetic collocation methods I shall here introduce a special Hilbert space for which it is possible to find an upper bound for $\|T\|$ and for which it is possible to calculate least-squares collocation approximations.

The actual gravitation field of the Earth is supposed to be generated by a mass distribution inside the Earth. The density of this mass distribution is non-negative and bounded by some constant K. For the normal gravitation field a mass distribution with the same properties can be found. Therefore T as the difference between these fields can also be regarded as generated by a mass distribution inside the Earth and having a bounded density so the integral of the square of this density is bounded.

Let ω denote the surface of the Earth, Ω_0 the interior and Ω the exterior of the Earth, and let D be the space $L_2(\Omega_0)$ i.e. Hilbert space of square integrable functions on Ω_0. We can regard the elements of D as densities of mass distributions. The gravitation field ϕ in Ω generated by $\rho \in D$ is given by

$$\phi(x) = \int_{\Omega_0} \frac{\rho(y)}{|x-y|}\, dy. \tag{13}$$

(13) defines a linear mapping

$$M\colon \rho \mapsto \phi \qquad \text{for } \rho \in D.$$

Let H be the image space consisting of $M\phi$ for all $\rho \in D$ such that

$$M\colon D \mapsto H, \qquad \rho \mapsto \phi.$$

H is a linear space of functions harmonic in Ω (regular at infinity) and we have just seen that $T \in H$. We shall prove that H can be made a Hilbert space.

By the definition of H the operator M is surjective, but it is not one to one: to a given harmonic function $\phi \in H$ there corresponds an infinite set of densities $\rho \in D$ such that $\phi = M\rho$. Let D_0 be the kernel of M, i.e. D_0 is the set of elements $\rho \in D$ such that $M\rho = 0$. D_0 is the set of square integrable densities on Ω_0 which generate potentials which are zero on Ω. If we can prove that D_0 is a closed linear subspace of D, then every element $\rho \in D$ can be written

$$\rho = \rho_0 + \rho_1$$

where $\rho_0 \in D_0$ and $\rho_1 \perp \rho_0$ (or $\rho_1 \in D_1$ where D_1 is the orthogonal complement to D_0 in D).

We have

$$M\rho = M\rho_0 + M\rho_1 = M\rho_1$$

and

$$\|\rho\|_D^2 = \|\rho_0\|_D^2 + \|\rho_1\|_D^2,$$

so that ρ_1—the orthogonal projection of ρ into D_1—is that element of D which has the smallest norm among those which generate the same outer potential as ρ.

It is now clear that if M_1 is the restriction of M to D_1:

$$M_1 \colon D_1 \mapsto H, \qquad \rho_1 \mapsto M\rho_1$$

then M_1 is linear, one to one, and onto: To every $\phi \in H$ there exists one $\rho \in D_1$ such that $M_1\rho = M\rho = \phi$, and we can define a norm $\|\cdot\|_H$ in H by

$$\|\phi\|_H = \|\rho\|_D \qquad \text{where } \phi \in H \text{ and } \rho \in D \text{ such that } \phi = M\rho$$

that is the norm of ϕ in H is the norm of that ρ which has the smallest norm of those densities which generate ϕ:

$$\|\phi\|_H = \min_{M\rho=\phi} \|\phi\|_D.$$

And we can prove that D_0 is closed! (That it is linear is trivial.) Define

$$k(x,y) = \frac{1}{|x-y|}, \qquad x \in \Omega, \quad y \in \Omega_0$$

then $k(x,\cdot)$ is continuous on $\overline{\Omega}_0$ and therefore $k(x,\cdot) \in D$ for all $x \in \Omega$ and we can write (13) as

$$\phi(x) = \big\langle \rho, k(x,\cdot) \big\rangle_D \tag{14}$$

i.e. for fixed $x \in \Omega$, $\phi(x)$ is a bounded linear functional on D, let us call it M_x:

$$M_x \colon D \mapsto R \colon \rho \mapsto \phi(x).$$

As a bounded linear functional M_x is continuous and therefore the set

$$S_x = \big\{\, \rho \in D \;\big|\; M_x = 0 \,\big\}$$

is closed. Clearly,

$$D_0 = \bigcap_{x \in \Omega} S_x$$

which is closed as an intersection of closed sets.

From (14) we can find a characterization of D_1, the orthogonal complement to D_0 in D. In fact (14) says that D_0 is the orthogonal complement in D to the linear subspace spanned by the elements $k(x,\cdot) \in D$, that is D_1 is exactly this subspace. It is at least plausible that D_1 consists exactly of those elements of D which are harmonic in Ω_0 and, in fact, it can be proved provided only that ω satisfies a very weak smoothness condition; it will not be proved or needed here, but we will return to the geodetic problem for a while.

In order to give the inequalities (11) and (12) a meaning we have found a Hilbert space H of functions harmonic outside the Earth such that T must be an element of this space and such that it is possible to find a number c such that

$$\|T\|_H < c. \tag{15}$$

In the next section we shall prove that the space H has a reproducing kernel K and that it is possible for $F, G \in H'$ to calculate $K(F,G)$ and so (11) and (12) have at least a meaning. Unfortunately, probably any constant c (in (15)) which can be calculated from our knowledge today about the mass distribution in the Earth will be much too great and so result in a much too pessimistic estimate for the approximation errors. c cannot be computed from satellite potentials because those ignore "higher order terms" i.e. the terms of T which decrease rapidly with the distance from the Earth's surface, but from satellite potentials we can find lower bounds for c.

There can be given three good reasons for our estimate being so pessimistic:

1. It lies in the very nature of maximum error estimates that they are more pessimistic than mean error estimates, they must therefore normally be looked upon with a reasonable large portion of optimism.
2. In the Gauss-distribution least-squares model the maximum error is infinite. Naturally it would be unfair already from the existence of finite maximum errors for this collocation model to deduce the invalidity of a Gauss-distribution model for geodetic collocation exactly as the fact that we know that geodetic observation errors in networks are bounded does not in practice preclude the application of the Gauss-distribution model.
3. Least-squares collocation is not fashioned with reference to the minimization of maximum errors. It would be possible to define a least-maximum error collocation and, more general, a least-norm collocation using Banach spaces with a not necessarily quadratic norm instead of Hilbert space. Then, however, the calculation would no longer be linear and I think the loss in simplicity would exceed the gain.

3. The Kernel

We have seen that $M_1 \colon D_1 \to H$ is an isometric operator:

$$\langle M_1\rho, M\sigma\rangle_H = \langle \rho, \sigma\rangle_D \qquad \text{for } \rho, \sigma \in D_1. \tag{16}$$

We want to find the reproducing kernel $K(x,y)$, $x, y \in \Omega$, for H. It is defined by the equation

$$\phi(x) = \big\langle \phi, K(x,\cdot)\big\rangle_H, \qquad x \in \Omega, \quad \phi \in H; \tag{17}$$

but from (14) and (16) follows:

$$\phi(x) = \big\langle \rho, k(x,\cdot)\big\rangle_D = \big\langle M_1\rho, k(x, M_1)\big\rangle_H = \big\langle \phi, k(x, M_1)\big\rangle_H \quad \text{for } \phi = M_1\rho,$$

therefore we must have

$$K(x,y) = k(x, M_1) = M_1\Big(\frac{1}{|x-z|}\Big) \qquad M_1 \text{ operates with respect to } z$$

$$= \int_{\Omega_0} \frac{dz}{|x-z|\,|y-z|} = \big\langle k(x,\cdot), k(y,\cdot)\big\rangle_D.$$

For given $x, y \in \Omega$ this can be found by numerical integration, but it is normally impossible to find a closed expression for it. But for $F, G \in H'$ we can find

$$K(F,G) = \big\langle k(F,\cdot), k(G,\cdot)\big\rangle_D$$

by numerical integration. $k(F,\cdot)$ is harmonic in Ω_0 and is called the Green's transform of the functional F, it is useful also for calculations in spaces with other norms.

There exist other Hilbert spaces for which the norm of T must be bounded. Some of them are more convenient. One could choose a Bjerhammar sphere Σ inside the Earth and define the norm for the mass density by

$$\|\rho\|^2 = \int_{\Sigma} w(r)\rho^2\,dx + \int_{\Omega_0-\Sigma} \rho^2\,dx$$

where $w(r) > 0$ is a continuous function of r, the distance from the centre of the Bjerhammar sphere. Also in this case such a norm defines a norm for harmonic functions in Ω as above; by a suitable choice of the weight function w that part of the reproducing kernel corresponding to the integral over Σ can be expressed by a closed expression such that the numerical integration has only to be performed over $\Omega_0 - \Sigma$.

4. Discussion

The fact that I in the foregoing section have entered into computational details concerning reproducing kernels for Hilbert spaces over Ω could perhaps make the reader suspect that I would recommend collocation "with respect to the surface of the Earth" instead of "with respect to a Bjerhammar sphere." This is *not* the case.

On the other hand, there is perhaps reason for expecting that the evolution of collocation will make a step or two in that direction.

I shall here give two possible reasons:

1. When the distance between points where observations of the same type are performed becomes small with respect to the distance to the Bjerhammar sphere then the normal equations have a tendency to become ill-conditioned. But where we are interested in a detailed picture of the local gravity field we have to use such neighbouring observations. Therefore we can be forced to lift up the Bjerhammar surface under certain circumstances so much that it is not a sphere any more. I have always regretted that the cheapest relevant source of information, the topography, is neglected by collocation. Here it comes in again, and I find it probable that a Bjerhammar surface approximating the physical surface roughly

from below in mountainous regions e.g. the Rheingraben can improve local results considerably. The topographical information is cheap but the numerical integration needed for its application for collocation is not cheap. But something can be done with suitable approximations and moreover any use of topography in physical geodesy costs a lot of calculations.

2. The field of applications of statistics is the unknown, but by using statistics on the field outside the Bjerhammar sphere we cut off the topography, the isostasy, etc. Was it not better to take advantage of what we know and apply statistics on that part of the mass distribution which is unknown? I do not want that the statistical speculations shall be buried, but I want them brought not only down to Earth but down under the surface of the Earth and this is quite different.

Bibliographical Notes

The Klein quotation is from [2], page 171. I have tried to translate it:

> Generally speaking geodesy is a brilliant illustration of the possibilities of application of mathematics and how to apply it. It is true that it gives only approximations, but as far as a geodetic investigation can be regarded as completed it always gives the measure for the approximation.

The quotation from myself—sans comparaison—is from [3]. This paper has been strongly influenced by the recent paper [1].

For slightly different aspects of the connection between collocation of the anomalous field and mass distributions in the Earth see [4] and [5].

References

[1] A. Dermanis. Probabilistic and deterministic aspects of linear estimation in geodesy. Technical report, Reports of the Department of Geodetic Science, 244, Columbus, Ohio, 1976.

[2] F. Klein. *Elementarmathematik vom höheren Standpunkt aus.* 3. edition, 1928.

[3] Torben Krarup. *A Contribution to the Mathematical Foundation of Physical Geodesy.* Meddelelse 44. Geodætisk Institut, København, 1969.

[4] S. L. Lauritzen. Random orthogonal set functions and stochastic models for the gravity potential of the Earth. Technical Report Preprint 2, Institute of Mathematical Statistics, University of Copenhagen, 1974.

[5] C. C. Tscherning. Some simple methods for the unique assignment of a density distribution to a harmonic function. Technical report, Reports of the Department of Geodetic Science, 213, Columbus, Ohio, 1974.

16

Apropos Some Recent Papers by Willi Freeden on a Class of Integral Formulas in the Mathematical Geodesy

I think that some geodesists after reading of one or two of the papers by Willi Freeden[1] would find it amusing to know if there is a relation between Freeden's method and the geodetic collocation method: is there an alternative to collocation or a complement to it here—or should the connections between the two theories be much more intimate as some formulas of Freeden's could make one suspect? The aim of this paper is to clear up this question.

I. Two Types of Approximation

In order to compare the two methods we will first define an abstract version of these methods, and in order to do that we will first formulate the problem they intend to solve.

The problem In a real Hilbert space $\mathcal{H}$ we concentrate on $N+1$ bounded linear functionals $\{f_i\}$, $i = 1, 2, \ldots, N$, and F. In the following the indices i and j will always have the range $1, 2, \ldots, N$. We suppose that these functionals are independent; that is, for any $\{v_i\} \in \mathbf{R}^N$ and $V \in \mathbf{R}$ there exists at least one element $\phi \in \mathcal{H}$ such that

$$f_i\phi = v_i \qquad \text{and} \qquad F\phi = V.$$

Now suppose that the values $\{v_i\}$ are given, from the independence of the $N+1$ functionals follows that V cannot be computed from $\{v_i\}$; the only linear functionals we can find from $\{f_i\}$ are those linearly dependent on $\{f_i\}$, i.e., the elements of the linear subspace of $\mathcal{H}'$ spanned by $\{f_i\}$. Let $\mathcal{N}$ be this N-dimensional subspace, and let us define:

Any function

$$A : \mathcal{H}' \to \mathcal{N} : F \mapsto \widetilde{F}$$

is called an approximation to F. If A is linear it will be called a linear approximation.

[1] [2], [3]

Clearly as both F and $\widetilde{F}$ are linear functionals, their difference

$$\widetilde{F} - F = E$$

is also a linear functional and we will call it the error functional for the approximation A. The numerical value of E for a given element $\phi \in \mathcal{H}$,

$$|E\phi| = |\widetilde{F}\phi - F\phi|,$$

will be called the error for A and ϕ. The $\mathcal{H}'$ norm of E, i.e., the smallest constant c such that

$$|E\phi| \leq c\|\phi\|,$$

where $\|\phi\|$ is the $\mathcal{H}$-norm of ϕ, is called the norm for the approximation. We will denote the $\mathcal{H}'$-norm by $\|\cdot\|'$, such that $\|E\|'$ is the norm of the approximation A.

We can now give the final formulation of the problem, namely by defining a *best approximation* in a reasonable way and stating a method for the computation of this best approximation. It shall be remarked that best approximation always is meant to be relative to the given structure of $\mathcal{H}$ as a Hilbert space, i.e., relative to the given norm $\|\cdot\|$.

The most obvious definition of best approximation is that approximation for which $\|E\|'$ is as small as possible. This definition gives the least-squares method and the result can be described as follows:

$\widetilde{F}$ is defined as the element in $\mathcal{N}$ for which $\|\widetilde{F} - F\|'$ is as small as possible; this means that $\widetilde{F}$ is the orthogonal projection of F into $\mathcal{N}$, so $\widetilde{F} - F$ must be orthogonal to every element of $\mathcal{N}$, and it is so if

$$\langle \widetilde{F} - F, f_i \rangle' = 0 \tag{1}$$

for every $\{f_i\}$, where $\langle \cdot, \cdot \rangle'$ is the scalar product in $\mathcal{H}'$. $\widetilde{F} \in \mathcal{N}$ so there exists a vector $a = \{a_i\}$ such that

$$\widetilde{F} = \sum_i a_i f_i,$$

and we can write (1) as

$$\sum_i a_i \langle f_i, f_j \rangle' = \langle F, f_j \rangle'.$$

The matrix

$$M = \{\langle f_i, f_j \rangle'\}$$

is non-singular because $\{f_i\}$ are independent; therefore

$$a = M^{-1} f_F,$$

where f_F means the vector

$$f_F = \{\langle F, f_i \rangle'\}.$$

If we pass from the functionals $\{f_i\}$ and F to their values $\{v_i\}$ and V, we may write the result as

$$\widetilde{V} = f_F^{\mathrm{T}} M^{-1} v, \tag{2}$$

where $v = \{v_i\}$.

We see that the least-squares approximation is linear. Another important property of this approximation which shall not be proved here is that it is consistent; that is given $\{f_i\}$ and $\{v_i\}$ there exists a unique element $\widetilde{\phi} \in \mathcal{H}$ such that for any $F \in \mathcal{H}'$,

$$\widetilde{V} = F\widetilde{\phi}.$$

Before we can define what we call the F-approximation and the F-method, we must give a few definitions.

We define the null space of the linear functional F as

$$\mathcal{H}_0 = \{\, \phi \in \mathcal{H} \mid F\phi = 0 \,\},$$

and we give $\mathcal{H}_0$ the structure as a Hilbert space induced on it as a subspace of $\mathcal{H}$. Its dual $\mathcal{H}_0'$ is the subspace of $\mathcal{H}'$ consisting of all elements of $\mathcal{H}'$ orthogonal to F.

The orthogonal complement to $\mathcal{H}_0$ in $\mathcal{H}$ will be called $\mathcal{G}$; it is one-dimensional and $\mathcal{G}'$ is the one-dimensional subspace of $\mathcal{G}'$ spanned by F. The orthogonal projector of $\mathcal{H}'$ onto $\mathcal{G}'$ is

$$Q = \frac{\langle F, \cdot \rangle'}{\langle F, F \rangle'} F,$$

and the orthogonal projector of $\mathcal{H}'$ onto $\mathcal{H}_0'$ is

$$P = I - \frac{\langle F, \cdot \rangle'}{\langle F, F \rangle'} F.$$

Now we can define the F-best approximation by the following three conditions:

1. $\widetilde{F} \in \mathcal{N}$
2. $Q\widetilde{F} = F$
3. $\widetilde{F}$ shall satisfy conditions 1 and 2 and have an error norm as small as possible.

Let $\mathcal{N}_F$ be the $N-1$ dimensional affine subspace of $\mathcal{H}'$ consisting of elements $\widetilde{F}$ of $\mathcal{N}$ for which $Q\widetilde{F} = F$, then 1 and 2 can be expressed as

$$\widetilde{F} \in \mathcal{N}_F, \tag{3}$$

and 3 can be expressed as the following: $\widetilde{F}$ is the orthogonal projection of F into $\mathcal{N}_F$.

1) means that there exists a vector $b = \{b_i\}$ such that

$$\widetilde{F} = \sum b_i f_i, \tag{4}$$

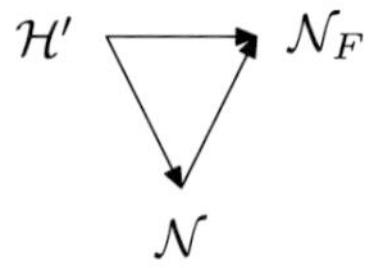

Fig. 16.1. Diagram of commuting orthogonal projections

and then 2) means that

$$\frac{\langle F, \widetilde{F}\rangle' F}{\langle F, F\rangle'} F = \sum b_i \langle F, f_i\rangle' \frac{F}{\langle F, F\rangle'} = F$$

or

$$\sum b_i \langle F, f_i\rangle' = \langle F, F\rangle' \qquad \text{or} \qquad f_F^{\mathrm{T}} b = \langle F, F\rangle', \tag{5}$$

so (3) means that there exists a vector b such that (4) and (5) are satisfied.

The orthogonal projection $\mathcal{H}' \to \mathcal{N}_F$ will now be effectuated by composition of the orthogonal projections $\mathcal{H}' \to \mathcal{N}$ and $\mathcal{N} \to \mathcal{N}_F$. This is possible because the diagram in Figure 16, where the arrows indicate orthogonal projections, commutes.

The first projection

$$\mathcal{H}' \to \mathcal{N} : F \mapsto F_{\mathcal{N}}$$

is effectuated by the least-squares method:

$$\begin{aligned} F_{\mathcal{N}} &= \sum a_i f_i \\ a &= M^{-1} f_F . \end{aligned}$$

The second projection is

$$\mathcal{N} \to \mathcal{N}_F : \sum a_i f_i \mapsto \sum b_i f_i ,$$

where b satisfies condition (5) is an orthogonal projection in an N-dimensional vector space onto an affine hyperplane in this space. Therefore

$$\sum b_i f_i - \sum a_i f_i$$

shall be orthogonal to the hyperplane in $\mathcal{N}$ determined by (5). A qualified guess is $b = \lambda a$, then

$$\sum b_i f_i - \sum a_i f_i = (\lambda - 1) \sum a_i f_i$$

and for $x = \{x_i\}$:

$$(\lambda-1)\Big\langle \sum a_i f_i , \sum x_i f_i \Big\rangle' = (\lambda-1) a^{\mathrm{T}} M x = (\lambda-1) f_F^{\mathrm{T}} M^{-1} M x = (\lambda-1) f_F^{\mathrm{T}} x ,$$

which in fact is zero for $\sum x_i f_i$ parallel to the hyperplane (5).

Putting $b = \lambda a$ in (5) gives

$$\lambda = \frac{\langle F, F\rangle'}{f_F^{\mathrm{T}} a} = \frac{\langle F, F\rangle'}{f_F^{\mathrm{T}} M^{-1} f_F}$$

and so the F-approximation is given by

$$\widetilde{F} = \sum b_i f_i,$$
$$b = \langle F, F\rangle' \frac{M^{-1} f_F}{f_F^{\mathrm{T}} M^{-1} f_F},$$

and

$$\widetilde{V} = \langle F, F\rangle' \frac{f_F^{\mathrm{T}} M^{-1} v}{f_F^{\mathrm{T}} M^{-1} f_F}. \tag{6}$$

This result is, as we shall see, apparently different from that of Freeden, which with our notations may be written as

$$\widetilde{V} = \langle F, F\rangle' \frac{f_F^{\mathrm{T}} M_0^{-1} v}{f_F^{\mathrm{T}} M_0^{-1} f_F}, \tag{7}$$

where

$$M_0 = \{\langle P f_i, P f_j\rangle'\}.$$

But

$$\langle P f_i, P f_j\rangle' = \langle P f_i, f_j\rangle' = \left\langle f_i - \frac{\langle F, f_i\rangle'}{\langle F, F\rangle'} F, f_j \right\rangle'$$
$$= \langle f_i, f_j\rangle' - \frac{\langle F, f_i\rangle' \langle F, f_j\rangle'}{\langle F, F\rangle'} = \langle f_i, f_j\rangle' - \frac{f_F f_F^{\mathrm{T}}}{\langle F, F\rangle'},$$

and

$$M_0^{-1} = \left\{ \langle f_i, f_j\rangle' - \frac{f_F f_F^{\mathrm{T}}}{\langle F, F\rangle'} \right\}^{-1} = M^{-1} + \frac{M^{-1} \frac{f_F f_F^{\mathrm{T}}}{\langle F, F\rangle'} M^{-1}}{1 - \frac{f_F^{\mathrm{T}} M^{-1} f_F}{\langle F, F\rangle'}}.$$

Therefore

$$f_F^{\mathrm{T}} M_0^{-1} v = f_F^{\mathrm{T}} M^{-1} v + \frac{f_F^{\mathrm{T}} M^{-1} f_F \frac{1}{\langle F, F\rangle'} f_F^{\mathrm{T}} M^{-1} v}{1 - \frac{f_F M^{-1} f_F}{\langle F, F\rangle'}}$$
$$= \left(1 + \frac{\frac{f_F^{\mathrm{T}} M^{-1} f_F}{\langle F, F\rangle'}}{1 - \frac{f_F^{\mathrm{T}} M^{-1} f_F}{\langle F, F\rangle'}} \right) f_F^{\mathrm{T}} M^{-1} v = \frac{f_F^{\mathrm{T}} M^{-1} v}{\langle F, F\rangle' - f_F^{\mathrm{T}} M^{-1} f_F},$$

and in the same way

$$f_F^{\mathrm{T}} M_0^{-1} f_F = \frac{f_F^{\mathrm{T}} M^{-1} f_F}{\langle F, F\rangle' - f_F^{\mathrm{T}} M^{-1} f_F},$$

so the expressions (6) and (7) are identical.

All these things found by calculations can in a moment be made transparent by regarding the plane in $\mathcal{H}'$ which contains the vector F and its projection into the span of $\{f_i\}$. This plane contains the four points O, F and the two approximating $\widetilde{F}$, namely the least-squares approximation and the F-approximation, and it becomes clear that the latter is $1/\cos^2\theta$ times the former, where θ is the angle between the vector F and (its projection into) span $\{f_i\}$.

Finally, we have three remarks:

1. The F-approximation is only defined for $\theta \neq \frac{\pi}{2}$;
2. The F-approximation is not linear;
3. The F-approximation is not consistent.

II. The Joy of Recognition

Perhaps it is not easy to recognize Freeden's Method in the F-approximation but just like reproducing kernels, the bridge from least squares to collocation it is also the bridge from the F-approximation back to Freeden's integral formulas. Let us therefore recall a few essential facts about Hilbert spaces with reproducing kernels (see [4] and [1]).

Let $\mathcal{H}$ be a Hilbert space of a class of functions on an open subset Ω of a Euclidean space. If the values of any element of $\mathcal{H}$ at points of Ω, evaluation functionals, are bounded linear functionals on $\mathcal{H}$, then there exists a reproducing kernel $K(\cdot,\cdot)$, a symmetric function on $\Omega \times \Omega$ such that for any

$$g \in \mathcal{H}' \quad \text{and} \quad h \in \mathcal{H}',$$

e.g., for evaluation functionals

$$g_x K(x,\cdot) \in \mathcal{H},$$

(we often write $K(g,\cdot)$ instead of $g_x K(x,\cdot)$) and for any $\phi \in \mathcal{H}$

$$g\phi = \big\langle K(g,\cdot), \phi \big\rangle,$$

where $\langle\cdot,\cdot\rangle$ is the scalar product in $\mathcal{H}$

$$g_x h_y K(x,y) = \langle g, h\rangle'$$

or as we also write it

$$K(g,h) = \langle g, h\rangle'.$$

Therefore if we know the reproducing kernel for our space $\mathcal{H}$, especially if this kernel is explicitly expressible in elementary functions, then the approximation problems are computationally simpler to solve if we in the formulas substitute $K(g,h)$ for $\langle g,h\rangle'$.

For Ω the unit sphere in $\mathbf{R}^3$ we have in Freeden's notations

$$f(\xi) = \frac{1}{(4\pi)^{2m}} \int_{\Omega} \left[(\Delta_\eta^* + \lambda) G^{(2m)}(\lambda; \xi, \eta)\right] (\Delta_\eta^* + \lambda) f(\eta)\, d\omega(\eta)$$

for $\lambda \neq n(n+1)$, i.e., $G^{(2m)}(\lambda; \xi, \eta)$ is the reproducing kernel for the Hilbert space $\mathcal{H}^{m,\lambda}$ consisting of functions on Ω for which the norm

$$\|f\|_{m,\lambda} = \left[\frac{1}{(4\pi)^{2m}} \int_{\Omega} \left[(\Delta_\eta^* + \lambda) f(\eta)\right]^2 d\omega(\eta)\right]^{1/2}$$

corresponding to the scalar product

$$\langle f, g\rangle_{m,\lambda} = \frac{1}{(4\pi)^{2m}} \int_{\Omega} \left[(\Delta_\eta^* + \lambda) f(\eta)\right] \left[(\Delta_\eta^* + \lambda) g(\eta)\right] d\omega(\eta) \tag{8}$$

is finite.

For

$$F(\phi) = \frac{1}{4\pi} \int_{\Omega} \phi(\eta)\, d\omega(\eta)$$

and $\lambda = 0$ we have that (8) is the scalar product on $\mathcal{H}_0$ for a suitable Hilbert space $\mathcal{H}$ of functions on Ω. The scalar product in $\mathcal{H}$ may be defined as

$$\langle f, g\rangle = \left[\frac{1}{4\pi} \int_{\Omega} f(\eta)\, d\omega(\eta)\right] \left[\frac{1}{4\pi} \int_{\Omega} g(\eta)\, d\omega(\eta)\right] + \frac{1}{(4\pi)^{2m}} \int_{\Omega} (\Delta^{*m} f)(\Delta^{*m} g)\, d\omega(\eta),$$

and the kernel for $\mathcal{H}$ then becomes

$$1 + G^{(2m)}(0; \xi, \eta).$$

Applying (7) in this situation we will find Freeden's result, remembering that $\{f_i\}$ are evaluation functionals and F is the mean value over the unit sphere, so $\langle F, F\rangle' = \langle F, f_i\rangle' = 1$.

The kernels $G^{(2m)}(0; \xi, \eta)$ are not expressible by elementary functions but instead of those one could use

$$K^{(2m)}(\xi, \eta) = \sum_{k=1}^{\infty} \frac{2k+1}{k(k+1)(k+2)\cdots(k+2m-1)} P_k(\xi, \eta),$$

which are explicitly expressible by elementary functions and are reproducing kernels in Hilbert spaces with norms equivalent to the norms in the spaces in which $G^{(2m)}$ are reproducing kernels; this is easily proved by decomposing the coefficients in (2) into partial fractions.

III. Instead of a Conclusion

Here I have tried to point out some connections between Freeden's formulas and collocation and to generalize Freeden's method in different directions, but I want to leave it to the reader to find out which one of the methods is preferred for his problems. I hope that the considerations above will help him in his choice.

I have observed that some geodesists in writing about applications of collocation have expressed surprise or perhaps complaint over the fact that the mean of an observed functional there in contrast to that calculated by Freeden's method is expressed by weights, the sum of which are less than one. Geodesists with such problems can perhaps by contemplating this paper find out where this is reasonable and where it is not.

References

[1] Philip J. Davis. *Interpolation and Approximation.* Blaisdell Publishing Company, Waltham, Massachusetts, 1963.

[2] Willi Freeden. An application of a summation formula to numerical computations of integrals over a sphere. *Bulletin Géodésique*, 52:165–175, 1978.

[3] Willi Freeden. *Über eine Klasse von Integralformeln der Mathematischen Geodäsie*, volume 27 of *Veröffentlichung des Geodätischen Instituts der Rheinisch-Westpfälzischen Technischen Hochschule Aachen.* 1979.

[4] Herbert Meschkowski. *Hilbertsche Räume mit Kernfunktion*, volume 113 of *Die Grundlehren der mathematischen Wissenschaften in Einzeldarstellungen mit besonderer Berücksichtigung der Anwendungsgebiete.* Springer, Berlin, 1962.

17

S-Transformation or How to Live Without the Generalized Inverse — Almost

I

Let

$$A\boldsymbol{x} = \boldsymbol{a} \tag{1}$$

be the normal equations of an adjustment problem in supernumerary parameters. Let n be the number of parameters, then A is an $n \times n$ matrix which is non negative definite but singular. Let its zero space $\mathcal{N}$: $\{\, \boldsymbol{y} \in \mathbf{R}^n \mid A\boldsymbol{y} = \mathbf{0} \,\}$ be d-dimensional, then the problem is said to have index d, and $\boldsymbol{x}$ is determined by (1) only modulo $\mathcal{N}$.

In order to make the problem uniquely solvable we add a suitable set of d fictitious observation equations

$$F\boldsymbol{x} = \boldsymbol{b} \tag{2}$$

where F is a $d \times n$ matrix. We suppose that these observation equations are weight normalized, such that the normal equations now become

$$(A + F^{\mathrm{T}}F)\boldsymbol{x} = \boldsymbol{a} + F^{\mathrm{T}}\boldsymbol{b}. \tag{3}$$

Such an addition to the problem is called a *soft postulation*, and the question which shall be treated here is how the inverse of the matrix

$$B = A + F^{\mathrm{T}}F$$

of the normal equations (3) depends on this soft postulation.

Let

$$O = \begin{bmatrix} M & N \end{bmatrix}$$

be an orthogonal $n \times n$ matrix where N is an $n \times d$ submatrix consisting of (d normalized and mutually orthogonal) column vectors in $\mathcal{N}$. Then M is an $n \times (n-d)$ submatrix consisting of $n-d$ normalized and mutually orthogonal column vectors in the subspace $\mathcal{M} \subset \mathbf{R}^n$ which is orthogonal to $\mathcal{N}$.

We find

$$O^{-1}BO = \begin{bmatrix} M^{\mathrm{T}} \\ N^{\mathrm{T}} \end{bmatrix} \begin{bmatrix} A + F^{\mathrm{T}}F \end{bmatrix} \begin{bmatrix} M & N \end{bmatrix}$$

$$= \begin{bmatrix} M^{\mathrm{T}}AM + M^{\mathrm{T}}F^{\mathrm{T}}FM & M^{\mathrm{T}}F^{\mathrm{T}}FN \\ N^{\mathrm{T}}F^{\mathrm{T}}FM & N^{\mathrm{T}}F^{\mathrm{T}}FN \end{bmatrix}$$

$$= \begin{bmatrix} I & (FM)^{\mathrm{T}} \\ \Theta & (FN)^{\mathrm{T}} \end{bmatrix} \begin{bmatrix} M^{\mathrm{T}}AM & \Theta \\ \Theta & I \end{bmatrix} \begin{bmatrix} I & \Theta \\ FM & FN \end{bmatrix}$$

where Θ means a zero matrix of suitable dimensions and I means a unity matrix. We will express the result as

$$O^{-1}BO = \begin{bmatrix} I & (FM)^{\mathrm{T}} \\ \Theta & (FN)^{\mathrm{T}} \end{bmatrix} \begin{bmatrix} M^{\mathrm{T}}AM & \Theta \\ \Theta & \lambda \end{bmatrix} \begin{bmatrix} I & \Theta \\ FM & FN \end{bmatrix}$$

$$\lambda = I. \tag{4}$$

In the following we shall treat λ as any scalar matrix not necessarily equal to I. Now

$$(O^{-1}BO)^{-1} = \begin{bmatrix} I & \Theta \\ FM & FN \end{bmatrix}^{-1} \begin{bmatrix} M^{\mathrm{T}}AM & \Theta \\ \Theta & \lambda \end{bmatrix}^{-1} \begin{bmatrix} I & (FM)^{\mathrm{T}} \\ \Theta & (FN)^{\mathrm{T}} \end{bmatrix}^{-1}. \tag{5}$$

The determinant of $\left[\begin{smallmatrix} I & \Theta \\ FM & FN \end{smallmatrix}\right]$ is equal to the determinant of FN. It is easy to see that (3) has a unique solution if

$$\det(FN) \neq 0. \tag{6}$$

(6) is the exact formulation of the condition that the observation equations (2) correspond to a suitable set of fictitious observations! We shall suppose that (6) is satisfied, and we find:

$$(O^{-1}BO)^{-1} = \begin{bmatrix} I & \Theta \\ -(FN)^{-1}FM & (FN)^{-1} \end{bmatrix} \begin{bmatrix} (M^{\mathrm{T}}AM)^{-1} & \Theta \\ \Theta & \lambda^{-1} \end{bmatrix} \times \begin{bmatrix} I & -(FM)^{\mathrm{T}}(FN)^{-\mathrm{T}} \\ \Theta & (FN)^{-\mathrm{T}} \end{bmatrix}. \tag{7}$$

The product of the three matrices is

$$\begin{bmatrix} (M^{\mathrm{T}}AM)^{-1} & -(M^{\mathrm{T}}AM)^{-1}(FM)^{\mathrm{T}}(FN)^{-\mathrm{T}} \\ -(FN)^{-1}FM(M^{\mathrm{T}}AM)^{-1} & \begin{Bmatrix} (FN)^{-1}FM(M^{\mathrm{T}}AM)^{-1}(FM)^{\mathrm{T}}(FN)^{-\mathrm{T}} \\ + \frac{1}{\lambda}\big((FN)^{\mathrm{T}}FN\big)^{-1} \end{Bmatrix} \end{bmatrix}.$$

We observe here that λ enters only at one place in the last expression and the result for $\lambda \to \infty$ is the same as if in the second matrix in (7), λ^{-1} had been replaced by Θ. Now $\lambda \to \infty$ corresponds to what could be called *hard postulation*, i.e., that the fictitious observations are given infinite weight or, equivalently, that the equations (2) are regarded as condition equations and not as observation equations. Writing Ω instead of B^{-1}, we may rewrite (7) as

$$O^{-1}\Omega O = \begin{bmatrix} I & \Theta \\ -(FN)^{-1}FM & (FN)^{-1} \end{bmatrix} \begin{bmatrix} (M^{\mathrm{T}}AM)^{-1} & \Theta \\ \Theta & \epsilon \end{bmatrix} \times \begin{bmatrix} I & -(FM)^{\mathrm{T}}(FN)^{-\mathrm{T}} \\ \Theta & (FN)^{-\mathrm{T}} \end{bmatrix}; \tag{8}$$

where

$$\epsilon = \begin{cases} I & \text{for soft postulation} \\ \Theta & \text{for hard postulation.} \end{cases}$$

We should find

$$\Omega = O \begin{bmatrix} I & \Theta \\ -(FN)^{-1}FM & (FN)^{-1} \end{bmatrix} O^{-1} O \begin{bmatrix} (M^{\mathrm{T}}AM)^{-1} & \Theta \\ \Theta & \epsilon \end{bmatrix} O^{-1} \times O \begin{bmatrix} I & -(FM)^{\mathrm{T}}(FN)^{-\mathrm{T}} \\ \Theta & (FN)^{-\mathrm{T}} \end{bmatrix} O^{-1}.$$

Now

$$\begin{aligned} O \begin{bmatrix} (M^{\mathrm{T}}AM)^{-1} & \Theta \\ \Theta & \epsilon \end{bmatrix} O^{-1} &= \begin{bmatrix} M & N \end{bmatrix} \begin{bmatrix} (M^{\mathrm{T}}AM)^{-1} & \Theta \\ \Theta & \epsilon \end{bmatrix} \begin{bmatrix} M^{\mathrm{T}} \\ N^{\mathrm{T}} \end{bmatrix} \\ &= \begin{bmatrix} M(M^{\mathrm{T}}AM)^{-1} & \epsilon N \end{bmatrix} \begin{bmatrix} M^{\mathrm{T}} \\ N^{\mathrm{T}} \end{bmatrix} \\ &= M(M^{\mathrm{T}}AM)^{-1}M^{\mathrm{T}} + \epsilon N N^{\mathrm{T}}. \end{aligned}$$

Here $NN^{\mathrm{T}} = \Pi_{\mathcal{N}}$, the orthogonal projector $\mathbf{R}^n \to \mathcal{N}$ and so independent of the choice of the d orthogonal vectors in $\mathcal{N}$. It is easily proved that $M(M^{\mathrm{T}}AM)^{-1}M^{\mathrm{T}}$ is also independent of the actual choice of the 'coordinate system' in $\mathcal{M}$. By the way,

$$M(M^{\mathrm{T}}AM)^{-1}M^{\mathrm{T}} = A^{+},$$

the generalized inverse of A.

Next

$$
\begin{aligned}
S_F &= O \begin{bmatrix} I & \Theta \\ -(FN)^{-1}FM & (FN)^{-1} \end{bmatrix} O^{-1} \\
&= \begin{bmatrix} M & N \end{bmatrix} \begin{bmatrix} I & \Theta \\ -(FN)^{-1}FM & (FN)^{-1} \end{bmatrix} \begin{bmatrix} M^{\mathrm{T}} \\ N^{\mathrm{T}} \end{bmatrix} \\
&= \begin{bmatrix} M - N(FN)^{-1}FM & N(FN)^{-1} \end{bmatrix} \begin{bmatrix} M^{\mathrm{T}} \\ N^{\mathrm{T}} \end{bmatrix} \\
&= MM^{\mathrm{T}} - N(FN)^{-1}FMM^{\mathrm{T}} + N(FN)^{-1}N^{\mathrm{T}}.
\end{aligned}
$$

Here $MM^{\mathrm{T}} = \mathit{\Pi}_{\mathcal{M}}$, the orthogonal projector $\mathbf{R}^n \to \mathcal{M}$, and $\mathit{\Pi}_{\mathcal{M}} = I - \mathit{\Pi}_{\mathcal{N}} = I - NN^{\mathrm{T}}$, so

$$
\begin{aligned}
S_F &= (I - NN^{\mathrm{T}}) - N(FN)^{-1}F(I - NN^{\mathrm{T}}) + N(FN)^{-1}N^{\mathrm{T}} \\
&= I - N(FN)^{-1}(F - N^{\mathrm{T}}),
\end{aligned} \tag{9}
$$

and

$$
\mathit{\Omega} = S_F\big(M(M^{\mathrm{T}}AM)^{-1}M^{\mathrm{T}} + \epsilon NN^{\mathrm{T}}\big)S_F^{\mathrm{T}}. \tag{10}
$$

A possible way to compute $M(M^{\mathrm{T}}AM)^{-1}M^{\mathrm{T}}$ is the following: Compute first $\mathit{\Omega}$ for soft postulation, then

$$
M(M^{\mathrm{T}}AM)^{-1} = S_F^{-1}\mathit{\Omega}S_F^{-\mathrm{T}} - NN^{\mathrm{T}},
$$

but

$$
\begin{aligned}
S_F^{-1} &= \left(O \begin{bmatrix} I & \Theta \\ -(FN)^{-1}FM & (FN)^{-1} \end{bmatrix} O^{-1} \right)^{-1} \\
&= O \begin{bmatrix} I & \Theta \\ FM & FN \end{bmatrix} O^{\mathrm{T}} \\
&= \begin{bmatrix} M & N \end{bmatrix} \begin{bmatrix} I & \Theta \\ FM & FN \end{bmatrix} \begin{bmatrix} M^{\mathrm{T}} \\ N^{\mathrm{T}} \end{bmatrix} \\
&= \begin{bmatrix} M + NFM & NFN \end{bmatrix} \begin{bmatrix} M^{\mathrm{T}} \\ N^{\mathrm{T}} \end{bmatrix}
\end{aligned}
$$

or

$$
\begin{aligned}
S_F^{-1} &= MM^{\mathrm{T}} + NFMM^{\mathrm{T}} + NFNN^{\mathrm{T}} \\
&= (I - NN^{\mathrm{T}}) + NF.
\end{aligned} \tag{11}
$$

More generally suppose that we have found Ω for the soft postulation corresponding to the equations (2), let us call it Ω_F, then we can find Ω_G; it is Ω for the soft postulation corresponding to the equations (2) with G instead of F in the formula

$$\Omega_G = S_G S_F^{-1} \Omega_F S_F^{-\mathrm{T}} S_G^{\mathrm{T}} = (S_G S_F^{-1}) \Omega_F (S_G S_F^{-1})^{\mathrm{T}}. \tag{12}$$

Such transformations $\Omega_F \to \Omega_G$ are called *S-transformations*; they evidently form a group.

I shall give an alternative proof of the result, a proof of a more statistical flavor.

If f is a linear functional of the parameters, then

$$\sigma^2(f) = f \Omega f^{\mathrm{T}}\, \sigma^2;$$

on the other hand if f_1 is a linear functional on $\mathcal{M}$, i.e., if f_1 is estimable, then from the definition of A^+ follows

$$\sigma^2(f_1) = f_1 A^+ f_1^{\mathrm{T}}\, \sigma^2,$$

and if $f_2 = hF$, where h is a d-dimensional row vector, i.e., if f_2 depends only on the postulation, then

$$\sigma^2(f_2) = \epsilon h h^{\mathrm{T}}\, \sigma^2.$$

But every f may be written as $f = f_1 + f_2$, with suitable f_1 and f_2 as above:

$$f = f_1 + hF, \qquad f_1 = f - hF, \qquad \Theta = f_1 N = fN - hFN$$

or

$$h = fN(FN)^{-1}, \qquad f_2 = fN(FN)^{-1}F, \qquad f_1 = f\big(I - N(FN)^{-1}F\big),$$

and as f_1 and f_2 are independent we have

$$\begin{aligned}\sigma^2(f) = \sigma^2(f_1) + \sigma^2(f_2) = f\big(I - N(FN)^{-1}F\big)A^+\big(I - N(FN)^{-1}F\big)^{\mathrm{T}} f^{\mathrm{T}} \sigma^2 \\ + \epsilon f N(FN)^{-1}\big(N(FN)^{-1}\big)^{\mathrm{T}} f^{\mathrm{T}}\, \sigma^2.\end{aligned}$$

This formula is even correct if the row vectors of N are not orthonormal! We have

$$N^{\mathrm{T}} N (N^{\mathrm{T}} N)^{-2} N^{\mathrm{T}} N = I,$$

so we can write

$$\begin{aligned}\sigma^2(f) &= f\big(I - N(FN)^{-1}F\big)A^+\big(I - N(FN)^{-1}F\big)^{\mathrm{T}} f^{\mathrm{T}} \sigma^2 \\ &\qquad + \epsilon f N(FN)^{-1} N^{\mathrm{T}}\big(N(N^{\mathrm{T}}N)^{-2}N^{\mathrm{T}}\big)\big(N(FN)^{-1}N^{\mathrm{T}}\big)^{\mathrm{T}} f^{\mathrm{T}}\, \sigma^2 \\ &= f S_F\big(A^+ + \epsilon N(N^{\mathrm{T}}N)^{-2}N^{\mathrm{T}}\big) S_F^{\mathrm{T}} f^{\mathrm{T}}\, \sigma^2.\end{aligned}$$

This is correct for all f; therefore we must have

$$\Omega = S_F\big(A^+ + \epsilon N(N^{\mathrm{T}}N)^{-2}N^{\mathrm{T}}\big) S_F^{\mathrm{T}}.$$

If $N^{\mathrm{T}} N = I$, i.e., if the row vectors of N are orthonormal this is formula (10), q.e.d.

II

As an example we shall regard a geodetic network in the Euclidean plane. Suppose there are m points and we have observed (sufficiently many) relative distances and/or directions, such that the coordinates can be estimated modulo the group of similarity transformations in $\mathbf{R}^2$. Let (1) be the normal equations in the coordinates and $n = 2m$. Then we will suppose that the coordinates x_1, y_1, x_2, and y_2 of two of the points are given (hard postulation) and that A^+ is found. We want to find S_F.

By (9)

$$S_F = I - N(FN)^{-1}F + N(FN)^{-1}N^{\mathrm{T}},$$

but as we are only interested in hard postulation the last addend is irrelevant, so we put

$$S_F = I - N(FN)^{-1}F. \tag{13}$$

Evidently

$$F = \left[\begin{array}{cccc|c} 1 & 0 & 0 & 0 & \\ 0 & 1 & 0 & 0 & \\ 0 & 0 & 1 & 0 & \Theta \\ 0 & 0 & 0 & 1 & \end{array}\right].$$

It is also evident that in order to find the reduced S_F in (13), it is not necessary to orthonormalize the vectors in N, so we may write

$$N = \begin{bmatrix} 1 & 0 & x_1 & -y_1 \\ 0 & 1 & y_1 & x_1 \\ 1 & 0 & x_2 & -y_2 \\ 0 & 1 & y_2 & x_2 \\ 1 & 0 & x_3 & -y_3 \\ 0 & 1 & y_3 & x_3 \\ & & \cdots & \end{bmatrix}.$$

In order to make the calculations simple, we will choose the coordinate system such that

$$x_1 = -\tfrac{1}{2}, \qquad y_1 = 0;$$
$$x_2 = \tfrac{1}{2}, \qquad y_2 = 0.$$

Then

$$FN = \begin{bmatrix} 1 & 0 & -\frac{1}{2} & 0 \\ 0 & 1 & 0 & -\frac{1}{2} \\ 1 & 0 & \frac{1}{2} & 0 \\ 0 & 1 & 0 & \frac{1}{2} \end{bmatrix},$$

and

$$(FN)^{-1} = \begin{bmatrix} \frac{1}{2} & 0 & \frac{1}{2} & 0 \\ 0 & \frac{1}{2} & 0 & \frac{1}{2} \\ -1 & 0 & 1 & 0 \\ 0 & -1 & 0 & 1 \end{bmatrix},$$

so

$$N(FN)^{-1} = \begin{bmatrix} 1 & 0 & 0 & 0 \\ 0 & 1 & 0 & 0 \\ 0 & 0 & 1 & 0 \\ 0 & 0 & 0 & 1 \\ \frac{1}{2} - x_3 & y_3 & \frac{1}{2} + x_3 & -y_3 \\ -y_3 & \frac{1}{2} - x_3 & y_3 & \frac{1}{2} + x_3 \\ \frac{1}{2} - x_4 & y_4 & \frac{1}{2} + x_4 & -y_4 \\ -y_4 & \frac{1}{2} - x_4 & y_4 & \frac{1}{2} + x_4 \\ & \dots & & \end{bmatrix}.$$

$N(FN)^{-1}F$ is then the square matrix:

$$N(FN)^{-1}F = \left[N(FN)^{-1} \mid \Theta \right].$$

Now

$$\begin{aligned} B = S_F A^+ S_F^{\mathrm{T}} &= A^+ - N(FN)^{-1}FA^+ - A^+F^{\mathrm{T}}(FN)^{-\mathrm{T}}N^{\mathrm{T}} \\ &\quad + N(FN)^{-1}(FA^+F^{\mathrm{T}})(FN)^{-\mathrm{T}}N^{\mathrm{T}} \\ &= A^+ - C - C^{\mathrm{T}} + D. \end{aligned}$$

Let us regard first the last addend

$$D = N(FN)^{-1}(FA^+F^{\mathrm{T}})(FN)^{-\mathrm{T}}N^{\mathrm{T}}.$$

Here FA^+F^{T} is a 4×4 submatrix of A^+. If the smallest eigenvalue of this submatrix is λ_1 and the largest is λ_4, then

$$0 \prec \lambda_1 I \prec FA^+F^{\mathrm{T}} \prec \lambda_4 I,$$

where for two symmetric matrices M_1 and M_2, $M_1 \prec M_2$ means that $M_2 - M_1$ is positive definite, and

$$\lambda_1 N(FN)^{-1}(FN)^{-\mathrm{T}}N^{\mathrm{T}} \prec D \prec \lambda_4 N(FN)^{-1}(FN)^{-\mathrm{T}}N^{\mathrm{T}}.$$

But

$$N(FN)^{-1}(FN)^{-\mathrm{T}}N^{\mathrm{T}} = \begin{bmatrix} A_{11} & A_{12} & \dots \\ A_{21} & A_{22} & \dots \\ & \dots & \end{bmatrix},$$

where the 2×2 submatrices A_{ij} are

$$A_{ij} = \begin{bmatrix} 2(x_i x_j + y_i y_j) + \frac{1}{2} & -2(x_i y_j - x_j y_i) \\ -2(x_i y_j - x_j y_i) & 2(x_i x_j + y_i y_j) + \frac{1}{2} \end{bmatrix},$$

and we see that the 'error-ellipses' of D have axes a_i and b_i which satisfy

$$\lambda_1\left(\tfrac{1}{2} + 2(x_i^2 + y_i^2)\right) < \begin{Bmatrix} a_i^2 \\ b_i^2 \end{Bmatrix} < \lambda_4\left(\tfrac{1}{2} + 2(x_i^2 + y_i^2)\right).$$

Remembering that the unit of coordinates is the distance between the two first points of the network we see, especially if these points are neighboring points near the boundary of the network, that these error ellipses can be very large compared with those of A^+.

The matrices C and C^{T} are of less importance, because they only grow linearly and they have changing signs.

The case where the coordinates are determined modulo translations is extremely simple. Here

$$F = \left[\begin{matrix} 1 & 0 \\ 0 & 1 \end{matrix} \,\middle|\, \Theta \right],$$

the coordinates of the first point are postulated. We find

$$B_{ij} = A_{ij}^+ - A_{1j}^+ - A_{i1}^+ + A_{11}^+,$$

where A_{ij}^+, (B_{ij}) are the 2×2 submatrices of A^+, (B) corresponding to the point with numbers i and j. The reader should try to find out why the diagonal elements of B has a tendency to grow with the distance of the corresponding point from point No. 1.

18

Integrated Geodesy

Summary

The paper ranges over the principles of Integrated Geodesy, characterizing its peculiar approach to geodetic problems like the adjustment and the combination of different measurements of geometric and gravimetric nature. Particular care is paid to the formulation of observation equations (distances, angles, etc.) including the description of the "local frame" by a suitable matrix formalism.

Sommario

Il lavoro esamina i principi della Geodesia Integrata, caratterizzandone il tipico approccio a problemi di combinazione di dati provenienti da misure di diversa natura (geometriche, gravimetriche, ecc.). Con particolare attenzione si illustra il metodo per la formazione di equazioni alle osservazioni (distanze, angoli, ecc.) includendo la trattazione del sistema locale di riferimento per mezzo di un opportuno formalismo matriciale.

The Leading Principles of Integrated Geodesy

Least-Squares Adjustment

The main framework for computation and discussion of classical geodetic networks is least-squares adjustment, which has proved to be very useful in combining different observations, so it is curious that geodesists in combining geometric and physical observations almost invariably use a more primitive technique. Collocation has made such an approach possible.

In such a framework the whole geodesy is formally reduced to three problems:

1. Forming of the observation equations.
2. Determination of the weights.
3. The adjustment procedure itself.

The Dynamic Aspect of Geodesy

Until recently great surveying projects had an overtone of eschatology: the computed coordinates were looked upon as being eternal and unalterable—a sort of domesday book. Today this has changed towards a more dynamic aspect, now a surveying institution regards as its treasure its continually increasing stock of observations rather than its file of computed coordinates, at any time new requirements can be covered from ad hoc adjustment from the relevant observations. We therefore want that integrated geodesy shall be flexible enough to handle both local and global coordinate computations.

Whether coordinates are measurable or not does not interest us: we have already coordinates, from new observations we shall be able to compute better coordinates, to glue two or more networks coordinated in different reference frames together into a common system of coordinates, and to take advantage of those observations concerning the gravity field, which make possible to find a reference frame with respect to the mass center of the Earth. Although it is contrary to all geodetic tradition I have proposed to use Cartesian coordinates in integrated geodesy because the passage from one frame of coordinates to another is most simple for Cartesian coordinates.

Consistent Mathematical Model

The observations we want to handle are those which are relevant to the determination of coordinates of points on (and outside) the surface of the Earth and of the gravity field of the Earth also on and outside the surface. The set of formulas which from a specification of such an observation gives the theoretical value of the result of the observation from the coordinates of the points involved in the observation and the mathematical expression for the gravity field outside the Earth.

A mathematical model will never be correct, but we want to use a model which is consistent, i.e. which would give unique and correct results under certain idealizing conditions, e.g. if the "real" geometry were Euclidean, if the instruments were correct, and if the refraction were zero.

We have to take into account that we do not live under these ideal conditions by applying *corrections* on the observations, that is we need also some—let us call them—peripheral mathematical models for the corrections in contrast to the proper mathematical model defined above.

Integrated geodesy differs from classical geometric geodetic methods mainly in the fact that the proper model includes the effects of the gravity field. One could say that what characterizes a geodetic method is where it places the dividing line between the proper mathematical model and the peripheral models.

I believe I have good reasons for placing refraction outside this dividing line and, at least for the time being, the effects of polar motion and plate tectonics, but I hope that in a not too far future integrated geodesy will be a sort of touchstone for theories in time dependent geodesy: can these new theories give us a better proper mathematical model or not.

Our proper mathematical model can shortly be described as follows:

The geometry is Euclidean, light propagates as in Newtonian vacuum, the Earth rotates with constant angular velocity around a fixed axis, in a coordinate system rigidly fixed to the thus rotating Earth, points at the surface of the Earth have unchangeable coordinates, the gravity field in this coordinate system is the sum of a centrifugal force part and a harmonic part satisfying certain well-known conditions at infinity, and both of these two parts are time independent.

The Reference

The starting point in integrated geodesy is the *reference*. Its analog in classical geodesy is the reference surface. Integrated geodesy has no reference surface but a *reference potential*, which is simply a mathematically given approximation to the gravity field. For historical reason the reference field will be called U. It plays almost the same role as the provisional coordinates which we therefore also call the *reference coordinates.*

We shall now see how an observation equation normally is found.

From the specification of the observation the mathematical model typically gives

$$\mathsf{o} = f\big(x, y, W(x), (\operatorname{grad} W)(x)\big) \tag{1}$$

where x and y are the coordinates of two points involved in the observation and W is the gravity potential. o is the value of the observation. If we in (1) for x, y, and W put the corresponding reference values x_0, y_0, and U we find

$$\mathsf{o}_0 = f\big(x_0, y_0, U(x_0), (\operatorname{grad} U)(x_0)\big). \tag{2}$$

o_0 is called the *reference value* of the observation.

By differentiation of (1) we find:

$$d\mathsf{o} = \sum_1^3 \frac{\partial f}{\partial x_i}\,dx_i + \sum_1^3 \frac{\partial f}{\partial y_i}\,dy_i + \frac{\partial f}{\partial W}\Big(\sum_1^3 \frac{\partial W}{\partial x_i}\,dx_i + T\Big) + \sum_{j=1}^3 \frac{\partial f}{\big(\frac{\partial W}{\partial x_j}\big)}\Big(\sum_1^3 \frac{\partial^2 W}{\partial x_i \partial x_j}\,dx_j + \frac{\partial T}{\partial x_j}\Big) \tag{3}$$

where dx_i and dy_i are the increments to the coordinates x_i and y_i, $i = 1, 2, 3$, and T is the increment to W.

If x and y are "near to" x_0 and y_0, and W is a "near to" U then we have the linearized observation equation (here we use the summation convention):

$$\Big(\frac{\partial f}{\partial x_i}\Big)_0 dx_i + \Big(\frac{\partial f}{\partial y_i}\Big)_0 dy_i + \Big(\frac{\partial f}{\partial W}\Big)_0 \bigg(\Big(\frac{\partial U}{\partial x_i}\Big)_0 dx_i + T(x_0)\bigg) \\ + \left(\frac{\partial f}{\frac{\partial W}{\partial x_j}}\right)_0 \bigg(M_{ij}\, dx_j + \Big(\frac{\partial T}{\partial x_j}\Big)_0\bigg) = \mathsf{o} - \mathsf{o}_0 + v_{\mathsf{o}}. \quad (4)$$

Here the index 0 means that the function shall be evaluated at $x = x_0$, $y = y_0$, and $W = U$. M is the matrix $\big[\big(\frac{\partial^2 U}{\partial x_i \partial x_j}\big)_0\big]$ and v_{o} is the correction to the observed value o. The meaning of (4) is that except from terms of the second (and higher) order in dx, dy, U, and $|\operatorname{grad} U|$, (4) is the condition for the compatibility of

$$\begin{gathered} \text{the coordinates } x_0 + dx \text{ and } y_0 + dy, \\ \text{the potential } U + T, \\ \text{the observation result } \mathsf{o} + v_{\mathsf{o}} \end{gathered} \quad (5)$$

with the mathematical model.

Sometimes it seems to be a help in the search for the observation equation to use a special coordinate system (or frame) $\mathcal{F}$ determined by the geodetic instrument and also a *reference frame* $\mathcal{F}_0$ which moreover depends on the reference. Naturally these frames must be "eliminated" so that no reference to them enters the final observation equation.

A set as (5) for each observation in question satisfying all the observation equations is called a *feasible (linear) solution.* It is clearly far from uniquely determined. A unique solution will normally be determined by the condition that it among all feasible solutions minimizes a given quadratic form—not only in the increments v_{o} to the observation, as in classical least-squares adjustment—but in v_{o} and T (least-squares collocation). The form for this quadratic form is an interesting problem which will not be discussed here.

It is clear that the use of the observation equations in the linearized form (4) introduces an error. With the optimism characteristic for an applied mathematician it is reasonable to expect that if the reference values are "good enough" then by iteration of the collocation procedure, i.e. by writing new observation equations, but every time with reference coordinates and reference potential equal to the already found coordinates and potential, i.e. writing

$$x_0 \qquad \text{for} \qquad x_0 + dx$$

and

$$U \qquad \text{for} \qquad U + T$$

etc. that by such an iteration it should be possible to find a set of values consistent with the mathematical model and satisfying the minimum condition. By a suitable definition of "good enough" this is even a tautology.

In this way it should be possible to come arbitrarily near to the ideal solution, i.e. that defined by the observations, the model, and the minimum principle. It is not so in classical geodetic calculation schemes, where already the use of a reference ellipsoid gives a bound for the precision of the approximation (spherical precision, ellipsoidal precision etc.).

Historical Notes

The name "integrated geodesy" has arisen in analogy to "integrated data processing" as the idea itself has arisen as a natural continuation of attempts at the integration of the geodetic data processing in the Danish Geodetic Institute at the same time as the idea of collocation. In fact collocation and integrated geodesy are two aspects of the same thing a natural generalization of least-squares methods in geodesy. It was in 1967.

In 1970 I gave it as a problem for a student to combine these ideas with the ideas in [1]. Dufour's ideas had interested me very much because he applied the classical French idea of "repère mobile" to geodesy. I suggested to the student that it could be done more elegantly using the well-known matrix expression of the Euclidean Group, [3], page 47. During the last few years the moving frame has gained a faithful community of geodesists through differential geometry.

Already the last year integrated geodesy was presented in a lecture at the Technical University in Berlin, but I deemed it too early to publish anything about it. In connection with the first summer school in Ramsau 1973 [2] was published, perhaps still too early, for among the many errors in that paper unfortunately some of them are not printing errors.

My own attitude to integrated geodesy has always been ambiguous. But as it seems that its ideas still have influence upon the evolution of the data processing philosophy at our institute I think that it may have a meaning to try to make colleagues outside our institute interested. And in these years where some geodesists of the older generation can have the impression that modern theoretical geodesy is emancipating itself from geodetic reality, this theory, which a few years ago I deemed perhaps too unrealistic, can even be regarded as a complementary movement.

The Observation Equations

Points, Vectors, and Frames in Euclidean Geometry

We will identify Euclidean n-space with $\mathbf{R}^n$, i.e. we suppose there has been selected a reference frame. We will denote the *point* with coordinates $x = \{x_i\}$, $i = 1, 2, \ldots, n$, by the $n+1$ dimensional column vector

$$\overline{x} = \begin{bmatrix} 1 \\ x \end{bmatrix} = \begin{bmatrix} 1 \\ x_1 \\ x_2 \\ \vdots \\ x_n \end{bmatrix} .$$

If $\overline{x}$ and $\overline{y}$ are points their difference

$$\overline{x} - \overline{y} = \begin{bmatrix} 0 \\ x_1 - y_1 \\ x_2 - y_2 \\ \vdots \\ x_n - y_n \end{bmatrix} = \begin{bmatrix} 0 \\ x - y \end{bmatrix}$$

is called a *vector*, so a vector describes the direction and the length of a line segment but not its position in the space. Vectors can be multiplied by scalars and added as usual to give new vectors and for two vectors x and y their scalar product $x^{\mathrm{T}}y = y^{\mathrm{T}}x$ is defined as usual. Moreover vectors may be added to points to give new points:

$$\overline{x} + \overline{y} = \begin{bmatrix} 1 \\ x + y \end{bmatrix},$$

but points can not be added, multiplied by scalars, etc.

A *coordinate frame* is determined by 1) a point $\overline{f}$ the zero point of the frame, and 2) n orthonormal vectors in the direction of the coordinate axes. We will combine the vectors in a matrix F, which will be an orthogonal matrix: $F^{-1} = F^{\mathrm{T}}$ and, if the coordinate frame is right-handed, which we shall say suppose here, then the determinant $\det F = 1$. We shall normally combine all this information on the frame into a single $(n+1) \times (n+1)$ matrix $\mathcal{F}$:

$$\mathcal{F} = \begin{bmatrix} 1 & 0 \\ f & F \end{bmatrix} = \begin{bmatrix} 1 & 0 & 0 & \dots & 0 \\ f_1 & f_{11} & f_{12} & \dots & f_{1n} \\ f_2 & f_{21} & f_{22} & \dots & f_{2n} \\ \vdots & \vdots & \vdots & & \vdots \\ f_n & f_{n1} & f_{n2} & \dots & f_{nn} \end{bmatrix} \tag{6}$$

and $\mathcal{F}$ shall mean the frame itself as well as this matrix. If $\mathcal{E}$ is the natural frame in $\mathbf{R}^n$ then $\mathcal{E} = 1_{n+1}$, the unit $(n+1) \times (n+1)$ matrix.

If a point $\overline{x}$ has the coordinate $x_{\mathcal{F}}$ in the frame $\mathcal{F}$ then we write

$$x = x_{\mathcal{F}}$$

and we find

$$\overline{x} = \mathcal{F}\overline{x}_{\mathcal{F}} = \begin{bmatrix} 1 & 0 \\ f & F \end{bmatrix} \begin{bmatrix} 1 \\ x_{\mathcal{F}} \end{bmatrix} = \begin{bmatrix} 1 \\ f + Fx_{\mathcal{F}} \end{bmatrix}.$$

In the same way for the vector x (the frame $\mathcal{G}$) with coordinates $x_{\mathcal{F}}$ ($\mathcal{G}_{\mathcal{F}}$) in the frame $\mathcal{F}$ we have

$$x = \mathcal{F}x_{\mathcal{F}} \qquad (\mathcal{G} = \mathcal{F}\mathcal{G}_{\mathcal{F}}). \tag{7}$$

The determinant of the matrix (6) is

$$\det \mathcal{F} = 1 \cdot \det F = 1,$$

therefore $\mathcal{F}^{-1}$ exists and is called the frame reciprocal to $\mathcal{F}$:

$$\mathcal{F}^{-1} = \begin{bmatrix} 1 & 0 \\ -F^{-1}f & F^{-1} \end{bmatrix} = \begin{bmatrix} 1 & 0 \\ -F^{\mathrm{T}}f & F^{\mathrm{T}} \end{bmatrix}. \tag{8}$$

Therefore from (7) follows

$$\mathcal{G}_{\mathcal{F}} = \mathcal{F}^{-1}\mathcal{G},$$

and we have in general:

If a point (a vector or a frame) α has the expressions

$$\alpha_{\mathcal{F}_i} \quad \text{in the frames } \mathcal{F}_i, \;\; i = 1,\, 2$$

then

$$\alpha = \mathcal{F}_i \alpha_{\mathcal{F}_i}$$

and

$$\alpha_{\mathcal{F}_2} = \mathcal{F}_2^{-1}\mathcal{F}_1\alpha_{\mathcal{F}_1}.$$

We shall use "infinitely small" orthogonal matrices in three dimensions: Let F be an orthogonal matrix; then we have

$$F^{\mathrm{T}}F = I,$$

$$F^{\mathrm{T}}\,dF + (dF)^{\mathrm{T}}F = 0,$$

or, if we write

$$S = F^{\mathrm{T}}dF = F^{-1}dF$$

$$S + S^{\mathrm{T}} = 0,$$

i.e. S is skew-symmetric and

$$dF = FS.$$

If we write S as

$$S = \begin{bmatrix} 0 & -S_3 & S_2 \\ S_3 & 0 & -S_1 \\ -S_2 & S_1 & 0 \end{bmatrix}$$

then S_i, $i = 1, 2, 3$, are the small Eulerian angles corresponding to rotations around the three coordinate axes.

The Observation Equations Illustrated by Examples

Distance Measurements

Suppose we have observed the logarithm of the distance between the two points $\overline{x}$ and $\overline{y}$ with reference coordinates $\overline{x}_0$ and $\overline{y}_0$. Let the observed value be l and let l_0 be the reference value:

$$l_0 = \ln|x_0 - y_0| = \tfrac{1}{2}\ln\sum_{i=1}^{3}(x_{0i} - y_{0i})^2.$$

We should have

$$l = \ln|x - y|.$$

Differentiation gives

$$\frac{\partial l}{\partial x_i} = \frac{x_i - y_i}{|x - y|}, \qquad \frac{\partial l}{\partial y_i} = \frac{y_i - x_i}{|x - y|}$$

and the equation for the increments Δx_i and Δy_i to the reference coordinates becomes:

$$\sum_{i=1}^{3} \frac{x_i - y_i}{|x - y|}(\Delta x_i - \Delta y_i) = l - l_0 + v_l$$

or

$$\frac{(x_0 - y_0)^{\mathrm{T}}(\Delta x - \Delta y)}{l_0} = l - l_0 + v_l.$$

Gravity Measurements

Let the gravity be measured in the point $\overline{x}$ with reference coordinates $\overline{x}_0$, the observed result is g. The reference value is

$$\gamma = |\operatorname{grad} U|_{\overline{x}_0}.$$

We have

$$dg = d|\operatorname{grad} W| = \frac{(\operatorname{grad} W)^{\mathrm{T}} d \operatorname{grad} W}{|\operatorname{grad} W|} \approx \frac{1}{\gamma}(\operatorname{grad} U)^{\mathrm{T}}(M\,dx + \operatorname{grad} T)$$

and the observation equation becomes:

$$\frac{1}{\gamma}(\operatorname{grad} U)^{\mathrm{T}}(M\,dx + \operatorname{grad} T) = g - \gamma + v_g.$$

Horizon Measurements By Theodolite

A theodolite determines a frame in the following way: The zero point is the center of the instrument, the first coordinate axis is the vertical axis orientated downwards, the second axis goes through the zero of the horizontal scale and the third axis through 90°. This is a right hand frame and will be called $\mathcal{F}$. We write

$$\mathcal{F} = \begin{bmatrix} 1 & 0 & 0 & 0 \\ x & \frac{1}{g}\operatorname{grad} W & \alpha & \beta \end{bmatrix}, \qquad F = \begin{bmatrix} \frac{1}{g}\operatorname{grad} W & \alpha & \beta \end{bmatrix}.$$

As we do not know x, W, and α, we must use a reference frame $\mathcal{F}_0$, namely

$$\mathcal{F}_0 = \begin{bmatrix} 1 & 0 & 0 & 0 \\ x_0 & \frac{1}{\gamma}\operatorname{grad} U & \alpha_0 & \beta_0 \end{bmatrix}, \qquad F_0 = \begin{bmatrix} \frac{1}{\gamma}\operatorname{grad} U & \alpha_0 & \beta_0 \end{bmatrix}$$

which can be calculated from reference coordinates, the reference potential, and one observation in $\overline{x}$.

In order to give relevant information the theodolite measurements at $\overline{x}$ shall consist of the values of the (horizontal) angles giving the directions to at least two points. To each angle measurement there will correspond an observation equation, and we shall now find that corresponding to a point $\overline{y} \neq \overline{x}$. $\overline{y}$ has the reference coordinates $\overline{y}_0$.

Let us find the coordinates of $\overline{y}$ in the frame $\mathcal{F}$ and the coordinates of $\overline{y}_0$ in the frame $\mathcal{F}_0$:

$$\overline{y}_{\mathcal{F}} = \mathcal{F}^{-1}\overline{y} = \begin{bmatrix} 1 & 0 \\ -F^{-1}x & F^{-1} \end{bmatrix} \begin{bmatrix} 1 \\ y \end{bmatrix} = \begin{bmatrix} 1 \\ F^{\mathrm{T}}(y-x) \end{bmatrix}$$

and

$$\overline{y}_{0\mathcal{F}_0} = \mathcal{F}_0^{-1}\overline{y}_0 = \begin{bmatrix} 1 & 0 \\ -F_0^{-1}x_0 & F_0^{-1} \end{bmatrix} \begin{bmatrix} 1 \\ y_0 \end{bmatrix} = \begin{bmatrix} 1 \\ F_0^{\mathrm{T}}(y_0-x_0) \end{bmatrix}.$$

We write

$$dy_{\mathcal{F}} = y_{\mathcal{F}} - y_{0\mathcal{F}_0} \qquad \text{etc.}$$

and ignore consequently terms of second and higher orders. Then

$$dy_{\mathcal{F}} = F_0^{\mathrm{T}}(dy-dx) + (dF^{\mathrm{T}})(y-x). \tag{9}$$

Now

$$dF^{\mathrm{T}} = (dF)^{\mathrm{T}}$$

and

$$dF = F_0 \begin{bmatrix} 0 & -S_3 & S_2 \\ S_3 & 0 & -S_1 \\ -S_2 & S_1 & 0 \end{bmatrix} = \begin{bmatrix} S_3\alpha_0 - S_2\beta_0 & -S_3\frac{\operatorname{grad} U}{\gamma} + S_1\beta_0 & S_2\frac{\operatorname{grad} U}{\gamma} - S_1\alpha_0 \end{bmatrix}. \tag{10}$$

As we are interested in angles we write with evident notation

$$\begin{aligned} y_{\mathcal{F}_1} &= r\sin h \\ y_{\mathcal{F}_2} &= r\cos h \cos\phi \\ y_{\mathcal{F}_3} &= r\cos h \sin\phi \end{aligned}$$

and find

$$\begin{aligned} dy_{\mathcal{F}_2} &= \cos\phi\, d(r\cos h) - r\cos h \sin\phi\, d\phi \\ dy_{\mathcal{F}_3} &= \sin\phi\, d(r\cos h) + r\cos h\cos\phi\, d\phi \end{aligned}$$

such that

$$r\cos h\, d\phi = \cos\phi\, dy_{\mathcal{F}_3} - \sin\phi\, dy_{\mathcal{F}_2}. \tag{11}$$

Equations (9), (10), and (11) give

$$\begin{aligned} r\cos h\, d\phi = {} & \big(\cos\phi\cdot\beta_0^{\mathrm{T}} - \sin\phi\cdot\alpha_0^{\mathrm{T}}\big)(dy-dx) - S_1\big(\cos\phi\cdot\alpha_0^{\mathrm{T}} + \sin\phi\cdot\beta_0^{\mathrm{T}}\big)(y-x) \\ & + \big(S_2\cos\phi + S_3\sin\phi\big)\frac{1}{\gamma}(\operatorname{grad} U)^{\mathrm{T}}(y-x). \end{aligned} \tag{12}$$

First consider the second term!

Here S_1 is the Eulerian angle corresponding to a rotation around the vertical axis, i.e. $S_1 = d\phi'$, the correction to the zero direction in the horizon

$$\cos\phi \cdot \alpha_0 + \sin\phi \cdot \beta_0 \tag{13}$$

is the unit vector in the direction of the orthogonal projection of $y - x$ into the horizontal plane, and so the second term is simply

$$-r\cos h\, d\phi'. \tag{14}$$

In the first term

$$\cos\phi \cdot \beta_0 - \sin\phi \cdot \alpha_0$$

is a horizontal unit vector orthogonal to (13), therefore it is proportional to the vector product

$$\frac{\operatorname{grad} U}{\gamma} \times (y - x),$$

the proportional factor being evidently

$$\frac{1}{r\cos h},$$

and so the first term must be:

$$\frac{1}{r\cos h}\left|\frac{\operatorname{grad} U}{\gamma} \quad y - x \quad dy - dx\right| \tag{15}$$

where $|\cdot|$ is the determinant.

In order to find S_2 and S_3 we use (10), or rather the first column of (10):

$$d\frac{\operatorname{grad} W}{g} = S_3\alpha_0 - S_2\beta_0$$

which gives

$$S_2 = -\beta_0^{\mathrm{T}} d\frac{\operatorname{grad} W}{g}$$

$$S_3 = \alpha_0^{\mathrm{T}} d\frac{\operatorname{grad} W}{g}.$$

Using the same reasoning as above, before (15), we find

$$S_2\cos\phi + S_3\cos\phi = -\frac{1}{r\cos h}\left|\frac{\operatorname{grad} U}{\gamma} \quad y - x \quad d\frac{\operatorname{grad} W}{g}\right|.$$

But

$$d\frac{\operatorname{grad} W}{g} = \frac{d\operatorname{grad} W}{\gamma} - \frac{\operatorname{grad} U}{\gamma^2}\, dg,$$

but the second term is proportional to $(\operatorname{grad} U)/\gamma$, so that

$$S_2 \cos\phi + S_3 \sin\phi = -\frac{1}{r\cos h}\left|\frac{\operatorname{grad} U}{\gamma} \quad y-x \quad d\frac{\operatorname{grad} W}{\gamma}\right|.$$

Here

$$\frac{d \operatorname{grad} W}{\gamma} = \frac{1}{\gamma}\big(\operatorname{grad} T + M\, dx\big),$$

and the third term in (12) becomes:

$$-\frac{\tan h}{\gamma^2}\left|\operatorname{grad} U \quad y-x \quad \operatorname{grad} T + M\, dx\right|. \tag{16}$$

From (12), (14), (15), and (16) follows:

$$d\phi = \frac{1}{r^2\cos^2 h}\left|\frac{\operatorname{grad} U}{\gamma} \quad y-x \quad dy-dx\right| \\ -\frac{\tan h}{\gamma^2 r\cos h}\left|\operatorname{grad} U \quad y-x \quad \operatorname{grad} T + M\, dx\right| - d\phi',$$

and if we write l for $r\cos h$ and h for $-h$, then h gets its "natural" orientation, and the observation equation becomes:

$$\frac{1}{\gamma l^2}\left|\operatorname{grad} U \quad y-x \quad dy-dx\right| + \frac{\tan h}{\gamma^2 l}\left|\operatorname{grad} U \quad y-x \quad \operatorname{grad} T + M\, dx\right| - d\phi' \\ = \phi - \phi_0 + v_\phi.$$

References

[1] H. M. Dufour. Toute la géodésie sans ellipsoïde. *Bulletin Géodésique*, 88:125–143, 1968.

[2] J. Eeg and T. Krarup. *Integrated Geodesy*, volume 13 of *Methoden und Verfahren der Mathematischen Physik*. Bibliographisches Institut Mannheim, 1975.

[3] Hermann Weyl. *The Classical Groups*. Princeton University Press, 1946.

19

A Measure for Local Redundancy —A Contribution to the Reliability Theory for Geodetic Networks

1. Motivation

It is a great merit of Baarda's to have drawn attention to the fact that besides the precision of a geodetic network, also its *reliability* is of great importance in practice. Here I regard reliability as an expression of the possibility of localizing gross errors in the observations.

Now it is clear that the possibility of localizing gross errors depends on the number of over-determinations in the network, the *redundancy* $r = N - m$, where N is the number of observations and m is the number of parameters in the model. Or rather it depends on the *relative redundancy* $\rho = r/N$. But that ρ does not always give a good picture of the reliability of a network is seen in two simple examples.

First, let the network $\mathcal{N}$ consist of two sub-networks $\mathcal{N}_1$ and $\mathcal{N}_2$ with no measurements between them. If the redundancy r_1 of $\mathcal{N}_1$ is very great and r_2 of $\mathcal{N}_2$ very small, then the redundancy $r_1 + r_2 = r$ of $\mathcal{N}$ gives a very poor picture of the reliability of $\mathcal{N}_1$.

That such unpleasant behavior is not restricted to unconnected networks is shown in the second example. Here a point P is only tied to the rest of the network by two distance measurements (from two different points, naturally). But no matter how great the redundancy of the network is, a gross error in one or both of these distance measurements cannot be localized.

In order to meet such difficulties, we shall define here the local redundancy, i.e., the redundancy ρ of every measurement in a network, such that

1. $0 \leq \rho < 1$;
2. The sum of the redundancies of all the measurements in the network is equal to the redundancy of the network as defined above.

It is clear that for any subset of measurements consisting of n measurements in the network, we can define the redundancy r as the sum of the redundancies of these n measurements and the relative redundancy $\rho = r/n$ and that for the single measurements the relative redundancy equals the redundancy.

2. The Model

Let the model space $\mathcal{M}$ be an m-dimensional vector space, and let $\{a_i\}$, $i = 1$, $2, \ldots, N$, be N linear functionals on $\mathcal{M}$ called the *measurements*. We suppose that there is at least one subset with n linearly independent measurements. We have also given N numbers, the *observations* $\{\mathsf{o}_i\}$, $i = 1, 2, \ldots, N$, the observed values of the measurement functionals, and the corresponding weights p_i, $i = 1, 2, \ldots, N$, which we put together so as to form a diagonal matrix P.

By the so-called weight normalization, i.e., the multiplication of the observation equations $a_i(x) = \mathsf{o}_i$, $x \in \mathcal{M}$, by $p_i^{1/2}$ we could suppose $P = 1$, and we shall do so eventually, but until then we must retain the $\{p_i\}$ in our formulas.

We organize the possible values of the observations into an N-dimensional vector space $\mathcal{O}$ in which the ith component is o_i, and so the actual observations determine the point $\mathsf{o} = (\mathsf{o}_1, \mathsf{o}_2, \ldots, \mathsf{o}_N)^{\mathrm{T}}$ and we define a metric in $\mathcal{O}$ such that the distance between point o^1 and o^2 is $\big((\mathsf{o}^1 - \mathsf{o}^2)^{\mathrm{T}} P (\mathsf{o}^1 - \mathsf{o}^2)\big)^{1/2}$.

The model space may be embedded in the observation space by defining the embedded $\mathcal{M}$ as the union of the points $\mathsf{o} \in \mathcal{O}$ such that

$$\mathsf{o}_i = a_i(x) \qquad \text{for } x \in \mathcal{M}.$$

For a given observation point $\mathsf{o} \in \mathcal{O}$, the least-squares solution $\hat{\mathsf{o}}$ may be defined as the orthogonal projection (with respect to the metric defined by P) onto the embedded $\mathcal{M}$ and for any reference system X on $\mathcal{M}$ the coordinates of $\hat{\mathsf{o}}$ are given by

$$A^{\mathrm{T}} P A x = A^{\mathrm{T}} P \mathsf{o},$$

where the $N \times m$ matrix A is the matrix with rows $\{a_i\}$, $i = 1, 2, \ldots, N$, expressed in the reference system X; that is

$$a_i(x) = a_{ij} x^i.$$

We shall denote the matrix of the normal equations by $N(P)$, where

$$N(P) = A^{\mathrm{T}} P A,$$

and we shall write

$$N = N(1).$$

We shall also write

$$\Delta(P) = \det N(P)$$

and

$$\Delta = \Delta(1) = \det N.$$

Evidently $\Delta(P)$ is a homogeneous polynomial in the $\{p_i\}$ of degree m, which depends on the choice of the reference system in $\mathcal{M}$ (the choice of the parameters). If we use another reference system Y such that

$$x = Sy,$$

where S is a regular $m \times m$ matrix, then

$$N(P)_Y = S^{\mathrm{T}} N(P)_X S$$

and

$$\Delta(P)_Y = (\det S)^2 \Delta(P)_X.$$

Two homogeneous polynomials f_1 and f_2 are called equivalent if and only if $f_1 = k f_2$ where k is a number different from zero, and we can say that the equivalence class of $\Delta(P)$ is independent of the reference system in $\mathcal{M}$.

Remark 19.1 If $I \colon \mathcal{M} \to \mathcal{O}$ is the embedding of $\mathcal{M}$ in $\mathcal{O}$ then $N(P)$ is the pull-back of the metric tensor P of $\mathcal{O}$:

$$N(P) = A^* P,$$

i.e., $N(P)$ is the metric tensor on $\mathcal{M}$ induced by the embedding in $\mathcal{O}$ or the metric tensor on $\mathcal{O}$ restricted to $\mathcal{M}$. It is clear that $N(P)$ is positive definite as P is; geometrically a Riemannian metric induces a Riemannian metric (rather than a pseudo-Riemannian) metric on a submanifold. It is curious and of some importance later on that $N(P)$ may also be positive definite in some cases where P is not, i.e., even if some of the weights are negative. The analog to this situation is the existence of space-like submanifolds of Einsteinian space-time.

$N \times N$ diagonal matrices constitute an N-dimensional vector space $\mathcal{P}$ with addition and multiplication with scalar defined in the evident way. For some such matrices P, $N(P)$ is positive definite, for others not. The first alternative is valid if all the diagonal elements in P are positive. We shall be interested in the subset $\mathcal{P}^+ \subset \mathcal{P}$ for which

$$P \in \mathcal{P}^+ \quad \Rightarrow \quad N(P) \text{ positive definite}$$

and which is connected and contains all $P \in \mathcal{P}$ with positive diagonal elements. At the boundary of $\mathcal{P}^+$, $\Delta(P)$ evidently vanishes. We shall determine later on that $\mathcal{P}^+$ is convex.

3.

Definition 19.1 *The redundancy ρ_i of the ith measurement a_i is defined by*

$$\rho_i = 1 - \left(\frac{1}{\Delta(P)} \frac{\partial \Delta(P)}{\partial p_i} \right)_{P=1}$$

First we observe that the multiplication of $\Delta(P)$ with a constant $c \neq 0$ does not change ρ_i.

Using Euler's theorem for homogeneous polynomials

$$\sum x_i \frac{\partial f}{\partial x_i} = nf,$$

for f a homogeneous polynomial of degree n in $\{x_i\}$ we find

$$\sum_{i=1}^{N} \rho_i = N - \sum_{i=1}^{N} \Bigl(\frac{1}{\Delta(P)} \frac{\partial \Delta(P)}{\partial p_i} \Bigr)_{P=1} = N - \frac{1}{\Delta} \sum_{i=1}^{N} \Bigl(\frac{\partial \Delta(P)}{\partial p_i} \Bigr)_{P=1}$$

$$= N - \frac{1}{\Delta} \Bigl(\sum_{i=1}^{N} p_i \frac{\partial \Delta(P)}{\partial p_i} \Bigr)_{P=1} = N - \frac{1}{\Delta} m \bigl(\Delta(P) \bigr)_{P=1} = N - m = r$$

and we have proved that the sum of the redundancies of all the measurements is equal to the redundancy of the network itself.

In order to find another expression for ρ_i, we shall use the following well-known formula:

$$\Bigl(\frac{d}{dt} \det(A + tB) \Bigr)_{t=0} = \det A \; \operatorname{tr}(A^{-1}B),$$

where A and B are $m \times m$ matrices and A is regular. This formula is proved as follows:

$$\Bigl(\frac{d}{dt} \det(A + tB) \Bigr)_{t=0} = \frac{d}{dt} \Bigl(\det A \; \det(1 + tA^{-1}B) \Bigr)_{t=0}$$

$$= \det A \Bigl(\frac{d}{dt} \det(1 + tA^{-1}B) \Bigr)_{t=0}$$

but as $\det(1 + tA^{-1}B)$ is a polynomial of degree m in t, $\frac{d}{dt} \det(1 + tA^{-1}B)_{t=0}$ is the coefficient to t in this polynomial which is the trace of $A^{-1}B$, q.e.d.

In case B is a matrix of rank 1, we can prove a little more. In fact in this case also $A^{-1}B$ is of rank 1 so all sub-determinants in $A^{-1}B$ of order 2, 3, ..., m are zero, and therefore all terms in the polynomial $\det(1 + tA^{-1}B)$ of degree higher than the first will vanish, and we have

$$\det(A + tB) = \det A \bigl(1 + t \operatorname{tr}(A^{-1}B) \bigr).$$

If and only if the rank of B is 1, then there exist two (column) vectors a and b such that

$$B = ab^{\mathrm{T}},$$

and

$$\operatorname{tr}(A^{-1}B) = \operatorname{tr}(A^{-1}ab^{\mathrm{T}}) = \operatorname{tr}(b^{\mathrm{T}}A^{-1}a) = b^{\mathrm{T}}A^{-1}a,$$

because the trace of a product is invariant with respect to a cyclic permutation of the factors, and they become $b^{\mathrm{T}}A^{-1}a$ which is a scalar and therefore equal to its trace. So we have

$$\frac{\det(A + tab^{\mathrm{T}})}{\det A} = 1 + tb^{\mathrm{T}}A^{-1}a.$$

Now put for a fixed j

$$p_i = \begin{cases} q_i > 0 & \text{for } i \neq j \\ q_j + t, q_j > 0 & \text{for } i = j, \end{cases}$$

then

$$N(P) = A^{\mathrm{T}} P A = \sum_{i=1}^{N} p_i a_i^{\mathrm{T}} a_i = t a_j^{\mathrm{T}} a_j + \sum_{i=1}^{N} q_i a_i^{\mathrm{T}} a_i = N(Q) + t a_j^{\mathrm{T}} a_j;$$

we remember that $\{a_i\}$ are row-vectors!

Then

$$\Delta(P) = \det\bigl(N(Q) + t a_j^{\mathrm{T}} a_j\bigr) = \Delta(Q)\bigl(1 + t a_j N(Q)^{-1} a_j^{\mathrm{T}}\bigr),$$

which shows us that for all p_i for $i \neq j$ fixed $\Delta(P)$ is an affine (i.e., nonhomogeneous linear) polynomial in p_j. As this is true for every choice of j it follows that all the terms in the polynomial $\Delta(P)$ are constants multiplied by the product of m different weights.

Putting $q_i = 1$, $i = 1, 2, \ldots, n$, i.e., all the weights p_i for $i \neq j$ is 1 gives

$$\Delta(P) = \Delta\bigl(1 + (p_j - 1) a_j N^{-1} a_j^{\mathrm{T}}\bigr),$$

so that

$$\left(\frac{1}{\Delta(P)} \frac{\partial \Delta(P)}{\partial p_j}\right)_{P=1} = a_j N^{-1} a_j^{\mathrm{T}}$$

or

$$\rho_j = 1 - a_j N^{-1} a_j^{\mathrm{T}}$$

which is an expression for the local redundancy which is easily computable.

Now $a_j N^{-1} a_j^{\mathrm{T}}$ is the a posteriori variance of the measurement a_j and the a priori variance of a_j is one, so we have the following:

The redundancy of a measurement is the difference between its a priori and its a posteriori variance.

From this also follows that $0 \leq \rho_j < 1$.

If the definition of local redundancy should be reasonable it should be so that if and only if the redundancy of a measurement is greater than zero, then the network should be determined also if that measurement is ignored, i.e., if it is given the weight 0 (all the other measurements having weight 1). Therefore from the continuity of $\Delta(P)$, it follows that if $-\mu_j$ is the greatest lower bound for p_j such that $\Delta(P) > 0$ for

$$p_i = \begin{cases} 1 & \text{for } j \neq i \\ p_j & \text{for } j = i \end{cases}$$

then $\rho_j = 0 \Rightarrow \mu_j = 0$ and $\rho_j > 0 \Rightarrow \mu_j > 0$, and if in fact it is so, we have

$$\Delta(P) = \Delta\bigl(1 + (p_j - 1) a_j N^{-1} a_j^{\mathrm{T}}\bigr) = \Delta\bigl(1 + (p_j - 1)(1 - \rho_j)\bigr).$$

If and only if $-\mu_j$ is the greatest lower bound for p_j such that $\Delta(P) > 0$, then $\Delta(P) = 0$ for $p_j = -\mu_j$, i.e.,

$$1 - (1 + \mu_j)(1 - \rho_j) = 0$$

$$1 + \mu_j = \frac{1}{1 - \rho_j}$$

$$\mu_j = -1 + \frac{1}{1 - \rho_j} = \frac{\rho_j}{1 - \rho_j}.$$

We can now take up again the second of the examples mentioned in the motivation section. Here a point was determined relative to the rest of the network by only two measurements, say a_1 and a_2 such that the network is not determined if we put p_1 and/or p_2 to zero, i.e., $\mu_1 = \mu_2 = 0$ and so $\rho_1 = \rho_2 = 0$, as it should be for ρ_i to be a reasonable measure for the local redundancy. By the way, it is easy to see that in this case p_1 and p_2 are factors in the polynomial $\Delta(P)$.

It is left to the reader to find out if our definition gives a reasonable result in connection with the first example.

The theory can be pushed much further, but we will proceed only one step more.

Can we find a condition for $\Delta(P)$ not to vanish if two given measurements, say a_1 and a_2, are given weights zero simultaneously (when all other weights are one)?

It would have been beautiful if the necessary and sufficient condition would have been $\rho_1 + \rho_2 > 1$. It is not that beautiful, but $\rho_1 + \rho_2 > 1$ is sufficient.

As pointed out in Section 2, $\mathcal{P}^+$ is a convex set. If we look at the intersection of $\mathcal{P}^+$ with the plane p_1, p_2 for $p_i = 1$ for $i = 3, 4, \ldots, N$, we find obviously a convex subset $\mathcal{P}^+_{12}$ of this plane where $\Delta(P) > 0$ and the points $(1, -\mu_1)$ and $(-\mu_2, 1)$ are situated at the boundary of $\mathcal{P}^+_{12}$. The line segment between these two boundary points must consist of points in the closure of $\mathcal{P}^+_{12}$, and a sufficient condition for the point $(0, 0)$ to be in $\mathcal{P}^+_{12}$ is therefore that this line segment passes under $(0, 0)$; that is that $\mu_1\mu_2 > 1$, i.e.,

$$\frac{\rho_1}{1 - \rho_1}\frac{\rho_2}{1 - \rho_2} > 1$$

$$\rho_1\rho_2 > 1 - \rho_1\rho_2 + \rho_1\rho_2$$

or

$$\rho_1 + \rho_2 > 1$$

q.e.d.

A closer analysis shows that if the a posteriori correlation between a_1 and a_2 is zero, then already $\rho_1 > 0$ and $\rho_2 > 0$ suffice, and it seems clear that for a further development of this theory the a posteriori correlations have to be taken into account.

Example 19.1 Linear regression for n equidistant points.

Here we have n observation equations of the form

$$x + py = \mathsf{o}_p$$

where

$$-\frac{n-1}{2} \leq p \leq \frac{n-1}{2},$$

and p is an integer if n is odd and a half-integer if n is even.

The weights are supposed to be 1, so the matrix of the normal equations is

$$\begin{bmatrix} n & 0 \\ 0 & \frac{(n-1)n(n+1)}{12} \end{bmatrix}$$

and the diagonal element labeled p of the covariance matrix for the adjusted observations is

$$\begin{bmatrix} 1 & p \end{bmatrix} \begin{bmatrix} \frac{1}{n} & 0 \\ 0 & \frac{12}{(n-1)n(n+1)} \end{bmatrix} \begin{bmatrix} 1 \\ p \end{bmatrix} = \frac{1}{n} + \frac{12p^2}{(n-1)n(n+1)}.$$

Finally, the redundancy for the observation labeled p is

$$\rho_p = 1 - \frac{1}{n} - \frac{12p^2}{(n-1)n(n+1)}.$$

If we instead of a line look for a polynomial of second degree, then with the notation introduced above we get the observation equations

$$x + py + p^2 z = \mathsf{o}_p$$

with the normal equation matrix

$$n(n^2-4) \begin{bmatrix} \frac{1}{n^2-1} & 0 & \frac{1}{12} \\ 0 & \frac{1}{12} & 0 \\ \frac{1}{12} & 0 & \frac{3n^2-7}{240} \end{bmatrix}$$

and variance-covariance matrix

$$\frac{1}{n(n^2-4)} \begin{bmatrix} \frac{9n^2-21}{4} & 0 & -15 \\ 0 & \frac{12(n^2-4)}{n^2-1} & 0 \\ -15 & 0 & \frac{180}{n^2-1} \end{bmatrix}$$

and the redundancy for the observation labeled p becomes

$$\rho_p = 1 - \frac{9n^2-21}{4n(n^2-4)} + \frac{18p^2}{n(n^2-4)} - \frac{180p^4}{n(n^2-4)(n^2-1)}.$$

20

A Convergence Problem in Collocation Theory

Summary

Collocation theory allows the approximation of the anomalous potential T, harmonic in a region Ω, by a smoother function $\hat{T}$ harmonic in a larger domain Σ and agreeing with measurements performed on T at discrete points. The smoothing least-squares collocation method is a part of collocation theory in which a hybrid norm is minimized, norm that depends upon a parameter λ that can be interpreted as the relative weight of the norm of $\hat{T}$ in Σ and in Ω.

The problem of the behaviour of $\hat{T}$ when the number of measurements tends to infinity and contemporarily $\lambda \to \infty$ is analyzed: the convergence to the correct solution is proved under suitable hypotheses.

Introduction

It has sometimes been claimed that using a discrete approach such as collocation to the solution of problems in physical geodesy should make the theoretical study of boundary value problems an idle sport for armchair geodesists. Naturally discrete methods can give results, and it is possible by analyzing them to have ad hoc evidence for their relevance, but it is equally obvious that the general evaluation of the relevance is only possible when the corresponding continuous boundary value problem has been studied.

That a discrete method is relevant must mean that if the number and the precision of the observations goes to infinity (in a way to be carefully specified), then the method will give results which converge to a solution of the boundary value problem defined by the observations, so if the boundary value problem has a unique solution then the result converges to the correct result.

To evaluate such a relevance must be to give formulas which estimate the errors of the results from specifications of the observations and the calculation methods. This is what I have tried to do in this paper for the *smoothing least-squares collocation method*. (The word smoothing seems to be important here!)

The method I use in order to find these estimates is to a given discrete collocation problem to find a continuous collocation problem to which the original problem may be looked at as an approximation, to estimate this approximation (in Section 2) and analyse the continuous problem, which is, in fact, a specially formulated boundary value problem (Section 1).

The condition for this procedure to work, i.e. for the proof for a solution to be relevant, is that the continuous collocation problem is what I call coercive, and this again is (almost) the same as saying that the corresponding boundary value problem is coercive, where this word now has the meaning defined in the theory for the variational approach to boundary value problems for elliptic partial differential equations.

1. Continuous Collocation

Let ω be a smooth (C^∞) closed surface outside the Earth (but possibly very near the surface of the Earth), which surrounds the Earth. Suppose that we have measured the T, or the free-air anomaly or such a thing at all points of ω. Mathematically we can formulate that by saying that we have given a linear differential operator D of finite order with coefficients which are smooth functions of the coordinates outside the Earth and that we have measured

$$D_x T = f$$

at all points $x \in \omega$. Here $D_x T$ means the value of DT at x and $f : \omega \to \mathbf{R}$. As ω is smooth and outside the Earth, the "true" value of f is a smooth function. We will suppose that the observed f is square integrable over ω.

We want to find an approximation T' to T such that

$$D_x T' = f + v \qquad \text{for all } x \in \omega \tag{1}$$

where $v : \omega \to \mathbf{R}$ and T' is harmonic outside a given Bjerhammar sphere σ. Then v is also square integrable. Let Σ be the subset of $\mathbf{R}^3$ outside the sphere σ and let H be a Hilbert space with norm $\|\cdot\|$ and scalar product $(\cdot,\cdot)$ such that 1) if ϕ is harmonic in any open set $\Omega \subset \mathbf{R}$ such that $\Sigma \subset \Omega$ then $\phi \in H$, and 2) every $\phi \in H$ is harmonic in Σ, and 3) strong convergence of ϕ_i in H involves uniform convergence of $\phi_i(X)$ on closed sets in Σ. Then H has a reproducing kernel $K(x, y)$.

The approximation T' we look for shall now be determined by

1. $T' \in H$,
2. T' and v satisfy (1),
3. $\|T'\|^2 + \lambda \int_\omega q v^2 \, d\omega$ shall be as small as possible,

here λ is a positive number and $q : \omega \to \mathbf{R}$ is a positive smooth function such that there exist two positive numbers c_1 and c_2 with

$$c_1 \le q(x) \le c_2 \qquad \text{for all } x \in \omega$$

q is the weight density of the observations.

We have now defined a smoothing continuous collocation problem, we want to find its solution and see what happens for $\lambda \to \infty$.

Instead of the clumsy expressions

$$D_x K(x, y) \qquad \text{etc.}$$

we will write

$$K(D_x, y) \qquad \text{etc.}$$

Then we can write the "normal equations" for our problem as

$$T'(x) + \lambda \left[\int_\omega K(x, D_z) q(z) \big(K(D_z, \cdot), T' \big)\, d\omega_z \right] = \lambda \int_\omega K(x, D_z) q(z) f(z)\, d\omega_z. \tag{2}$$

By applying the operator D_x on both sides of (2) and remembering that

$$D_x T' = \big(K(D_x, \cdot), T' \big)$$

we find

$$D_x T' + \lambda \int_\omega K(D_x, D_z) q(z) D_z T'\, d\omega_z = \lambda \int_\omega K(D_x, D_z) q(z) f(z)\, d\omega_z.$$

Substitution of (1) in this equation gives

$$v(x) + \lambda \int_\omega K(D_x, D_z) q(z) v(z)\, d\omega_z + f(x) = 0. \tag{3}$$

This is an integral equation for v with a kernel

$$K(D_x, D_z) q(z),$$

which is not symmetric, but if we write

$$q(x)^{1/2} v(x) = w(x)$$

we find:

$$w(x) + \lambda \int_\omega q(x)^{1/2} K(D_x, D_z) q(z)^{1/2} w(z)\, d\omega_z + q(x)^{1/2} f(x) = 0 \tag{4}$$

and this is an integral equation for w with the symmetric kernel

$$H(x, y) = q(x)^{1/2} K(D_x, D_y) q(y)^{1/2}.$$

H is smooth and positive definite (not necessarily strictly positive definite), so (4) is a Fredholm integral equation with positive eigenvalues λ_i, $i = 1, 2, \ldots$, and it has a unique solution for every positive value of λ.

By substitution of (1) in (2) we find analogously:

$$T'(x) = -\lambda \int_\omega K(x, D_y) q(y)^{1/2} w(y)\, d\omega_y \tag{5}$$

so when w is found T' can be found by (5), and it is easy to see that for all $0 < \lambda < \infty$, $T' \in H$.

This solution is a curious one: T' is in (5) calculated directly from w, the (weight normalized) correction to the observations, that is, in a way from the measuring errors. But Nature often seems to like paradoxes!

We want to find out what happens for $\lambda \to \infty$. It is most amusing to do this using the eigenfunctions and eigenvalues defined by

$$\phi_i(x) - \lambda_i \int_\omega H(x, y) \phi_i(y)\, d\omega_y = 0 \qquad i = 1, 2, 3, \ldots \tag{6}$$

where $\phi : \omega \to \mathbf{R}$.

From the theory of Fredholm integral equations we know that (6) has only positive eigenvalues, that they have no point of accumulation and that the eigenfunctions can be chosen so as to be orthonormal in $L^2(\omega)$. Let us call the Hilbert Space $L^2(\omega)$ H_ω, the subspace spanned by $\{\phi_i\}$ H'_ω, and let the orthogonal complement to H'_ω in H_ω be called H''_ω. H''_ω may be the zero space. $H(x, y)$ can be represented as

$$H(x, y) = \sum \frac{\phi_i(x)\phi_i(y)}{\lambda_i}.$$

Define

$$\Phi_i(x) = \lambda_i^{1/2} \int_\omega K(x, D_y) q(y)^{1/2} \phi_i(y)\, d\omega_y \tag{7}$$

then

$$\Phi_i : \Sigma \to \mathbf{R}$$

and Φ_i is harmonic in Σ.

Clearly

$$(\Phi_i, \Phi_j) = \lambda_i^{1/2} \lambda_j^{1/2} \int_\omega \int_\omega H(x, y) \phi_i(x) \phi_j(y)\, d\omega_x\, d\omega_y = \begin{cases} 1 & \text{for } i = j, \\ 0 & \text{for } i \neq j, \end{cases}$$

so $\{\Phi_i\}$ is a complete orthonormal system for (at least) a subspace $H' \subset H$. Let H'' be the orthogonal complement to H' in H and $\{\psi_i\}$ be an orthonormal system in H''.

Then

$$K(x, y) = \sum_i \Phi_i(x) \Phi_i(y) + \sum_j \psi_j(x) \psi_j(y).$$

From (7) it follows also that

$$K(x, D_y)q(y)^{1/2} = \sum_i \frac{\Phi_i(x)\phi_i(y)}{\lambda_i^{1/2}},$$

so that

$$q(x)^{1/2} D_x \Phi_i = \frac{\phi_i(x)}{\lambda_i^{1/2}}$$

and

$$q(x)^{1/2} D_x \psi_j = 0,$$

and it follows that H'' consists of those $u \in H$ for which $D_x u = 0$ for all $x \in \omega$.

Let us write

$$q^{1/2} f = f' + f'',$$

where f' and f'' are the projections of $q^{1/2} f$ into the subspaces H'_ω and H''_ω. Then f' can be expressed as

$$f' = \sum_i f'_i \phi_i,$$

where

$$f'_i = \int_\omega f(x)\phi(x)\, d\omega,$$

and

$$\|f'\|^2_\omega = \int_\omega \left(f'(x)\right)^2 d\omega = \sum_i f'^2_i.$$

If $\{\psi_j\}$ is a complete orthonormal system for H''_ω, then $f'' = \sum_j f''_j \psi_j$ and

$$q^{1/2} f = \sum_i f'_i \phi_i + \sum_j f''_j \psi_j. \tag{8}$$

Putting (8) and the analogous representation of w:

$$w = \sum_i w'_i \phi_i + \sum_j w''_j \psi_j$$

in (4) we find

$$\sum_i w'_i \phi_i + \sum_j w''_j \psi_j + \lambda \sum_i \frac{w'_i}{\lambda_i} \phi_i + \sum_i f'_i \phi_i + \sum_j f''_j \psi_j = 0,$$

or

$$\left(1 + \frac{\lambda}{\lambda_i}\right) w'_i + f'_i = 0,$$

$$w''_j + f''_j = 0;$$

that is

$$w = -\sum_i \frac{\lambda_i}{\lambda_i + \lambda} f'_i \phi_i - \sum_j f''_j \psi_j, \tag{9}$$

or

$$f + v = q^{-1/2} \sum_i \frac{\lambda}{\lambda_i + \lambda} f'_i \phi_i. \tag{10}$$

$f + v$ is the "corrected observations" and so (10) shows

1. that $\|f + v\|_\omega \leq \|f\|_\omega$ for all $\lambda > 0$
2. that $\lim_{\lambda \to \infty}(f + v) = q^{-1/2} f'$.

On the other hand putting (9) into (5) and using (7) gives:

$$T'(x) = \sum_i \frac{\lambda \lambda_i^{1/2}}{\lambda_i + \lambda} f'_i \Phi_i(x),$$

thus $T' \subset H$ for all $0 < \lambda < \infty$ but for $\lambda \to \infty$ one finds formally

$$T' \sim \sum_i \lambda_i^{1/2} f'_i \Phi_i,$$

that is T' converges in H for $\lambda \to \infty$, if and only if $\sum_i \lambda_i f'^2_i < \infty$, in fact $\lim_{\lambda \to \infty} T' \in H$ if and only if (1) or

$$D_x T' = q^{-1/2} f'$$

has a solution in H.

We can now conclude:

> The continuous collocation problem as we have defined it here has always a solution for $0 < \lambda < \infty$, i.e. if it is a smoothing collocation problem, but the corresponding not smoothing problem ($\lambda = \infty$) has normally no solution.

Before we try to repair this situation we will look at a few examples. First let $D_x T = T(x)$ for $x \in \omega$ and let Ω be the points of $\mathbf{R}^3$ outside ω. Now

$$\|T\|^2_\Omega = \int_\omega q^2 T^2 \, d\omega$$

is a norm for a Hilbert space of functions harmonic in Ω which contains all functions harmonic in any open set containing $\bar{\Omega}$, and this space has a reproducing kernel. (These simple facts will not be proved here).

Therefore if we use the collocation theory as described above and let $\lambda \to \infty$ then even if T does not converge in Σ it converges in Ω and it converges to a potential which does not depend on the norm we choose in Σ, namely to that regular potential in Ω which has the observed $T(x)$, $x \in \omega$ as boundary values (in a certain general sense).

Next let ω be a sphere (still outside the Earth) concentric with the Bjerhammar sphere and

$$D_x T = \left(\frac{\partial T}{\partial r} + \frac{2}{r} T\right)(x).$$

Under the *Symmetry assumption:* q is constant and the norm $\|\cdot\|$ is invariant with respect to a rotation about 0, the eigenfunctions ϕ_i, Φ_i, ψ_j are spherical harmonics and I think it would be very instructive for the reader to make the calculations above for this special case. If he does so he will find for $\lambda \to \infty$:

1. That $f + v$ converges to a function on ω for which the boundary problem (the Stokes' problem) has a solution.
2. That T converges (in an appropriate norm) in Ω to the well-known solution of Stokes' problem.

If we drop the symmetry assumption the result 1 is still true but 2 must be changed slightly: From the theory of boundary problems it follows that

$$\|T\|_{\Omega}^2 = \int_{\omega} \left(\frac{\partial T}{\partial r} + \frac{2}{r} T\right)^2 d\omega$$

is a norm for the space H', i.e. this norm defines a pre-Hilbert space with the norm $\|\cdot\|'_{\Omega}$ and the elements equal to the elements of H'. By completion we get a Hilbert space $H'_{\Omega} \supset H'$ which consists of functions harmonic in Ω. In this case for $\lambda \to \infty$ the component of T in the space H'_{Ω} converges to a solution of the Stokes' problem in Ω and the component of T in H'' is 0. But because H' is defined as the orthogonal complement to H'' in H it is *not* independent of the norm $\|\cdot\|$ of H; in fact it is only independent modulo H'', that is the difference of solutions found by using different norms is an outer spherical harmonic of degree 1.

After these examples we can formulate the final result. First define:

A continuous collocation problem as defined above is *coercive* if

$$\|T\|_{\Omega}^{\prime 2} = \int_{\omega} (D_x T)^2 \, d\omega,$$

which is a semi-norm defined for all elements $T \in H$, has the following property: The completion of H with respect to the semi-norm $\|\cdot\|'_{\Omega}$ consists of functions harmonic in Ω.

If the collocation problem is coercive then for $\lambda \to \infty$ T_λ converges in Ω (in the semi-norm $\|\cdot\|'_{\Omega}$) to a solution of the boundary value problem:

$$\begin{aligned} \Delta T &= 0 \qquad && \text{in } \Omega, \\ D_x T &= f + v \qquad && \text{on } \omega, \end{aligned}$$

where v depends on the weight density q but not on the norm in Σ (that is v does not depend on the reproducing kernel); and therefore modulo a solution to the homogeneous boundary value problem:

$$\begin{aligned} \Delta T &= 0 \qquad \text{in } \Omega, \\ D_x T &= 0 \qquad \text{on } \omega. \end{aligned}$$

T does not depend on the choice of the reproducing kernel.

2. Discrete Collocation

The notations from Section 1 remain valid also in this section but we have to introduce some more notations.

In the discrete case we have measured

$$D_x T = f$$

only at N points $x_n \in \omega$, $n = 1, 2, \ldots, N$. Writing D_n for D_{x_n} we want to find an approximation T' to T such that

$$D_n T' = f_n + v_n,$$

with $T' \in H$ such that

$$\|T'\| + \lambda \sum_{n=1}^{N} p_n v_n^2$$

is as small as possible where $\{p_n\}$ are positive weights.

We want to investigate what happens when $N \to \infty$ using the results from the continuous case. The method we shall use consists in regarding the discrete case as a "finite element" approximation to the continuous case for fixed N and then afterwards letting $N \to \infty$.

First we make a subdivision of ω into N open sets $\omega_n \subset \omega$ such that

1. $x_n \in \omega_n$ for $n = 1, 2, \ldots, N$,
2. $\omega = \cup_{n=1}^{N} \bar{\omega}_n$
3. the boundary of ω_n is a closed piecewise smooth curve which is star shaped with respect to x_n for $n = 1, 2, \ldots, N$.

Functions which are only defined in the points x_n, $n = 1, 2, \ldots, N$, will now be extended to almost all points of ω by defining

$$\tilde{f}(x) = f_n, \qquad \tilde{v}(x) = v_n \quad \text{for } x \in \omega_n.$$

The points of ω where these functions are not defined are on the boundary curves of the sets ω_n.

If f is a continuously differentiable function on ω and we put $f_n = f(x_n)$, then $\tilde{f}$ can be regarded as an approximation to f. How good is this approximation?

For $y \in \omega_n$

$$f(y) - \tilde{f}(y) = f(y) - f(x_n) = \int_{x_n}^{y} \operatorname{grad} f \, d\bar{s},$$

where the curve integral is taken along a geodesic c_n on ω from x_n to y. Write $|c_n|$ for the length of c_n. We have then:

$$\left| f(y) - \tilde{f}(y) \right| \leq |c_n| \sup_{\omega_n} |\operatorname{grad} f|,$$

and

$$\begin{aligned} \|f(y) - \tilde{f}(y)\|_{\omega}^{2} &= \int_{\omega} q(y) \big(f(y) - \tilde{f}(y)\big)^2 \, d\omega \\ &\leq \sum_n \int_{\omega_n} q(y) \big(c_n(y)\big)^2 \sup_{\omega_n} |\operatorname{grad} f|^2 \, d\omega_n \\ &\leq \sum_n \sup_{\omega_n} |\operatorname{grad} f|^2 \int_{\omega_n} q(y) c_n(y)^2 \, d\omega_n . \end{aligned}$$

If we write

$$\int_{\omega_n} q c_n^2 \, d\omega_n = \rho_n^2 \int_{\omega_n} q \, d\omega_n = p_n \rho_n^2$$

where

$$p_n = \int_{\omega_n} q \, d\omega_n \tag{11}$$

then

$$\|f - \tilde{f}\|_{\omega}^{2} \leq \sum_n \sup_{\omega_n} |\operatorname{grad} f|^2 p_n \rho_n^2 .$$

Here $\sup_{\omega_n} |\operatorname{grad} f|^2$ exists because $\operatorname{grad} f$ is continuous and ω is compact. Putting

$$\rho = \max_n c_n$$

we get at last

$$\|f - \tilde{f}\|_{\omega}^{2} \leq \rho^2 \sum_n p_n \sup_{\omega_n} |\operatorname{grad} f|^2 . \tag{12}$$

We will now in (3) approximate $v(x)$ and $K(D_x, D_z)$ in the same way as we approximate $f(x)$. The smooth function q is not approximated but we suppose the connection between q and p_n given by (11). (3) in approximated form becomes

$$\tilde{v}(x) + \lambda \int_{\omega} \tilde{K}(D_x, D_z) q(z) \tilde{v}(z) \, d\omega_z + \tilde{f}(x) = 0. \tag{13}$$

Subtracting (13) from (3) now gives

$$(v-\tilde{v})+\lambda\int\limits_{\omega} K(D_x,D_z)q(v-\tilde{v})\,d\omega+(f-\tilde{f})+\lambda\int\limits_{\omega}(K-\tilde{K})q\tilde{v}\,d\omega=0,$$

or if we write

$$\tilde{f}=(f-\tilde{f})+\lambda\int\limits_{\omega}(K-\tilde{K})q\tilde{v}\,d\omega,$$

we have

$$(v-\tilde{v})+\lambda\int\limits_{\omega} K(D_x,D_z)q(z)\big(v(z)-\tilde{v}(z)\big)\,d\omega_z+\tilde{f}=0,$$

i.e. $v-\tilde{v}$ is a solution of an integral equation of the same form as (3) and with the same kernel, and therefore it follows from result of Section 1 especially equation (9) that

$$\|v-\tilde{v}\|_\omega \le \|\bar{f}\|_\omega. \tag{14}$$

But

$$\bar{f}=\bar{f}_1+\bar{f}_2+\bar{f}_3$$

where

$$\bar{f}_1=f-\tilde{f},$$

$$\bar{f}_2=\lambda\int\limits_{\omega}\big(K(D_x,D_z)-K(D_x,\tilde{D}_z)\big)q(z)\tilde{v}(z)\,d\omega_z,$$

$$\bar{f}_3=\lambda\int\limits_{\omega}\big(K(D_x,\tilde{D}_z)-K(\tilde{D}_x,\tilde{D}_z)\big)q(z)\tilde{v}(z)\,d\omega_z,$$

so that from (14) it follows

$$\|v-\tilde{v}\|\le\|\bar{f}_1\|_\omega+\|\bar{f}_2\|_\omega+\|\bar{f}_3\|_\omega. \tag{15}$$

From (12)

$$\|\bar{f}_1\|_\omega=\rho\Big(\sum_n p_n\sup_{\omega_n}|\operatorname{grad} f|^2\Big)^{1/2}.$$

Using a standard routine from the theory of integral equations we find:

$$\begin{aligned}\|\bar{f}_2\|_\omega^2 &\le \lambda^2\int\limits_{\omega}\int\limits_{\omega}\big(K(D_x,D_y)-K(D_x,\tilde{D}_y)\big)^2 q(x)q(y)\,d\omega_x d\omega_y\int\limits_{\omega}\tilde{v}(x)^2q(x)\,d\omega_x\\ &=\lambda^2\sum_{n=1}^{N}\sum_{m=1}^{N}\int\limits_{\omega_n}\int\limits_{\omega_m}\big(K(D_x,D_y)-K(D_x,D_{y_m})\big)^2 q(x)q(y)\,d\omega_x d\omega_y\\ &\qquad\times\sum_{n=1}^{N}\int\limits_{\omega_n} v_n^2 q(x)\,d\omega_x\end{aligned}$$

$$\leq \lambda^2 \sum_{m=1}^{N} \int_{\omega_n} \rho^2 \Big(\sum_{m=1}^{N} p_m \sup_{y\in\omega_m} \big|K(D_x, \operatorname{grad} D_y)\big|^2 q(x) \Big)\, d\omega_n \sum_{n=1}^{N} v_n^2 p_n$$

$$\leq \lambda^2 \rho^2 \sum_{n=1}^{N} \sum_{m=1}^{N} \sup_{\substack{x\in\omega_n\\ y\in\omega_m}} \big|K(D_x, \operatorname{grad} D_y)\big|^2 p_n p_m \sum_{n=1}^{N} v_n^2 p_n.$$

For $\|\bar{f}_3\|_\omega^2$ we find the same estimate as for $\|\bar{f}_2\|_\omega^2$, and we have:

$$\|v - \tilde{v}\|_\omega \leq \rho \Big(\sum_{n=1}^{N} p_n \sup_{\omega_n} |\operatorname{grad} f|^2 \Big)^{1/2}$$

$$+ 2\rho\lambda \Big(\sum_{n=1}^{N} \sum_{m=1}^{N} \sup_{\substack{x\in\omega_n\\ y\in\omega_m}} \big|K(D_x, \operatorname{grad} D_y)\big|^2 p_n p_m \sum_{n=1}^{N} v_n^2 p_n \Big)^{1/2}. \quad (16)$$

The perspicacious reader, seeing that the complicated part and, as we shall see soon, the rather unpleasant part of the estimate (16) stems from the fact that we have used an approximation to the kernel instead of using the kernel itself may wonder why I do the task so complicated for myself and for him, so I shall give the explanation here and now.

Let us go back to (13) writing it as

$$v_n + \lambda \sum_{m=1}^{N} K(D_n, D_m) p_m v_m + f_n = 0, \qquad n = 1, 2, \ldots, N,$$

but this is together with

$$T(x) = -\lambda \sum_{n=1}^{N} K(x, D_n) p_n v_n$$

the expression for the solution of the discrete collocation problem (written in an asymmetrical form) and therefore (16) gives an estimate for the difference between the result of a continuous collocation problem and a discrete one.

It is possible to continue this procedure and apply it to (5) or rather to

$$T'(x) = -\lambda \int_\omega K(x, D_y) q(y) v(y)\, d\omega_y,$$

and its $\sim$-approximation

$$T''(x) = -\lambda \int_\omega K(x, \tilde{D}_y) q(y) \tilde{v}(y)\, d\omega_y,$$

in order to estimate $\|T' - T''\|$, the norm of the difference between T', the solution to the continuous problem, and T'', the solution to the discrete problem. I will not do that here—the interested reader should be able to do it without difficulties using the same method as above—but it is relevant to use the results from Section 1 under the assumption of coercivity. In this case

$\|T' - T''\|'_{\Omega} \leq \|v - \tilde{v}\|_{\omega}$ and (16) give directly an estimate for $T' - T''$ outside ω which alone is physically relevant.

For $\lambda \to \infty$ also the last member of (16) $\to \infty$, therefore (16) only gives an estimate for smoothing collocation. This is perhaps because the estimation method is not strong enough, but I do not think so. At any case this question deserves further investigations. On the other hand, as I have already mentioned immediately after (16) it is possible to find from a finite set of observations an approximation to the solution of the continuous collocation problem which remains finite also for $\lambda \to \infty$, although this method demands more calculation, I think it should be further investigated also!

We have now arrived at our essential problem namely what happens when the number of observations goes to infinity? We must first, however, thoroughly define the limiting process.

Instead of one set of discrete observations we have a sequence of finite sets of observations, one for every ν, $\nu = 1, 2, \ldots, \infty$, such that:

1. $N_\nu \to \infty$, i.e. the number of observations goes to infinity with ν
2. $f_{n\nu} = f(x_{n\nu})$, i.e. all the observation results are compatible with the same continuously differentiable function f. (This hypothesis could easily be replaced by
$$|f_{n\nu} - f(x_{n\nu})| \leq l_\nu,$$
where $l_\nu \to 0$, the price would be an extra addend in the resulting estimate.)
3. $\rho_\nu \to 0$, a necessary (but not sufficient) condition for this is that the areas of $\omega_{n\nu} \to 0$.
4. $p_{n\nu} = \int\limits_{\omega_{n\nu}} q \, d\omega$, from this follows
$$\sum_{n=1}^{N_\nu} p_{n\nu} = \int\limits_{\omega} q \, d\omega,$$
that is the weights are normalized in such a way that the sum of the weights for the νth set is independent of ν, and
$$p_{n\nu} \to 0 \qquad \text{for } \nu \to \infty,$$
it is important to remember this normalization when you try to interpret the final result especially the meaning of the parameter λ.

The first addend on the right hand side of (16) is ρ multiplied by a factor which is finite for all ν and

$$\lim_{\nu \to \infty} \left(\sum_{n=1}^{N_\nu} p_{n\nu} \sup_{\omega_{n\nu}} |\operatorname{grad} f|^2 \right)^{1/2} = \left(\int\limits_{\omega} q |\operatorname{grad} f|^2 \, d\omega \right)^{1/2},$$

which is finite because f is continuously differentiable on ω, therefore there exists a constant A (normally between, say, 2 and 5) such that

$$\left(\sum_{n=1}^{N_\nu} p_{n\nu} \sup_{\omega_{n\nu}} |\operatorname{grad} f|^2\right)^{1/2} \leq A \int_\omega q |\operatorname{grad} f|^2 \, d\omega .$$

The second addend is $2\rho\lambda$ multiplied by two factors, the first of which is

$$\left(\sum_{n=1}^{N_\nu} \sum_{m=1}^{N_\nu} \sup_{\substack{x\in\omega_{n\nu}\\ y\in\omega_{m\nu}}} \left|K(D_x, \operatorname{grad} D_y)\right|^2 p_{n\nu} p_{m\nu}\right)^{1/2},$$

it is also finite and it converges for $\nu \to \infty$ to

$$\left(\int_\omega \int_\omega \left|K(D_x, \operatorname{grad} D_y)\right|^2 q(x) q(y) \, d\omega_x \, d\omega_y\right)^{1/2}.$$

The second factor is

$$\left(\sum_{n=1}^{N_\nu} v_{n\nu}^2 p_{n\nu}\right)^{1/2} \leq \left(\sum_{n=1}^{N_\nu} f_{n\nu}^2 p_{n\nu}\right)^{1/2}.$$

This again is finite and it converges to $\|f\|_\omega$. For the second addend we can find a factor B similar to A and find:

$$\begin{aligned}\|v - \tilde{v}\|_\omega \leq \rho_\nu A & \left(\int_\omega q |\operatorname{grad} f|^2 \, d\omega\right)^{1/2} \\ & + 2\rho_\nu \lambda B \|f\|_\omega \left(\int_\omega \int_\omega \left|K(D_x, \operatorname{grad} D_y)\right|^2 q(x) q(y) \, d\omega_x \, d\omega_y\right)^{1/2}.\end{aligned}$$

This inequality shows not only that $\tilde{v}$ converges to v under our assumptions 1–4 for λ a finite constant, but also, in connection with Section 1, that again under the assumptions 1–4 together with

5. $\lim_{\lambda\nu\to\infty} \rho_\nu \lambda = 0$, and
6. the continuous collocation problem is coercive,

that is under the assumptions 1–6, T converges outside ω to a solution of the boundary value problem.

For observations with equal precision we have

$$\frac{1}{\rho_\nu} = O\left(\frac{1}{N_\nu}\right)$$

and it would be natural to choose $\lambda \sim cN_\nu$, c a constant, but in this case $\lambda\rho$ would not go to zero so e.g.

$$\lambda \sim cN_\nu^{1/2}$$

would be better.

Remarks

For the understanding of the collocation formulas used in this paper it would be useful to read Sections II,2 and III,3 in [2], the continuous collocation is treated only in that book. A short introduction to the variational theory of boundary value problems is given in [3] Section II,9. For the theory of integral equations see [1], Chapter III.

It is not usual to apply different weights for the observations in collocation, when I have used weights in this paper it was in order to obtain a better fitting between discrete and continuous collocation. It is possible that the use of weights proportional to the areas of ω_i can improve the numerical behaviour of collocation methods in regions with dense observations.

References

[1] R. Courant and D. Hilbert. *Methoden der Mathematischen Physik.* Springer, Berlin, 3rd edition, 1968.

[2] Torben Krarup. *A Contribution to the Mathematical Foundation of Physical Geodesy.* Meddelelse 44. Geodætisk Institut, København, 1969.

[3] Jacques-Louis Lions and Enrico Magenes. *Non-Homogeneous Boundary Value Problems and Applications*, volume 181 of *Grundlehren der mathematischen Wissenschaften.* Springer, Berlin, 1972.

21

Non-Linear Adjustment and Curvature

Most of the adjustment problems in geodesy are non-linear. We have learned that we should 'linearize' the problems by differentiation of the equations of observation, and after doing so we normally forget about the non-linearity of the original problem and the relations between the solution of this and the solution of the linearized one.

This paper will treat one single aspect of non-linear adjustment viz. the curvature, which is a measure of the non-linearity of the problem, and which I regard as essential for our understanding of it.

The first section describes a model example which is extremely simple but illustrates the impact of curvature for adjustment in a nutshell and should make the understanding of the following section easy.

In Section 2 I try to give a geometric intuitive exposition of non-linear adjustment under the aspect of curvature, and in Section 3 I describe a geodetic problem—it could be called pseudo-free adjustment on an ellipsoid—for which the curvature is very large, and which could hardly be solved by linear methods.

I must emphasize that this is a first report on a complex of problems which has occupied me for some years, and that it has been necessary here to focus on a very small part of this complex.

1

Suppose we know that a point P is situated on the circle with radius r and center at the origin of a coordinate system in the plane, and that we have uncorrelated observations with equal weights of its coordinates x and y. We want to find P by adjustment according to the rules.

We express P by its coordinates $(r\cos t, r\sin t)$ where t is the parameter to be determined. Put $t = t_0 + \Delta t$, where t_0 is our guessed value for t, then the 'linearized' equations of observation are:

$$-r \sin t_0 \, \Delta t \approx x - r \cos t_0;$$

$$r \cos t_0 \, \Delta t \approx y - r \sin t_0;$$

and the normal equation is

$$r^2 \, \Delta t = r(y \cos t_0 - x \sin t_0).$$

Putting

$$x = R \cos \phi, \quad y = R \sin \phi, \qquad R > 0$$

we find

$$\Delta t = \tfrac{R}{r} \sin(\phi - t_0).$$

If our guessed value for t is rather good, i.e.,

$$\sin(\phi - t_0) \approx \phi - t_0,$$

then we get according to the rules

$$\Delta t \approx \tfrac{R}{r}(\phi - t_0)$$

and not

$$\Delta t = \phi - t_0$$

as it should be.

If $\frac{R}{r} \approx 1$ then we may expect that iteration of our linear adjustment will converge to the correct result, but if $R > 2r$ this expectation is normally frustrated.

But why is it so?

In order to understand this we will generalize the problem slightly. Instead of the circle we suppose given a two times continuously differentiable plane curve C; let the observation point be Q and let B be the distance from C to Q and let δ be the distance from C to Q such that δ is in this case the estimated mean error of the observations after the adjustment. Let P be a point on C such that $PQ = \delta$ and let P_0 on C be our guess for the adjustment result. We will use s, the arc-length along C, as the parameter in the adjustment and suppose that Δs, the distance between P and P_0, is so small that we in practice may identify the distance between P and P_0 along the chord and along the curve, i.e., we ignore quantities of the second order in Δs.

Now the effect of our 'linearized' adjustment procedure is exactly the orthogonal projection P' of Q into the tangent, but—and this is essential—not the tangent at P but the tangent at P_0, and the angle between these two tangents is Δs multiplied by c, the curvature of C in some point between P_0 and P, and therefore the distance between P_0 and P' is not Δs but this value changed by the amount of

$$\delta \tan(c\Delta s) \approx \delta c \Delta s,$$

still ignoring second order quantities in Δs. I leave it to the reader to discuss the sign of this discrepancy, i.e., to see how the result depends on how Q and the relevant center of curvature are positioned with respect to C.

In any case, we see that the unpleasant discrepancy is proportional to δ divided by a certain radius of curvature, and we shall see this is true for adjustment problems in higher dimensions, also. The reader may argue that as geodetic observations are normally very precise it would be unlikely that for any reasonable geodetic problem the mean error should be great with respect to this radius of curvature. To that I should remark that, as far as I know, geodesists have not been so much interested in these curvatures as they are in estimating their magnitude relative to the mean errors, and later in this paper I shall discuss a geodetic problem with a very small radius of curvature. Whether that problem is a reasonable one—that is another question.

2

It is well known that linear adjustment problems may be looked at from the geometrical point of view as orthogonal projections from points in a Euclidean N-space $\mathcal{O}$ (N is the number of observations) into a linear subspace. In generalizing this idea to non-linear problems we must be a little sophisticated in defining the relevant manifolds and mappings.

The space of observations $\mathcal{O}$ is, as in the linear case, a Euclidean space. If the observations are uncorrelated and of equal weight, then $\mathcal{O}$ has the usual $\mathbf{R}^N$-metric:

$$d^2 = \sum_{i=1}^{N} \Delta x_i^2 .$$

The range of the solution of the problem is described by the parameter manifold $\mathcal{Q}$. The meaning of this is best understood through an example.

Suppose we want to find ten points on a given sphere. Locally these points are determined by twenty coordinates, but no coordinate system exists on the sphere which gives a global one-to-one correspondence between the position of a point on the sphere and pairs of coordinates. Therefore we must use the concept of a differentiable manifold which just rests on the idea of local coordinate systems. In this case, the manifold will be the product manifold $S^2 \times S^2 \times \cdots \times S^2$ of ten spheres.

For any point of $\mathcal{Q}$ there corresponds a set of values of the N observations, so we have a mapping

$$F : \mathcal{Q} \longmapsto \mathcal{O};$$

the observation equations are the expression of this mapping in local coordinates. Here we will suppose that the mapping F is C^∞, i.e., arbitrarily often continuously differentiable. The image of F will be called $\mathcal{Q}'$ and is a differentiable submanifold of $\mathcal{O}$.

Let Q be a point of $\mathcal{Q}$ with coordinates

$$q = \{q^i\}, \qquad i = 1, 2, 3, \ldots, m$$

in a local coordinate system in a neighborhood of Q, m is the dimension of $\mathcal{Q}$, and define the matrix

$$A(q) = \left\{ \frac{\partial \mathrm{o}^\alpha}{\partial q^i} \right\}, \qquad \alpha = 1, 2, 3, \ldots, N.$$

A is the left-hand matrix of the observation equations, and $\{\mathrm{o}^\alpha\}$ are the values of the observations.

If rank $A(q) = N$ for all points $Q \in \mathcal{Q}$ (this condition is evidently independent of the choice of coordinate system), then the mapping F is an *immersion.* If F is an immersion then for any $Q \in \mathcal{Q}$ there exists a neighborhood of Q such that F is one-to-one with a C^∞ inverse, but we cannot deduce that F is globally one-to-one, as we can if F is linear. In the linear case rank $A = N$ is necessary and sufficient for Q to be estimable, while in the non-linear case it is only necessary.

As a differentiable manifold imbedded in $\mathcal{O}$, $\mathcal{Q}'$ is a Riemannian manifold with metric induced by the imbedding in a Euclidean space. If F is an immersion, this Riemannian structure can be pulled back to $\mathcal{Q}$ and the matrix of the first fundamental form g is given by

$$g = (A^{\mathrm{T}} A)^{-1} \qquad \text{or} \qquad g_{ij} = \sum_{\alpha=1}^{N} \left(\frac{\partial \mathrm{o}^\alpha}{\partial q^i} \frac{\partial \mathrm{o}^\alpha}{\partial q^j} \right)^{-1}.$$

If $T \in \mathcal{O}$ corresponds to a set of observations, then we will define $Q \in \mathcal{Q}$ as a solution of the adjustment problem if the distance δ in $\mathcal{Q}$ between T and $Q' = FQ$ is as small as possible:

$$\delta = \inf_{Q \in \mathcal{Q}} \|T - FQ\|.$$

If $\mathcal{Q}$—and thus $\mathcal{Q}'$—is compact, then we are sure that the problem

$$\delta = \inf_{Q' \in \mathcal{Q}'} \|T - Q'\|$$

has at least one solution—but it may have more than one, possibly infinitely many solutions and, as F is not necessarily one-to-one, to each of these points Q' there may correspond more than one point $Q \in \mathcal{Q}$.

It is even more complicated. A necessary condition for δ to have a local minimum at Q' is that the line $Q'T$ is orthogonal to the tangent space at Q' and it is exactly such points we are looking for using the normal procedure for adjustment; we do not know if we have a global minimum or even perhaps a maximum etc. In normal geodetic praxis we know approximately where to look for the solution (the starting point Q_0), but it may be that at least in exceptional occasions we find a 'solution' Q which does not correspond to a global minimum.

Now let Q be a 'stationary point,' i.e., such that the line $Q'T$ is orthogonal to the tangent space at Q' ($Q' = FQ$) and let Q_0 be 'close to' Q such that $Q'_0 = FQ_0$ is close to Q' and let us start the traditional adjustment procedure at Q_0. Exactly as in the example in Section 1 we find Q' as the orthogonal projection of T into the tangent space at Q_0' so the 'angle' between this tangent space and that at Q' is important for the behavior of the method, and again this 'angle' has to do with the curvature of the manifold $\mathcal{Q}'$ near Q'. Here, I shall shortly refer the result of a closer analysis of the behavior—the reader who has our first example in mind will at least not be surprised by the result.

The general problem has two complications with respect to that in two dimensions: First, the normal space at Q' has more than one dimension, and second, the tangent space at Q' has also more than one dimension.

The first of these complications is evaded by projecting the line $Q'T$ and a neighborhood in $\mathcal{Q}'$ of Q' into the linear space spanned by T and the tangent space at Q'. Let $\mathcal{Q}''$ be this projection of (part of) $\mathcal{Q}'$. Generally in Q', m lines of curvature of $\mathcal{Q}''$ will meet orthogonally and the corresponding centers of curvature (for Q') will be situated on the normal $Q'T$.

Now for Q' to be a point of *local* minimum for δ, it is necessary that none of these centers of curvature lies between Q' and T and it suffices if, moreover, T is not such a center.

Moreover, it is clear that if $\delta = Q'T$ is not small with respect to all these radii of curvature in Q' then we will have difficulties with the convergence if we simply iterate the adjustment procedure if the line $Q'Q_0'$ should not happen to be orthogonal to all the lines of curvature corresponding to small radii of curvature through Q'.

I should like to state here that the discussion above is coordinate-free so the results are independent of the way we choose the coordinates to determine. On the other hand, everything rests on the supposition that Q_0 is close to Q; if this is not the case new difficulties may very well occur which depend on the choice of the local coordinate system. But it is so to speak the most favorable situation we have treated.

As a conclusion of this section we can say that a measure for the non-linearity of an adjustment problem, i.e., a measure for the difficulties caused by the non-linearity of such a problem, is the greatest principal curvature of the manifold $\mathcal{Q}'$ as imbedded in $\mathcal{O}$ in the region where the solution point can be. For a numerically given adjustment problem it is relatively easy to compute this curvature (at least after the problem itself has been solved!), but for a given class of adjustment problems it can be very difficult. In the next section we shall give a *rough* estimate for the curvature for a certain class of adjustment problems, which should be of some geodetic relevance.

3

Now we shall consider free adjustment of distance networks on a sphere. Suppose we have $m/2$ points on a given sphere; here m is an even integer greater than 4, and that we have measured the distance (or rather squares of the distances) between N pairs of these points, sufficiently many to determine the configuration, that is $N > m - 3$, and we suppose that these squares of distances are measured with equal weights. This sounds abominable in the ears of a geodesist, but I promise to repair the situation later!

Now the parameter manifold $\mathcal{Q}$ is the product manifold of $m/2$ spheres, i.e., there is a one-to-one correspondence between points $Q \in \mathcal{Q}$ and $\frac{m}{2}$-tuples of points on the sphere. The mapping $F : \mathcal{Q} \longmapsto \mathcal{O}$ is differentiable, also for the case where two or more of the $m/2$ points of the sphere coincide—it was exactly in order to obtain this differentiability that we choose that unnatural weighting. But in this case, F is not an immersion, because if we rotate the sphere with the $m/2$ points, then the distances between pairs of them are the same; that is all points $Q \in \mathcal{Q}$ corresponding to rotations of a given configuration of $m/2$ points on the sphere correspond to the same point $Q' \in \mathcal{Q}'$. As the orthogonal group in three-dimensional space is three-dimensional, the rank of $A^{\mathrm{T}}A$ is at most $m-3$. It follows from what we have supposed that this rank is exactly $m-3$, and the dimension of $\mathcal{Q}'$ is also $m-3$ while that of $\mathcal{Q}$ is m. But nevertheless the manifold $\mathcal{Q}' \subset \mathcal{O}$ is well defined and so is the problem to find a point of $\mathcal{Q}$ with the smallest distance to T. Our geodetic experience and optimism even promises us that this problem, which is properly the adjustment problem, should not be complicated by large curvatures. The problem to find $Q \in \mathcal{Q}$ corresponding to $Q' \in \mathcal{Q}' \subset \mathcal{O}$ is then to find the inverse image of $\mathcal{Q}'$ by F. This is why I usually say that the difficulty of free adjustment is not in the problem of adjustment but only in the problem of inverse mapping.

If we consider distance networks on an ellipsoid instead of on a sphere, the situation changes. Suppose the $m/2$ points on an ellipsoid E with semi-axes $a > b > c$:

$$E = \Big\{ (x, y, z) \in \mathbf{R}^3 \;\Big|\; \tfrac{x^2}{a^2} + \tfrac{y^2}{b^2} + \tfrac{z^2}{c^2} = 1 \Big\}.$$

The only transformations of E onto itself which are isometric is the finite group generated by reflections with respect to the coordinate axes. Therefore generally a configuration of $m/2$ points on E cannot be displaced with a small amount without a change in the distances between them, i.e., generally the position of a distance network on E is locally determined. Clearly there are exceptions, e.g., the number of points must at least be 5, but we may suppose that the number of points and the number of distances suffice.

In the elliptic case, $\mathcal{Q}_E$ is the product manifold of $m/2$ ellipsoids. We have equipped $\mathcal{Q}$ with the suffix E in order to distinguish it from the corresponding manifold in the spherical case, which from now on will be denoted by $\mathcal{Q}_S$. Its dimension is still m, but the dimension of $\mathcal{Q}'_E$ is now also m (and not $m-3$ as for the sphere) and the mapping $F_E : \mathcal{Q}_E \to \mathcal{Q}'_E$ is here an immersion.

In order to be able to compare adjustment on the sphere and ellipsoid we will choose a mapping G of the ellipsoid E onto the sphere S with radius r and center at 0. r will later be fixed to $\sqrt{ac}$.

$$G : E \to S; \qquad (x, y, z) \longmapsto \big(\tfrac{a}{r}x, \tfrac{b}{r}y, \tfrac{c}{r}z\big).$$

This mapping is evidently differentiable and one-to-one, and its inverse $G^{-1} : S \to E$ is also differentiable.

We can now define a 'pseudo-rotation' R^* on the ellipsoid E as

$$R^* = G^{-1}RG,$$

where R is an orthogonal transformation of the sphere S. This transformation R^* is evidently differentiable and one-to-one.

A given configuration of $m/2$ points on E determines as above the points $Q_E \in \mathcal{Q}_E$ and $Q'_E \in \mathcal{Q}'_E \subset \mathcal{O}$, but applying the transformation G to each of the points we find corresponding points $Q_S \in \mathcal{Q}_S$ and $Q'_S \in \mathcal{Q}'_S \subset \mathcal{O}$.

If we apply a pseudo-rotation R^* on this configuration, the corresponding point, let us call it $R^*Q'_E$, is in general different from Q'_E, but RQ'_S will be equal to Q'_S, and we see that each point $Q'_S \in \mathcal{Q}'_S$ will correspond to a three-dimensional manifold of points in $\mathcal{Q}'_E$ as R varies over the orthogonal group in three dimensions.

If a/c is 'near to one' then the ratio between distances between corresponding pairs of points on E and on S are also near to one, and therefore the distances in $\mathcal{O}$ between points Q'_S and the corresponding points Q'_E are small. On account of the continuity of the mappings involved, for any $\epsilon > 0$ it is possible to find a three-axed ellipsoid such that for every $Q'_S \in \mathcal{Q}'_S$ and the corresponding points $Q'_E \in \mathcal{Q}'_E$ the distance $(Q'_S Q'_E) < \epsilon$, i.e., the whole m-dimensional manifold $\mathcal{Q}'_E$ is in the interior of a 'tubular' neighborhood defined as those points in $\mathcal{O}$ which have a distance less than ϵ from the manifold $\mathcal{Q}'_S$.

The 'tube' itself, i.e., the set of points $P \in \mathcal{O}$ such that $\text{dist}(\mathcal{Q}_S P) = \epsilon$ is an $N-1$-dimensional submanifold of $\mathcal{O}$ which at each point has $(N-1)-(m-3) = N-m+2$ orthogonal tangents such that the radii of curvature in the directions of these tangents are equal to ϵ. Now $\mathcal{Q}_E$ is an m-dimensional closed manifold which is situated inside this tube.

Let $\mathcal{C}$ be a closed curve in 3-space such that the curvature of $\mathcal{C}$ is less than $1/\epsilon$, and let $\mathcal{T}$ be the tube consisting of the points with distance ϵ from $\mathcal{C}$. Then it is clear that $\mathcal{T}$ at every point has a circle of curvature with radius ϵ (and center on $\mathcal{C}$). If $\mathcal{V}$ is a differentiable surface inside $\mathcal{T}$ and the greatest curvature at a point $P \in \mathcal{V}$ is $c(P)$, then it seems plausible that the mean value of $c(P)$ over $\mathcal{V}$ is greater than $1/\epsilon$, but I must confess that I have not yet found a proof for it. If the reader will admit this conjecture as a fact, then he will perhaps go a step further and admit that the three greatest curvatures of $\mathcal{Q}_E$ at a point for most points of $\mathcal{Q}_E$ are greater than $1/\epsilon$. But if this is so he will also understand that for the adjustment problem we are discussing, the curvature can be arbitrarily great if only a/c is sufficiently near to one.

Finally we will find an estimation of ϵ as a function of $a/c - 1$, but first we must release ourselves from the unnatural weight-convention with which we have worked above; in fact we will suppose that it is the logarithms of the distances which are observed with equal weights. The price we have for such an adaption of the mathematical model is that $\mathcal{Q}_S$ and $\mathcal{Q}_E$ are not differentiable manifolds any longer: The observations—the logarithms of the distances—are even not continuous when the distances go to zero. But for every network there is a positive number such that distances between points less than this number will never be actual; therefore it seems as we may ignore this blemish.

First we have to compare distances between pairs of points on the ellipsoid E and distances between the corresponding pairs of points on the sphere S, when the correspondence is given by the mapping G.

For the scale μ of the mapping G we have evidently

$$\sqrt{\tfrac{c}{a}} \le \mu \le \sqrt{\tfrac{a}{c}},$$

so if C is a curve on S and the length of C is, say λ_S, then for the length of $G^{-1}(C)$ on E, say λ_E, we have

$$\sqrt{\tfrac{c}{a}}\,\lambda \le \lambda_E \le \sqrt{\tfrac{a}{c}}\,\lambda_S$$

and

$$\sqrt{\tfrac{c}{a}}\,\lambda \le \lambda_E \le \sqrt{\tfrac{a}{c}}\,\lambda_E.$$

Now let C be a (minimal) geodesic connecting the points A and $B \in E$, then

$$\operatorname{dist}(A,B) = \lambda_E \ge \sqrt{\tfrac{c}{a}}\,\lambda_S \ge \sqrt{\tfrac{c}{a}}\,\operatorname{dist}(GA,GB).$$

On the other hand, let C' be a shortest great circle arc connecting the points GA and GB, then

$$\operatorname{dist}(GA,GB) = \lambda'_S \ge \sqrt{\tfrac{c}{a}}\,\lambda_E \ge \sqrt{\tfrac{c}{a}}\,\operatorname{dist}(A,B),$$

i.e.,

$$\sqrt{\tfrac{c}{a}} \le \frac{\operatorname{dist}(A,B)}{\operatorname{dist}(GA,GB)} \le \sqrt{\tfrac{a}{c}}$$

or

$$\left|\log\operatorname{dist}(A,B) - \log\operatorname{dist}(GA,GB)\right| \le \tfrac{1}{2}\log\tfrac{a}{c} \le \tfrac{a-c}{2c}.$$

From this and the definition of N and ϵ follows

$$\epsilon \le \tfrac{a-c}{2c}\sqrt{N}.$$

Epilogue

Indeed the position of a distance network on a three-axed ellipsoid is theoretically estimable, but no geodesist would find it in this way, for it is only poorly estimable. Nevertheless I have analyzed this example because I think it helps us to understand the non-linear nature of some adjustment problems better, and also because it emphasizes the strength of the geometrical reasoning in connection with such problems at a time where the statistical aspects of adjustment almost control the market. The depth of an object is best perceived by changing the point of view. In another—not yet published—paper, I have discussed a more down-to-earth problem, the Helmert transformation in three dimensions; from the point of view of non-linearity. Probably nobody has observed this non-linearity—simply because they have not tried to iterate the linearized adjustment procedure.

We have now arrived at the question: How can we make sure that the results of our adjustments are not invalidated from effects of non-linearity?

As far as I see, the simplest and most effective way is to iterate the linear adjustment procedure from scratch, i.e., to recalculate the linearized observation equations with the result from the foregoing adjustment as provisional value, compute the normal equations, etc. Also for purely linear problems, this strategy will have a favorable effect, so—in my opinion—an adjustment can not be looked upon as finished if it has not been verified or improved by such an iteration.

In this connection I should like to mention a concept which has been used constantly at the Danish Geodetic Institute for 15 years as an indicator mainly for the numerical behavior of adjustments. For obvious reasons we call it ENP, exponent of numerical precision.

ENP is calculated as

$$-\log_{10}\sqrt{\frac{\sum c_j^2}{\sum p_i v_i^2}},$$

where $\{c_j\}$ are the constants on the right-hand sides of the Cholesky-reduced normal equations, and it is an estimate for the number of *numerically* significant decimals of the adjustment calculations. The idea behind ENP is the philosophy that the measure for the numerical precision of an adjustment calculation is the angle under which the distance from the calculated adjustment result to the exact adjustment result is seen from the observation point in observation space. This angle cannot be computed because the exact result is unknown, but ENP can, and it gives valuable information. Especially when curvature behavior of an adjustment is dangerous, the ENP will reach a reasonable value only after many iterations or perhaps not at all.

So my first advice to geodesists who have seen that curvature may be dangerous: Use iteration from scratch and ENP!

I can give no references to publications treating adjustment from the aspect of this paper but I can mention that most of the ideas used here can be traced back to the book [1].

References

[1] John Milnor. *Morse Theory*, volume 51 of *Annals of Mathematics Studies*. Princeton University Press, 1963.

22

Mechanics of Adjustment

The tendency of looking for structures common for different branches of mathematics and applications of mathematics has been very pronounced in the last fifty years. Clearly the establishment of such common structures is advantageous for the economy of reasoning but also for the understanding of different branches of the science for the possibility of illuminating problems and methods in one branch by corresponding problems and methods in other branches.

It is well-known that part of adjustment theory may be expressed by a geometric structure—namely orthogonal projection, it is also known that there is an analogy between adjustment and statics, in fact the notion of forces is often used heuristically in the discussion of geodetic networks.[1]

On the other hand modern presentations of analytical dynamics are founded upon symplectic geometry i. e. the geometrical structure of phase space, and it is proved that this structure is also inherent in statics, therefore I have found that in a time where analytical dynamics has attracted the attention of many geodesists it could perhaps be of interest to try to find the corresponding structure in adjustment theory. In fact I shall prove that there is a common geometrical structure in mechanics, in non-linear adjustment, and in Sansò's approach to Molodenskiy's problem.

Non-Linear Adjustment

We map the values of n observations in the usual way to a point $\boldsymbol{u}_0$ in $\mathcal{U}$, the n-dimensional observation space, which is a Euclidean space with the constant metric tensor P, the weight matrix of observations. The coordinate axes correspond in a one-to-one way to the single observations. The square of the distance from any point $\boldsymbol{u} \in \mathcal{U}$ to the observation point is then

$$2E = (\boldsymbol{u} - \boldsymbol{u}_0)^{\mathrm{T}} P(\boldsymbol{u} - \boldsymbol{u}_0).$$

E is a quadratic polynomial in the coordinates $\{\, \boldsymbol{u}_\alpha \mid \alpha = 1, 2, \ldots, n \,\}$ of $\boldsymbol{u}$.

We shall call E *the energy function* in $\mathcal{U}$:

$$E : \mathcal{U} \to \mathbf{R}.$$

The energy is non-negative and zero only for $\boldsymbol{u} = \boldsymbol{u}_0$.

$$dE = \sum_\alpha \frac{\partial E}{\partial \boldsymbol{u}_\alpha}\, d\boldsymbol{u}_\alpha$$

is a differential one-form on $\mathcal{U}$, which we shall call the *force form* on $\mathcal{U}$, and the vector

$$\boldsymbol{\tau} = \{\boldsymbol{\tau}_\alpha\} = \left\{ \frac{\partial E}{\partial \boldsymbol{u}_\alpha} \right\} \in \mathcal{U}' \qquad \text{(the dual of } \mathcal{U}\text{)}$$

is called the *force*. We can look upon $\boldsymbol{\tau}$ as an elastic force directed more or less towards the observation point. If P is not a scalar multiple of the unit matrix there is a 'plumb-line declination.' Exactly as when we treat the gravity field of the Earth we here give the wrong sign to the force!

Let the purpose of our observations be to determine a point $\boldsymbol{x} \in \mathcal{A}$, where $\mathcal{A}$ is an m-dimensional differential manifold ($m < n$), for which we have given a mapping

$$f : \mathcal{A} \to \mathcal{U}.$$

f is a vector-valued function such that

$$f(\boldsymbol{x}) = \{\, f_\alpha(\boldsymbol{x}) \mid \alpha = 1, 2, \ldots, n \,\}$$

where $f_\alpha(\boldsymbol{x})$ is the theoretical value of the i'th observation corresponding to the point $\boldsymbol{x} \in \mathcal{A}$. We suppose that f is (sufficiently many times) differentiable and that the mapping f is an immersion, i. e. for any admissible coordinate system $\{\, \boldsymbol{x}^i \mid i = 1, 2, \ldots, m \,\}$ the matrix

$$A = \{A_{\alpha i}\} = \left\{ \frac{\partial f_\alpha}{\partial \boldsymbol{x}^i} \right\}$$

has rank $= m$. Then the mapping of $\mathcal{A}$ to its image in $\mathcal{U}$ is locally one-to-one. Under this condition $\mathcal{A}$ becomes a Riemannian space with the metric tensor

$$g = \{g_{ij}\} = \left\{ \sum_\alpha \frac{\partial f_\alpha}{\partial \boldsymbol{x}^i} \frac{\partial f_\alpha}{\partial \boldsymbol{x}^j} \right\} \tag{1}$$

which defines the metric on the image of $\mathcal{A}$ as a submanifold of the Euclidean space $\mathcal{U}$.

We can also pull the energy and the force form back to $\mathcal{A}$:

We define $E_\mathcal{A}$, the energy function on $\mathcal{A}$, as

$$E_\mathcal{A}(\boldsymbol{x}) = E\big(f(\boldsymbol{x})\big),$$

and we call the differential one-form $E_\mathcal{A}$ *the force form on* $\mathcal{A}$.

In any given admissible coordinate system $\{x_i\}$ on $\mathcal{A}$ we define the m-dimensional vector $\boldsymbol{\sigma} = \{\, \boldsymbol{\sigma}^i \mid i = 1, 2, \ldots, m \,\}$ by

$$dE_{\mathcal{A}} = \sum_i \boldsymbol{\sigma}^i \, d\boldsymbol{x}^i,$$

i. e.

$$\boldsymbol{\sigma}^i = \frac{\partial E_{\mathcal{A}}}{\partial \boldsymbol{x}^i}$$

or, for short,

$$\boldsymbol{\sigma} = \frac{\partial E_{\mathcal{A}}}{\partial \boldsymbol{x}}. \tag{2}$$

We shall say that a point $\boldsymbol{x}_0 \in \mathcal{A}$ is a solution point of the adjustment problem if the distance between $\boldsymbol{u}_0$ and $f(\boldsymbol{x})$ for $\boldsymbol{x} \in \mathcal{A}$ attains a local minimum at $\boldsymbol{x} = \boldsymbol{x}_0$. Evidently this means that $E_{\mathcal{A}}(\boldsymbol{x})$ attains a minimum at $\boldsymbol{x}_0$ and a necessary condition for this is that the force vanishes at $\boldsymbol{x}_0$, that is $\boldsymbol{x}_0$ is a point for which the right hand member of (2) vanishes.

If

$$\det\left(\frac{\partial^2 E_{\mathcal{A}}}{\partial \boldsymbol{x}^i \partial \boldsymbol{x}^j}\right) \neq 0 \qquad i, j = 1,\, 2,\, \ldots,\, m$$

for $\boldsymbol{x} = \boldsymbol{x}_1$, say, then from the implicit function theorem follows that if

$$\boldsymbol{\sigma}_1 = \left(\frac{\partial E_{\mathcal{A}}}{\partial \boldsymbol{x}}\right)_{\boldsymbol{x}=\boldsymbol{x}_1},$$

then in a neighborhood of $\boldsymbol{\sigma}_1$ the equation (2) has a solution $\boldsymbol{x}$. By a well-known reasoning[2] there exists a function F from the cotangent space of $\mathcal{A}$ to the reals such that this solution can be expressed by

$$-\boldsymbol{x}^i = \frac{\partial F}{\partial \boldsymbol{\sigma}^i},$$

or

$$-\boldsymbol{x} = \frac{\partial F}{\partial \boldsymbol{\sigma}}, \tag{3}$$

and

$$E_{\mathcal{A}} - F - \sum \boldsymbol{\sigma}^i \boldsymbol{x}^i = 0.$$

Evidently

$$\frac{\partial \boldsymbol{\sigma}^i}{\partial \boldsymbol{x}^j} = \frac{\partial^2 E_{\mathcal{A}}}{\partial \boldsymbol{x}^i \partial \boldsymbol{x}^j},$$

and

$$-\frac{\partial \boldsymbol{x}^j}{\partial \boldsymbol{\sigma}^i} = \frac{\partial^2 F}{\partial \boldsymbol{\sigma}^i \partial \boldsymbol{\sigma}^j},$$

therefore the matrices

$$-\left[\frac{\partial^2 E_{\mathcal{A}}}{\partial \boldsymbol{x}^i \partial \boldsymbol{x}^j}\right] \qquad \text{and} \qquad \left[\frac{\partial^2 F}{\partial \boldsymbol{\sigma}^i \partial \boldsymbol{\sigma}^j}\right]$$

are inverse to one-another, for corresponding values of $\boldsymbol{x}$ and $\boldsymbol{\sigma}$.

It should be noticed that $\left[\frac{\partial^2 E_{\mathcal{A}}}{\partial \boldsymbol{x}^i \partial \boldsymbol{x}^j}\right]$ is not equal to g in (1), the coefficient matrix of the so-called linearized normal equations, it depends also on the curvature behaviour of the image of $\mathcal{A}$ in $\mathcal{U}$.

The forces are as differential one-forms elements of the cotangent spaces of $\mathcal{A}$ as are the momentums in analytical dynamics, and as in dynamics we will combine the space $\mathcal{A}$, the configuration space, and the cotangent spaces of $\mathcal{A}$ (they are all congruent to $\mathbf{R}^m$), to an m-dimensional manifold $\mathcal{S}$, the phase-space which is the same as the cotangent bundle $T^*\mathcal{A}$, of $\mathcal{A}$. Let

$$\theta = \sum_i \boldsymbol{\sigma}^i \, d\boldsymbol{x}^i$$

and

$$\omega = d\theta = \sum_i d\boldsymbol{\sigma}^i \wedge d\boldsymbol{\sigma}^i$$

be the canonical 1- respective 2-forms on $\mathcal{S} = T^*\mathcal{A}$, such that $(\mathcal{S}, \omega)$ is a symplectic space.

The transformation

$$\begin{bmatrix} \boldsymbol{x} \\ \boldsymbol{\sigma} \end{bmatrix} \to \begin{bmatrix} \boldsymbol{\sigma} \\ -\boldsymbol{x} \end{bmatrix}$$

is a canonical transformation, and we may express part of the result of this section by saying that the execution of this transformation corresponds to solving the normal equations.

Partial Adjustment

In this section we suppose that the m-dimensional manifold $\mathcal{A}$ introduced in the foregoing section may be regarded as a product manifold of a p-dimensional manifold $\mathcal{B}$ and a q-dimensional manifold $\mathcal{C}$, i. e. $\mathcal{A}$ is diffeomorphic to the set of ordered pairs $(\boldsymbol{x}, \boldsymbol{y})$, with $\boldsymbol{x} \in \mathcal{B}$ and $\boldsymbol{y} \in \mathcal{C}$. Naturally $p + q = m$.

The mapping

$$F : \mathcal{A} = \mathcal{B} \times \mathcal{C} \to \mathcal{U}$$

is now a function of the two variables $\boldsymbol{x}$ and $\boldsymbol{y}$, and also the energy $E_{\mathcal{A}}$ on $\mathcal{A}$ is a function of these variables.

The force form $dE_{\mathcal{A}}$ may be decomposed into $\mathcal{B}$- and $\mathcal{C}$-components; or rather $T^*\mathcal{B}$- and $T^*\mathcal{C}$-components:

$$dE_{\mathcal{A}} = \left\langle \frac{\partial E}{\partial \boldsymbol{x}}, d\boldsymbol{x} \right\rangle + \left\langle \frac{\partial E}{\partial \boldsymbol{y}}, d\boldsymbol{y} \right\rangle = dE_{\mathcal{B}} + dE_{\mathcal{C}},$$

but it is important to remember that $dE_{\mathcal{B}}$ as well as $dE_{\mathcal{C}}$ depend on both $\boldsymbol{x}$ and $\boldsymbol{y}$ in analogy with the way partial differential coefficients do! It is also important to observe that $dE_{\mathcal{B}}$ and $dE_{\mathcal{C}}$ are independent of the choice of coordinates in $\mathcal{B}$ and $\mathcal{C}$.

In an evident way we can define the $\mathcal{B}$- respective $\mathcal{C}$-phase spaces $\mathcal{S}_{\mathcal{B}}$ respective $\mathcal{S}_{\mathcal{C}}$ with the respective canonical two-forms $\omega_{\mathcal{B}}$ and $\omega_{\mathcal{C}}$, with

$$\omega = \omega_{\mathcal{B}} + \omega_{\mathcal{C}},$$

and perhaps more correctly

$$\omega = (\omega_{\mathcal{B}}, 0) + (0, \omega_{\mathcal{C}}).$$

Now we can define partial adjustment:

> Let the point $\boldsymbol{y} \in \mathcal{C}$ be given, find a point $\boldsymbol{x} \in \mathcal{B}$ such that the energy attains a local minimum, and find the force form $dE_{\mathcal{C}}$.
> A necessary condition is that the force form $\omega_{\mathcal{B}}$ vanishes.

By introducing local coordinate systems $\boldsymbol{x} = \{\, \boldsymbol{x}^i \mid i = 1, 2, \dots, p\}$ and $\boldsymbol{y} = \{\, \boldsymbol{y}^k \mid k = 1, 2, \dots, q\}$, we can write the force vectors in $\mathcal{B}$ and $\mathcal{C}$ as

$$\boldsymbol{\tau}^i = \frac{\partial E_{\mathcal{A}}}{\partial \boldsymbol{x}^i} \qquad \text{and} \qquad \boldsymbol{\lambda}^k = \frac{\partial E_{\mathcal{A}}}{\partial \boldsymbol{y}^k};$$

and we have

$$d\boldsymbol{\tau}^i = \sum_j \frac{\partial^2 E_{\mathcal{A}}}{\partial \boldsymbol{x}^i \partial \boldsymbol{x}^j}\, d\boldsymbol{x}^j + \sum_l \frac{\partial^2 E_{\mathcal{A}}}{\partial \boldsymbol{x}^i \partial \boldsymbol{y}^l}\, d\boldsymbol{y}^l;$$

$$d\boldsymbol{\lambda}^k = \sum_j \frac{\partial^2 E_{\mathcal{A}}}{\partial \boldsymbol{y}^k \partial \boldsymbol{x}^j}\, d\boldsymbol{x}^j + \sum_l \frac{\partial^2 E_{\mathcal{A}}}{\partial \boldsymbol{y}^k \partial \boldsymbol{y}^l}\, d\boldsymbol{y}^l;$$

or, using matrices:

$$\begin{aligned} d\boldsymbol{\tau} &= A\, d\boldsymbol{x} + B\, d\boldsymbol{y}; \\ d\boldsymbol{\lambda} &= B^{\mathrm{T}}\, d\boldsymbol{x} + C\, d\boldsymbol{y}. \end{aligned} \tag{4}$$

If $\det A \neq 0$ then the system can be solved locally with respect to $-d\boldsymbol{x}$ and $d\boldsymbol{\lambda}$:

$$\begin{aligned} -d\boldsymbol{x} &= -A^{-1}\, d\boldsymbol{\tau} + A^{-1} B\, d\boldsymbol{y}; \\ d\boldsymbol{\lambda} &= B^{\mathrm{T}} A^{-1}\, d\boldsymbol{\tau} + (C - B^{\mathrm{T}} A^{-1} B)\, d\boldsymbol{y}. \end{aligned}$$

The fact that the matrix of the coefficients of this system, i. e.

$$\begin{bmatrix} -A^{-1} & A^{-1}B \\ B^{\mathrm{T}} A^{-1} & C - B^{\mathrm{T}} A^{-1} B \end{bmatrix}$$

is symmetric makes it plausible that there exists a function $F(\boldsymbol{\tau}, \boldsymbol{y})$ such that

$$-\boldsymbol{x} = \frac{\partial F}{\partial \boldsymbol{\tau}} \qquad \text{and} \qquad \boldsymbol{\lambda} = \frac{\partial F}{\partial \boldsymbol{y}};$$

or

$$-\sum_i \boldsymbol{x}^i\, d\boldsymbol{\tau}^i + \sum_j \boldsymbol{\lambda}^j\, d\boldsymbol{y}^j = dF. \tag{5}$$

In fact subtracting (5) from

$$\sum_i \boldsymbol{\tau}^i \, d\boldsymbol{x}^i + \sum_j \boldsymbol{\lambda}^j \, d\boldsymbol{y}^j = dE$$

gives

$$\sum_i (\boldsymbol{\tau}^i \, d\boldsymbol{x}^i + \boldsymbol{x}^i \, d\boldsymbol{\tau}^i) = dE - dF$$

or

$$d(E - F) = d\left(\sum_i \boldsymbol{\tau}^i \boldsymbol{x}^i\right),$$

so

$$F = E - \sum_i \boldsymbol{\tau}^i \boldsymbol{x}^i$$

will do, and we see that solving the partial adjustment problem corresponds to using the canonical transformation defined by

$$\begin{bmatrix} \boldsymbol{x} \\ \boldsymbol{y} \\ \boldsymbol{\sigma} \\ \boldsymbol{\lambda} \end{bmatrix} \to \begin{bmatrix} \boldsymbol{\sigma} \\ \boldsymbol{y} \\ -\boldsymbol{x} \\ \boldsymbol{\lambda} \end{bmatrix}.$$

As mentioned above the condition for the local solution of the problem is that $\det A \neq 0$, that is that $\boldsymbol{x}$ is determined by the observations and the fixed value of $\boldsymbol{y}$, therefore the method can be applied in the typical geodetic case of a (more or less) free network where at least sufficiently many parameters are fixed.

Conclusion

We have seen that certain canonical transformations play a role in adjustment, namely what could be called interchange transformations. It is natural to ask if other—or perhaps all—canonical transformations of the phase space should have obvious interpretations also. We shall see here that there are two subgroups of all the canonical transformations which have:

1. A canonical transformation which maps the configuration space onto itself and which also maps the force space onto itself is called a point transformation; it is simply a diffeomorphism of the configuration space and the contragradient diffeomorphism of the force space. In adjustment theory it corresponds simply to a coordinate transformation in the admissible subspace $\mathcal{A}$.
2. A canonical transformation which maps every point of the configuration space in itself can be defined in coordinates by

$$\begin{bmatrix} \boldsymbol{x}^i \\ \boldsymbol{\sigma}^i \end{bmatrix} \to \begin{bmatrix} \boldsymbol{x}^i \\ \boldsymbol{\sigma}^i + \frac{\partial f}{\partial \boldsymbol{x}^i} \end{bmatrix}$$

where f is any differentiable function from configuration space to $\mathbf{R}$. But this is exactly the way the observation data is introduced in the phase space.

Transformations of the three kinds mentioned here generate the whole group of canonical transformations, but I still miss a good idea for a natural interpretation of this whole group in adjustment theory.

In the note[2] I mentioned the Legendre transformation, so it is natural to compare the transformation theory here with Sansò's transformation of the Molodenskiy problem, [3]. This transformation is the same as that used here in the section non-linear adjustment, but it is commonly called the Legendre transformation, which can be justified because the configuration space there is (part of) a Euclidean space, and a Euclidean space is canonically isomorph to its tangent space at every point. Nevertheless I think it would be more fruitful to regard it as a canonical transformation. There exist canonical transformations which near the surface of the Earth are equal to Sansò's transformation but which far from the surface are equal to the identity. Using such a transformation it would perhaps be possible to generalize Sansò's theory to rotating planets also.

Notes

[1]An actual example of such reasonings is on pp. 104 and 118 in Peter Meissl's formidable work [2].

[2]The canonical transformation we introduce here has a striking likeness to the Legendre transformation, but it is not the same thing. The Legendre transformation is a fibre mapping between the tangent bundle of the phase space to the cotangent bundle, the mapping here is between the phase space itself and the cotangent bundle (the moment space—or as it should be called in connection with adjustment—the force space). For a beautiful introduction to the modern treatment of analytical dynamics and especially for the definition of the Legendre transformation, see [1]. Moreover it will help the understanding of this transformation to confront it with its generalization in the next section.

References

[1] Saunders Mac Lane. Hamiltonian mechanics and geometry. *American Mathematical Monthly*, 77:570–586, 1970.

[2] Peter Meissl. *A Priori Prediction of Roundoff Error Accumulation in the Solution of a Super-Large Geodetic Normal Equation System.* Professional Paper 12. National Oceanic and Atmospheric Administration, 1980.

[3] Helmut Moritz. *Advanced Physical Geodesy.* Wichmann, Karlsruhe, 1980.

23

Angelica Returning or The Importance of a Title

Shortly after he had become president for a study group on mathematical problems in physical geodesy in 1967 Professor Moritz wrote a circular letter to the members of this study group, where he pointed out some fundamental problems, which he hoped would be taken up by the members. This letter was a great encouragement to continue my work seriously, for now I saw that I would have at least one reader, for whom these problems really mattered.

Some months later I wrote a letter to the study group where I sketched the lines along which I had planned to attack the problems pointed out by Moritz.

Again a few months later, when I was trying to combine what is now known as collocation with Moritz's method for interpolation of gravity anomalies, I found that the first thing (collocation) was simply a generalization of the second. This discovery fascinated me so much that I made it the starting point in [2] in the hope that it would be easier to understand starting from well-known facts and generalizing a well established statistical method.

Considering the fact that I have always lacked the right feeling for statistics this was a bad idea. I did not consider the fact and I have often had occasion to regret it afterwards, but I had the satisfaction that my colleagues have judged the result rather by what is right than by what is wrong in it, and I am very grateful to them for that.

It is curious, however, that it seems as geodesists have read the mentioned booklet as if its title were: 'Collocation a new method for calculation of the gravity field' and not as a contribution to the mathematical foundation of physical geodesy as the title really is. Naturally I was not uninterested in using collocation for practical computations but my main purpose was to use it as an important link in the answer to the fundamental problem in physical geodesy e.g. as formulated in Moritz's circular letter.

I shall try here, avoiding technical details, to summarize my thoughts about the foundation of physical geodesy, this paper may at least be a key to [2] and [3] for the reader who does not wish to make his discovering himself using hints which are not necessarily placed exactly where he would expect to find them.

1

The fundamental problem of physical geodesy may be formulated as follows: how may we from a finite set of observations concerning the potential of the Earth—or, mathematically from a finite set of observations of linear functionals of the potential—calculate a useful approximation to the potential itself.

It is very difficult to work with functions which are harmonic only outside surfaces of complicated form, such as the surface of the Earth, therefore it is important to know that any function which is harmonic outside the Earth may be approximated arbitrarily well by a function which is harmonic overall outside any given nonvoid closed point set in the interior of the Earth, e.g. down to any Bjerhammar sphere (i.e. a sphere inside the Earth) or even down to (but not including) the center of the Earth. (The precise meaning of the expression that a function outside the Earth may be approximated arbitrarily well is that for any closed point set outside the Earth the absolute value of the error of the approximation can be made less than any $\epsilon > 0$). This result—the Runge theorem—was well known for elliptic differential equations in the sixties.

From Runge's theorem many important facts may be deduced but certainly not, that one may treat series in spherical harmonics as if they were convergent overall where the potential they represent is regular. The following example is illustrative in this direction.

In [2], pages 52–53, it is demonstrated how a harmonic function given by its expansion in a series in spherical harmonics convergent outside a sphere with radius R may be approximated by another harmonic function expressed by a series in spherical harmonics converging outside another sphere concentric with the first one, but with radius $R_1 < R$. But this approximation is generally a good one only outside the first sphere. So if the first sphere contains the Earth and the second one is contained in the Earth (i.e. is a Bjerhammar sphere) then we can not find an approximation to a function harmonic outside the Earth which is valid down to the surface of the Earth. To the reader who wants to use Runge's theorem I recommend thorough study of this example.

Then how can we find a Runge-approximation down to the Earth?

In order to answer this question we must first quantitatively describe what it is we want.

Let ϕ be the known function which is harmonic in the (open) domain $\mathcal{D}$ then we want to find a function ϕ_1, which is harmonic in a greater domain $\mathcal{D}_1$ such that $\bar{\mathcal{D}}_1 \supset \mathcal{D}$, where $\bar{\mathcal{D}}$ is the closure of $\mathcal{D}_1$ and such that $|\phi_1 - \phi|$ is 'small' in $\mathcal{D}$. In order to define what we mean by $|\phi_1 - \phi|$ is small we define a norm for harmonic functions in $\mathcal{D}$ such that this norm of ϕ is finite. But even by prescribing the norm of $\phi_1 - \phi$ the approximating function ϕ, is not determined. We can introduce also a norm for harmonic functions in $\mathcal{D}_1$ and call the two norms $\|\cdot\|$ and $\|\cdot\|_1$ respectively, then it is reasonable to look for a ϕ_1 such that both $\|\phi_1\|_1$ and $\|\phi_1 - \phi\|$ are small. If ϕ is bounded in $\mathcal{D}$ the most obvious choice for the norm would be $\|\phi\| = \sup_{x \in \mathcal{D}} |\phi(x)|$ such that $\|\phi_1 - \phi\| = \sup_{x \in \mathcal{D}} |\phi_1(x) - \phi(x)|$. It is possible to build up a theory using these norms but if we choose quadratic norms (Hilbert-norms) instead, all the problems will reduce to linear problems analogous to least-squares problems, therefore we shall in the following suppose that all norms used are quadratic norms.

Given such norms we will look for a ϕ, which minimizes the expression $\|\phi_1\|_1^2 + \lambda\|\phi_1 - \phi\|^2$ where λ is a given positive number. Clearly if it can be proved—as in fact it can—that this minimum problem has a solution for every $\lambda > 0$ then for increasing values of λ and corresponding solution ϕ_1, $\|\phi_1\|_1$ will be increasing and $\|\phi_1 - \phi\|$ decreasing. From Runge's theorem it then follows that for λ going to infinity $\|\phi_1 - \phi\|$ converges to zero, but ϕ_1 does not converge unless ϕ is extendable to $\mathcal{D}_1$, in which case ϕ_1 does in fact converge to the analytic continuation of ϕ to $\mathcal{D}_1$. If we regard the restriction of ϕ_1 to the domain $\mathcal{D}$ however, then this function converges to ϕ in all cases.

I have here sketched how a Runge approximation is found for an explicitly given harmonic function, an effective proof can be based upon simple facts from potential theory and the theory for compact operators in Hilbert space. But as the harmonic function in our case is not explicitly given but is only partly determined by a finite set of observations this result does not help us very much.

2

An obvious adaption of this method would be among all harmonic functions defined in $\mathcal{D}_1$ and compatible with the, say n, observations to look for that for which $\|\cdot\|_1$ is minimum, that is to look for that ϕ_1 which minimizes $\|\phi_1\|_1$ under n given conditions. If these conditions are independent then this problem has exactly one solution, and this solution is exactly the collocation solution. We may also, as above, look for harmonic functions in $\mathcal{D}_1$ which minimize the square of $\|\cdot\|_1$ plus λ times a weighted square sum of the residuals of the observations. Then we have smoothing collocation. For $\lambda \to \infty$ we get the solution of the non-smoothing collocation problem if and only if the observations are independent. It can undoubtedly be supposed that readers of this book know how the solution of these minimum problems are found by solving system of linear equations derived by use of the theory for Hilbert-spaces with reproducing kernels, and also that the norms $\|\cdot\|_1$ used in practice are invariant with respect to rotations around the center of the Earth such that the region $\mathcal{D}_1$ becomes the part of the space outside a Bjerhammar sphere, and also that users of the collocation method agree in the choice of the norm $\|\cdot\|_1$ (or equivalently of reproducing kernel) in consideration of quasi statistical principles (covariance functions). In my opinion there has been made amazingly little research in the norm problem under other aspects also.

It should here be mentioned too that in practice the potential to which an approximation is looked for is not the potential of the Earth itself but the anomalous potential or rather the correction to the 'so-far-best-known' approximation to the potential itself, for it is naturally this correction which should be minimized.

3

What the collocation method gives us is a method for computing a harmonic function in $\mathcal{D}$, which is compatible with the observations (or in the smoothing case with a reasonable choice of λ is reasonably compatible with the observations taking their precision into account). Our problem is then in which respect our solution gives a good approximation to the Earth's potential.

Another formulation of this problem is that of a 'convergence problem of collocation':

If we have given an infinite sequence of sets of observations of functionals of the potential, such that the errors of the observations in the mth set goes to zero for $m \to \infty$ and points of the observations become dense on a certain smooth surface ω around the Earth and near its surface, under which conditions does the result of the collocation applied to the mth set of observations converge to the correct result for $m \to \infty$?

Clearly the best thing we may expect is that the restriction of ϕ_1 to the domain $\mathcal{D}$ (the part of the space outside the surface ω) converges to the (anomalous) potential of the Earth, for this potential is in general not extendable to $\mathcal{D}$ (cf. the result under 1).

Instead of attacking the convergence problem here directly let us regard first some variations of it.

First suppose that ω is known and that the value of the potential is known and finite at all points of ω. If we try to use smoothing collocation instead of n linear equations we have to solve an integral equation of Fredholm type. By analyzing this equation we would find that for $\lambda \to \infty$ the restriction of ϕ_1 to $\mathcal{D}$ in fact converges to the solution of Dirichlet's problem as we want and that the limit is independent of the norm $\|\cdot\|_1$ selected in $\mathcal{D}_1$. This follows from the unique solvability of the boundary value problem, which here is Dirichlet's problem.

There is another way however to regard this problem.

The expression which should be minimized is

$$\|\phi_1\|_1^2 + \lambda \int_\omega (\phi_1 - \phi)^2 \, d\omega,$$

but it can be proved (in fact it follows easily from some (unpublished) papers of mine of Hilbert spaces of harmonic functions) that

$$\|\phi\|_{\mathcal{D}}^2 = \int_\omega \phi(x)^2 \, d\omega$$

defines a norm for functions harmonic in $\mathcal{D}$ and so we are back to the problem treated under 1.

We can intuitively look at the limiting process as a swelling of $\|\phi_1\|_1$—and therefore also of ϕ_1—as λ increases but this swelling takes place only inside ω (in $\mathcal{D}_1 \setminus \bar{\mathcal{D}}$) exactly because $\|\phi_1 - \phi\|_{\mathcal{D}}$ is a norm and it enters with increasing weight for increasing λ. That this norm converges to 0 and therefore $\phi_1 \to \phi$ in $\mathcal{D}$ follows from Runge's theorem, so we have here an interplay between boundary value problem theory, Runge's theorem, and the theory for Hilbert spaces of harmonic functions.

This example may naturally be generalized to other boundary value problems e.g. to Molodenskiy's problem, but here it is important to notice that the boundary value problems which really are of interest in geodesy are rather complicated, because there are observed different kinds of boundary value data on overlapping regions of ω, e.g. different kinds of data over continents and oceans. There is here a great challenge for those working with boundary value problems to prove that such actual observations define norms of harmonic functions in the outer space, but this is a slight generalization of the so-called coersivity problem in the theory of boundary value problems.

We have not yet arrived at the goal. We are faced with that set of problems which in my opinion is what makes mathematical geodesy so fascinating namely those of the interplay between continuous and discrete aspects, for still we have only finitely many observations.

In my Assisi lecture I have bridged the gap between these two aspects by using a method of interpolation of the observations to achieve functions on ω. Today I should prefer a more direct method which I hope could give practical results in an easier way.

So we have again to look at our problem with only a finite sets of observations and a corresponding smoothing collocation problem. Clearly by increasing the value of λ or by increasing the number n of observations the resulting ϕ_1 becomes more 'short waved.' On the other hand the more long waved ϕ_1 is the more 'feels' ϕ_1 the observations acting on it through the factor to λ, the weighted square sums of the residuals as (the square of) a norm. When n and the value of λ approach infinity the restriction of ϕ_1 to $\mathcal{D}$ will only be bounded if the 'ideal boundary value problem' corresponding to the observations is coersive, so we see also here the relevance of the study of boundary value problems. Under assumption of coercivity in my Assisi lecture I succeeded in proving the convergence for the number of observations and λ going to infinity only if λ increases sufficiently slowly with respect to the number of observations and I made the conjecture that this was essential, such that it should be impossible to prove convergence of the non-smoothing collocation. I have not yet succeeded in proving or disproving this conjecture.

The traditional mathematical proof of convergence is build upon a manipulation of ϵ and δ. So also here. It has been said that only the result, not the ϵ δ manipulation is of interest for the user of mathematics. So it is not here. The limit we find, if it exists, is independent of e.g. the norm $\|\cdot\|_1$ but the velocity of the convergence is not. For the practical problems we are interested very much in how δ depends on ϵ and the dependance of this dependance of the kind of observations, the distance between them, of the radius of the Bjerhammar sphere etc. A deeper study of the convergence problem as sketched here should have good chances for throwing light on the practical problems in physical geodesy and perhaps not only on the collocation method. And from where should really good and new practical results come if not from a solid mathematical foundation of physical geodesy.

The second title of this paper may have given some reader the desire for an explication of the first title. Here it is.

Angelica was a Chinese princess who plays an important role in [1]. But at a certain point of time the author looses interest in her, and concentrates on other parts of the action which brings the reader all around the Earth even with excursions to the underworld and to the moon. Ariosto leaves Angelica writing that perhaps some more clever poet would write about her later adventures on her way back to China. I do not think that Ariosto were expressing a sincere hope concerning a continuation of his work, and certainly he had no particular author in mind when scribing so, but I had when I closed my 'contribution' quoting this passage from Ariosto. Relatively soon later this hope of mine was fulfilled beyond any expectation, subsequent work especially of Moritz has eminently contributed to the acceptance of the method of collocation in the geodetic community, and I have the feeling that the sporadic resistance against it has rather deepened the understanding and contributed to this acceptance.

But still .... Well, I do not pretend to have written something which can be read with pleasure even 450 years later but as I expressed it in common words in the introduction of this paper, I had hoped—and still I hope it—that not only a ramification of the plot but chiefly the 'grand tour around the world' should serve as a starting point for fruitful investigation by my colleagues.

Perhaps one should never make explicit what one has written between the lines. But I have done it here in order to counterbalance my habit of writing too much between the lines, which may have to do with the fact that I regard writing not as a means to propagate my meanings but as a means to make my colleagues and myself think.

For years I have been dreaming of writing a paper without formulas, as you see, I have not completely succeeded. I know that a paper without references would hardly be accepted, so here they are!

References

[1] Ludovico Ariosto. *Orlando Furioso.* Venezia, 1532.

[2] Torben Krarup. *A Contribution to the Mathematical Foundation of Physical Geodesy.* Meddelelse 44. Geodætisk Institut, København, 1969.

[3] Torben Krarup. A convergence problem in collocation theory. *Bollettino di Geodesia e Scienze Affini*, XL:225–240, 1981.

24

Evaluation of Isotropic Covariance Functions of Torsion Balance Observations

Abstract

Torsion balance observations in spherical approximation may be expressed as second-order partial derivatives of the anomalous (gravity) potential T:

$$T_{13} = \frac{\partial^2 T}{\partial x_1 \partial x_3}, \quad T_{23} = \frac{\partial^2 T}{\partial x_2 \partial x_3}, \quad T_{12} = \frac{\partial^2 T}{\partial x_1 \partial x_2}, \quad T_{\Delta} = \frac{\partial^2 T}{\partial x_1^2} - \frac{\partial^2 T}{\partial x_2^2}$$

where x_1, x_2, and x_3 are local coordinates with x_1 "east," x_2 "north," and x_3 "up." Auto- and cross-covariances for these quantities derived from an isotropic covariance function for the anomalous potential will depend on the directions between the observation points. However, the expressions for the covariances may be derived in a simple manner from isotropic covariance functions of torsion balance measurements. These functions are obtained by transforming the torsion balance observations in the points to local (orthogonal) horizontal coordinate systems with first axis in the direction to the other observation point. If the azimuth of the direction from one point to the other point is α, then the result of this transformation may be obtained by rotating the vectors

$$\begin{bmatrix} T_{13} \\ T_{23} \end{bmatrix} \quad \text{and} \quad \begin{bmatrix} T_{\Delta} \\ 2T_{12} \end{bmatrix}$$

the angles $\alpha - 90°$ and $2(\alpha - 90°)$ respectively.

The reverse rotations applied on the 2×2 matrices of covariances of these quantities will produce all the direction dependent covariances of the original quantities.

1. Introduction

Let $K(P, Q)$ be a rotational invariant reproducing kernel of a Hilbert space of harmonic functions, or equivalently a so-called empirical covariance function of the anomalous gravitational potential T. It is a function of two variables

P, Q, both points in $\mathbf{R}^3$ outside some sphere with radius R_B, the Bjerhammar sphere, bounding the set of harmonicity for the functions in the Hilbert space. Then

$$K(P,Q) = \sum_{i=2}^{\infty} \sigma_i \left(\frac{R_B^2}{rr'} \right)^{i+1} P_i(\cos\psi), \tag{1}$$

where ψ is the spherical distance between P and Q, r and r' the radial distances of P, Q from the origin respectively, and P_i are Legendre polynomials. σ_i are positive constants, the so-called degree-variances.

The inner product of two linear functionals L_1 and L_2 or equivalently their covariance, is obtained by applying these functionals on K

$$L_1(L_2(K(P,Q))) := K(L_1, L_2). \tag{2}$$

Expressions for such quantities where L_1 and L_2 are linear functionals associated with zero-, first-, or second-order derivatives of T are given, e.g., in [8]. Similar expressions have been derived in [2] for harmonic functions given in a half space $\{\, (x_1, x_2, x_3) \mid x_3 > 0 \,\}$.

If we want to derive the quantities $K(L_1, L_2)$ for cases where L_1 or L_2 are linear functionals associated with torsion balance (or gravity gradiometer) observations, then the equations given in [8] cannot be used directly. Certain linear combinations of the equations must be used.

The covariance functions will depend on the directions between P and Q. This makes the direct derivation of expressions for these functions rather involved. We will show here that the expressions may be derived in a very straightforward manner from *isotropic* covariance functions of torsion balance observations. Such functions are also useful in cases where empirical covariances have to be estimated in a situation where few observations are available. (This is because sampling of products of pairs of observations will have to be done only with respect to the distance between the observation points.)

Isotropic covariances are derived in Section 4 following some mathematical preparations in Sections 2 and 3. Finally in Section 5 the isotropic covariances are used to derive not only the general expressions for the covariances of torsion balance observations, but also expressions for covariances of several other gravity field quantities.

2. Basic equations

Suppose we have a usual Cartesian coordinate system with axes x, y, z so that the last one coincides with the rotation axis of the Earth. All points not located on the z-axis will then have the usual spherical coordinates ϕ, latitude, λ, longitude, and r, the distance from the origin. For these points we may also define a local coordinate system with coordinates (x_1, x_2, x_3) so that the first axis points east, the second north, and the third in the direction of the radius vector. We may keep the origin or move it to the point, as we like.

In spherical approximation we may then express the torsion balance observations as follows:

$$T_{13} = \frac{\partial^2 T}{\partial x_1 \partial x_3} = \frac{\partial}{\partial r}\Big(\frac{1}{r\cos\phi}\frac{\partial T}{\partial\lambda}\Big) \tag{3}$$

$$T_{23} = \frac{\partial^2 T}{\partial x_2 \partial x_3} = \frac{\partial}{\partial r}\Big(\frac{1}{r}\frac{\partial T}{\partial\phi}\Big) \tag{4}$$

$$T_\Delta = \frac{\partial^2 T}{\partial x_1^2} - \frac{\partial^2 T}{\partial x_2^2} = -\frac{1}{r^2}\Big(\frac{1}{\cos\phi}\frac{\partial}{\partial\phi}\Big(\cos\phi\frac{\partial}{\partial\phi}T\Big) - \frac{1}{\cos^2\phi}\frac{\partial^2 T}{\partial^2\lambda}\Big) \tag{5}$$

$$2T_{12} = 2\frac{\partial^2 T}{\partial x_1 \partial x_2} = \frac{2}{r}\frac{\partial}{\partial\phi}\Big(\frac{1}{r\cos\phi}\frac{\partial T}{\partial\lambda}\Big) \tag{6}$$

see [4] and [7], Equation (58).

For the sake of completeness, we write the equations associated with some other important types of observations, the height anomaly ζ, the gravity anomaly Δg, and the deflections of the vertical (η, ξ). Then, with γ equal to the normal gravity, and still using spherical approximation, we have

$$\zeta = T/\gamma$$

$$\Delta g = -\frac{\partial T}{\partial x_3} - \frac{2}{r}T = -\frac{\partial T}{\partial r} - \frac{2}{r}T$$

$$\eta = -\frac{1}{\gamma}\frac{\partial T}{\partial x_1} = -\frac{1}{\gamma r\cos\phi}\frac{\partial T}{\partial\lambda}$$

$$\xi = -\frac{1}{\gamma}\frac{\partial T}{\partial x_2} = -\frac{1}{\gamma r}\frac{\partial T}{\partial\phi}.$$

It is obvious that the evaluation functional or the functional $\frac{\partial}{\partial x_3}\big|_P$ will result in quantities which are still isotropic and of the form given in Equation (7); except for factors r^k or $(r')^k$, $k = -1$ or -2. T_{13} and T_{23} will in this respect be treated as if they were deflections of the vertical (η, ξ). In order to see this

$$\frac{\partial}{\partial r}\Big(\frac{1}{r}\frac{\partial}{\partial\phi}T\Big) = \frac{1}{r}\frac{\partial}{\partial\phi}\Big(\frac{\partial T}{\partial r} - \frac{1}{r}T\Big)$$

$$\frac{\partial}{\partial r}\Big(\frac{1}{r\cos\phi}\frac{\partial}{\partial\lambda}T\Big) = \frac{1}{r\cos\phi}\frac{\partial}{\partial\lambda}\Big(\frac{\partial T}{\partial r} - \frac{1}{r}T\Big).$$

Then with $i, j = 1$ or 2

$$K\Big(\frac{\partial^2}{\partial x_j \partial x_3}\Big|_P, Q\Big) = \frac{\partial}{\partial x_j}\Big|_P\Big(\frac{\partial K}{\partial r} - \frac{1}{r}K\Big)$$

$$K\Big(P, \frac{\partial^2}{\partial x_j \partial X_3}\Big|_Q\Big) = \frac{\partial}{\partial x_j}\Big|_Q\Big(\frac{\partial K}{\partial r'} - \frac{1}{r'}K\Big)$$

$$K\Big(\frac{\partial^2}{\partial x_j \partial x_3}\Big|_P, \frac{\partial^2}{\partial x_i \partial x_3}\Big|_Q\Big) = \frac{\partial}{\partial x_j}\Big|_P \frac{\partial}{\partial x_i}\Big|_Q\Big(\frac{\partial^2 K}{\partial r \partial r'} - \frac{1}{r}\frac{\partial K}{\partial r'} - \frac{1}{r'}\frac{\partial K}{\partial r} + \frac{1}{rr'}K\Big).$$

In order to facilitate the use of these quantities in the subroutine COVAX [8] we introduce the variable $s = R_B^2/(rr')$ and compute the derivatives using s. (Note, that we later use the term "s" for another quantity.)

Because

$$\frac{\partial}{\partial r} = \frac{\partial s}{\partial r}\frac{\partial}{\partial s} = -\frac{s}{r}\frac{\partial}{\partial s}, \qquad \frac{\partial}{\partial r'} = \frac{\partial s}{\partial r'}\frac{\partial}{\partial s} = -\frac{s}{r'}\frac{\partial}{\partial s}$$

we get

$$\frac{\partial T}{\partial r} - \frac{1}{r}T = -\frac{1}{r}\Big(s\frac{\partial K}{\partial s} + K\Big)$$

$$\frac{\partial T}{\partial r'} - \frac{1}{r'}T = -\frac{1}{r'}\Big(s\frac{\partial K}{\partial s} + K\Big)$$

$$\begin{aligned}\frac{\partial^2 K}{\partial r\partial r'} - \frac{1}{r}\frac{\partial K}{\partial r'} - \frac{1}{r'}\frac{\partial K}{\partial r} + \frac{1}{rr'}K &= \frac{s}{rr'}\frac{\partial}{\partial s}\Big(s\frac{\partial K}{\partial s} + K\Big) + \frac{1}{rr'}\Big(s\frac{\partial K}{\partial s} + K\Big)\\ &= \frac{1}{rr'}\Big(s^2\frac{\partial^2 K}{\partial s^2} + 3s\frac{\partial K}{\partial s} + K\Big).\end{aligned}$$

With

$$D = s\frac{\partial K}{\partial s} + K$$

$$E = \Big(s^2\frac{\partial^2 K}{\partial s^2} + 3s\frac{\partial K}{\partial s} + K\Big)s/R_B^2$$

we have

$$K\Big(\frac{\partial^2}{\partial x_j\partial x_3}\Big|_P, Q\Big) = \frac{\partial}{\partial x_j}\Big|_P (D)/(-r) \tag{7}$$

$$K\Big(P, \frac{\partial^2}{\partial x_i\partial x_3}\Big|_Q\Big) = \frac{\partial}{\partial x_i}\Big|_Q (D)/(-r') \tag{8}$$

$$K\Big(\frac{\partial^2}{\partial x_i\partial x_3}\Big|_P, \frac{\partial^2}{\partial x_j\partial x_3}\Big|_Q\Big) = \frac{\partial}{\partial x_i}\Big|_P \frac{\partial}{\partial x_j}\Big|_Q (E). \tag{9}$$

Note that

$$D = \sum_{i=2}^{\infty} \sigma_i (i+2) s^{i+1} P_i(\cos\psi)$$

and

$$E = \sum_{i=2}^{\infty} \frac{\sigma_i}{R_B^2}(i+2)^2 s^{i+2} P_i(\cos\psi).$$

3. Change of covariances caused by a rotation of the local coordinate system

It is well known that isotropic covariance functions for deflections of the vertical are obtained by using the longitudinal and transversal components of a

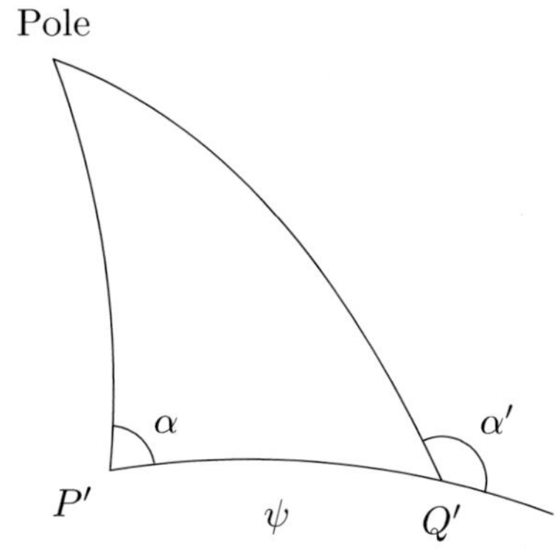

Fig. 24.1. P' and Q' are the projections of P, Q respectively on the unit sphere

pair of deflections in a point P relative to a point Q, see [1], [3], [5], and [6]. These components arise as first-order partial derivatives with respect to the coordinates of a local coordinate system obtained by rotating the local coordinate system the angle $\alpha - 90°$ with axes pointing east and north, where α is the azimuth from the point P to the point Q. Another angle α' will obviously have to be used in the point Q, see Figure 24.1.

Let us then compute the change in the second-order horizontal derivatives following a rotation β and let us denote the new coordinates z_1 and z_2. Then

$$\begin{bmatrix} z_1 \\ z_2 \end{bmatrix} = R(\beta) \begin{bmatrix} x_1 \\ x_2 \end{bmatrix} \tag{10}$$

with

$$R(\beta) = \begin{bmatrix} \cos\beta & -\sin\beta \\ \sin\beta & \cos\beta \end{bmatrix}. \tag{11}$$

For the first-order derivatives we have

$$\begin{bmatrix} \dfrac{\partial}{\partial z_1} \\ \dfrac{\partial}{\partial z_2} \end{bmatrix} = R(\beta) \begin{bmatrix} \dfrac{\partial}{\partial x_1} \\ \dfrac{\partial}{\partial x_2} \end{bmatrix} \tag{12}$$

and for the second-order derivatives

$$\begin{bmatrix} \dfrac{\partial^2}{\partial z_1^2} & \dfrac{\partial^2}{\partial z_1 \partial z_2} \\ \dfrac{\partial^2}{\partial z_1 \partial z_2} & \dfrac{\partial^2}{\partial z_2^2} \end{bmatrix} = R(\beta) \begin{bmatrix} \dfrac{\partial^2}{\partial x_1^2} & \dfrac{\partial^2}{\partial x_1 \partial x_2} \\ \dfrac{\partial^2}{\partial x_1 \partial x_2} & \dfrac{\partial^2}{\partial x_2^2} \end{bmatrix} R(-\beta). \tag{13}$$

The functionals related to the torsion balance observations $2T_{12}$ and T_Δ are then transformed as follows:

$$\frac{\partial^2}{\partial z_1^2} - \frac{\partial^2}{\partial z_2^2} = \left(\cos^2\beta - \sin^2\beta\right)\frac{\partial^2}{\partial x_1^2} - \left(\cos^2\beta - \sin^2\beta\right)\frac{\partial^2}{\partial x_2^2} - 2\sin\beta\cos\beta\, 2\frac{\partial^2}{\partial x_1 \partial x_2}$$

$$2\frac{\partial^2}{\partial z_1 \partial z_2} = 2\cos\beta\sin\beta\left(\frac{\partial^2}{\partial x_1^2} - \frac{\partial^2}{\partial x_2^2}\right) + 2\left(\cos^2\beta - \sin^2\beta\right) 2\frac{\partial^2}{\partial x_1 \partial x_2}.$$

Hence

$$\begin{bmatrix} \dfrac{\partial^2}{\partial z_1^2} - \dfrac{\partial^2}{\partial z_2^2} \\ 2\dfrac{\partial^2}{\partial z_1 \partial z_2} \end{bmatrix} = \begin{bmatrix} \cos 2\beta & -\sin 2\beta \\ \sin 2\beta & \cos 2\beta \end{bmatrix} \begin{bmatrix} \dfrac{\partial^2}{\partial x_1^2} - \dfrac{\partial^2}{\partial x_2^2} \\ 2\dfrac{\partial^2}{\partial x_1 \partial x_2} \end{bmatrix}. \tag{14}$$

The vector containing $(T_\Delta, 2T_{12})$ is then transformed by a rotation the angle 2β while the vector (T_{13}, T_{23}) is transformed by a rotation the angle β, just as if it were a vector containing the negative deflection components $(-\eta, -\xi)$.

Because of the rotational invariance of $K(P, Q)$ we may now choose the local coordinate systems in a convenient way, so that the derivative with respect to the first coordinate variable corresponds to the directional derivatives between the points. We will then work with two systems, so that the first makes the evaluation of the linear functionals associated with P easy and so that the second makes the evaluations in Q easy. The original derivatives in northern and eastern directions may then be obtained by executing the rotations $90° - \alpha$ in P and $270° - \alpha'$ in Q.

4. Evaluation of isotropic covariances

We now introduce the two coordinate systems. One is associated with P having coordinates (x_1, x_2, x_3) and the other with Q having coordinates (y_1, y_2, y_3). They are obtained from the above described local coordinate systems by rotations $\alpha - 90°$ and $\alpha' - 270°$ respectively.

The third axis will in both systems coincide with the radius vector passing through the point, and the first axis will be in the plane spanned by the two radius vectors. In both cases the direction of the first axis is selected so that the new azimuth is 0. The second axis is selected so that the three axes span a right-handed Cartesian coordinate system. Here $P = (0, 0, r)$ in the x-system and $Q = (0, 0, r')$ in the y-system. Let the angle between the radius vectors be $\psi =$ the spherical distance and denote the coordinates of a point x in the y-system by $\widetilde{x}$ and vice versa. Then we have the following coordinate transformation:

$$\begin{aligned} \widetilde{x}_1 &= -x_1 \cos\psi + x_3 \sin\psi = -x_1 t + x_3 s \\ \widetilde{x}_2 &= -x_2 \\ \widetilde{x}_3 &= x_1 \sin\psi + x_3 \cos\psi = x_1 s + x_3 t. \end{aligned} \tag{15}$$

Note that this transformation is its own inverse, i.e., $\widetilde{\widetilde{x}} = x$, and that it is also valid for $\widetilde{\widetilde{y}}$. We now put $t = \cos\psi$ and $s = \sin\psi$.

We will now express the covariance function by some more convenient parameters

$$u = (x, \widetilde{y}) = (\widetilde{x}, y) = \sum_{i=1}^{3} x_i \widetilde{y}_i = \sum_{j=1}^{3} \widetilde{x}_j y_j \tag{16}$$

$$v = \tfrac{1}{2}|x|^2|y|^2 \tag{17}$$

so that

$$K(x, y) := f(u, v).$$

However, we may here use D or E instead of K, see Section 1.

Because of our choice of coordinate system, we may now compute the derivatives in these two new systems, and subsequently evaluate the result in P and Q. The derivatives may then be transformed back to the old coordinate system using Equations (12) and (14).

We denote the derivatives of f with respect to u and v using subscripts, so that

$$\frac{\partial^m f}{\partial u^i \partial v^j} := f_{ij}, \qquad m = i + j.$$

We will need the derivatives of u and v with respect to the coordinates

$$\frac{\partial u}{\partial x_i} = \widetilde{y}_i, \qquad \frac{\partial v}{\partial x_i} = x_i|\widetilde{y}|^2 = x_i|y|^2$$

and also

$$\frac{\partial \widetilde{y}_1}{\partial y_1} = -t, \qquad \frac{\partial \widetilde{y}_1}{\partial y_3} = s, \qquad \frac{\partial \widetilde{y}_2}{\partial y_2} = -1.$$

Then

$$\frac{\partial}{\partial x_i} K = f_{10}\widetilde{y}_i + f_{01}x_i|y|^2$$

and

$$\frac{\partial^2}{\partial x_i \partial x_j} K = f_{20}\widetilde{y}_i\widetilde{y}_j + f_{11}(\widetilde{y}_i x_j + \widetilde{y}_j x_i)|y|^2 + f_{02}x_i x_j|y|^4 + \frac{\partial x_i}{\partial x_j} f_{01}|y|^2.$$

When evaluating the derivatives in P, and then subsequently in Q following $\lozenge$, we may after each *completed* differentiation insert the actual values of the coordinates. This gives us the following simple equations:

$$\frac{\partial}{\partial x_1} K = f_{10}\widetilde{y}_1 \lozenge f_{01}r's \tag{18}$$

$$\frac{\partial}{\partial x_2} K = f_{10}\widetilde{y}_2 \lozenge 0 \tag{19}$$

$$F := \frac{\partial^2 K}{\partial x_1^2} - \frac{\partial^2 K}{\partial x_2^2} = f_{20}(\widetilde{y}_1^2 - \widetilde{y}_2^2) \diamondsuit f_{20}(r's)^2 \tag{20}$$

$$G := \frac{\partial^2 K}{\partial x_1 \partial x_2} = f_{20}\widetilde{y}_1\widetilde{y}_2 \diamondsuit 0. \tag{21}$$

We must then differentiate these quantities with respect to coordinates y_i, using $P = (0, 0, r)$ in x-coordinates:

$$\frac{\partial^2 K}{\partial y_j \partial x_i} = \widetilde{y}_i\big(f_{20}\widetilde{x}_j + f_{11}y_j|x|^2\big) + \frac{\partial \widetilde{y}_i}{\partial y_j}(f_{10}) \\ + x_i\big(|y|^2(f_{11}\widetilde{x}_j + f_{02}y_j|x|^2) + 2y_j|y|f_{01}\big)$$

and then

$$\frac{\partial^2 K}{\partial y_1 \partial x_1} = \widetilde{y}_1(f_{20}\widetilde{x}_1 + f_{11}y_1|x|^2) - tf_{10} \diamondsuit s^2 rr' f_{20} - tf_{10} \tag{22}$$

$$\frac{\partial^2 K}{\partial y_2 \partial x_1} = \widetilde{y}_1(f_{20}\widetilde{x}_2 + f_{11}y_2|x|^2) \diamondsuit \widetilde{y}_1\widetilde{x}_2 f_{20} = 0 \tag{23}$$

$$\frac{\partial^2 K}{\partial y_1 \partial x_2} = \widetilde{y}_2(f_{20}\widetilde{x}_1 + f_{11}y_1|x|^2) \diamondsuit 0 \tag{24}$$

$$\frac{\partial^2 K}{\partial y_2 \partial x_2} = \widetilde{y}_2(f_{20}\widetilde{x}_2 + f_{11}y_2|x|^2) - f_{10} \diamondsuit -f_{10} \tag{25}$$

$$\frac{\partial F}{\partial y_1} = \frac{\partial}{\partial y_1}(\widetilde{y}_1^2)f_{20} + (\widetilde{y}_1^2 - \widetilde{y}_2^2)(f_{30}\widetilde{x}_1 + f_{21}y_1|x|^2) \diamondsuit -2tsr'f_{20} + (sr')^2 rsf_{30} \tag{26}$$

$$\frac{\partial F}{\partial y_2} = -2y_2 f_{20} + (\widetilde{y}_1^2 - \widetilde{y}_2^2)(f_{30}\widetilde{x}_2 + f_{21}y_2|x|^2) \diamondsuit 0 \tag{27}$$

$$\frac{\partial^2 F}{\partial y_1 \partial y_2} = -2y_2\frac{\partial}{\partial y_1}(f_{20}) + y_2|x|^2\frac{\partial}{\partial y_1}\big((y_1^2 - y_2^2)f_{21}\big) \diamondsuit 0 \tag{28}$$

$$\begin{aligned}\frac{\partial^2 F}{\partial y_1^2} &= 2t^2 f_{20} - 2t\widetilde{y}_1(f_{30}\widetilde{x}_1 + f_{21}y_1|x|^2) - 2t\widetilde{y}_1(f_{30}\widetilde{x}_1 + f_{21}y_1|x|^2) \\ &\quad + (\widetilde{y}_1^2 - \widetilde{y}_2^2)\left[\widetilde{x}_1(f_{40}\widetilde{x}_1 + f_{31}y_1|x|^2) + |x|^2\left(f_{21} + y_1\frac{\partial}{\partial y_1}(f_{21})\right)\right] \\ &\quad \diamondsuit 2t^2 f_{20} - 4ts^2 rr' f_{30} + (r's)^2\big((rs)^2 f_{40} + r^2 f_{21}\big)\end{aligned} \tag{29}$$

$$\begin{aligned}\frac{\partial^2 F}{\partial y_2^2} &= -2f_{20} - 2y_2\frac{\partial}{\partial y_2}(f_{20}) + (\widetilde{y}_1^2 - \widetilde{y}_2^2)|x|^2\left(f_{21} + y_2\frac{\partial}{\partial y_2}(f_{21})\right) \\ &\quad + 2\widetilde{y}_2\frac{\partial}{\partial y_2}(f_{21}y_2|x|^2) \diamondsuit -2f_{20} + (rr's)^2 f_{21}.\end{aligned} \tag{30}$$

Hence

$$\Big(\frac{\partial^2}{\partial y_1^2}-\frac{\partial^2}{\partial y_2^2}\Big)F = 2(t^2+1)f_{20}-4ts^2rr'f_{30}+(rr')^2s^4f_{40} \tag{31}$$

$$\frac{\partial G}{\partial y_1} = \widetilde{y}_2\Big(\frac{\partial \widetilde{y}_1}{\partial y_1}(f_{20})+\widetilde{y}_1(f_{30}\widetilde{x}_1+f_{21}y_1|x|^2)\Big)\diamondsuit\, 0 \tag{32}$$

$$\frac{\partial G}{\partial y_2} = \widetilde{y}_1\big(-f_{20}+\widetilde{y}_2(f_{30}\widetilde{x}_2+f_{21}y_2|x|^2)\big)\diamondsuit -r's f_{20} \tag{33}$$

$$\frac{\partial^2 G}{\partial y_2\partial y_1} = tf_{20}-\widetilde{y}_1\big(f_{30}\widetilde{x}_1+f_{21}y_1|x|^2\big)+\widetilde{y}_2\frac{\partial}{\partial y_2}(\dots)\diamondsuit\, tf_{20}-rr's^2f_{30}. \tag{34}$$

These expressions may be slightly simplified by introducing the derivatives of K, D, and E with respect to $t=\cos\psi$.

Because

$$\frac{\partial}{\partial u}=\frac{\partial t}{\partial u}\frac{\partial}{\partial t}=\frac{1}{rr'}\frac{\partial}{\partial t}$$

then with

$$K_n=\frac{\partial^n}{\partial t^n}K, \qquad \text{and similarly for } D \text{ and } E$$

we have the following non-zero expressions

$$\frac{\partial K}{\partial x_1}=\frac{s}{r}K_1 \tag{35}$$

$$\frac{\partial^2 K}{\partial x_1^2}-\frac{\partial^2 K}{\partial x_2^2}=\frac{s^2}{r^2}K_2 \tag{36}$$

$$\frac{\partial^2 K}{\partial y_1\partial x_1}=(s^2K_2-tK_1)/(rr') \tag{37}$$

$$\frac{\partial^2 K}{\partial y_2\partial x_2}=-K_1/(rr') \tag{38}$$

$$\frac{\partial}{\partial y_2}\Big(\frac{\partial^2 K}{\partial x_1\partial x_2}\Big)=-sK_2/(r^2r') \tag{39}$$

$$\frac{\partial}{\partial y_1}\Big(\frac{\partial^2 K}{\partial x_1^2}-\frac{\partial^2 K}{\partial x_2^2}\Big)=(s^3K_3-2tsK_2)/(r^2r') \tag{40}$$

$$\frac{\partial^2}{\partial y_2\partial y_1}\frac{\partial^2 K}{\partial x_2\partial x_1}=(tK_2-s^2K_3)/(rr')^2 \tag{41}$$

$$\Big(\frac{\partial^2}{\partial y_1^2}-\frac{\partial^2}{\partial y_2^2}\Big)\Big(\frac{\partial^2 K}{\partial x_1^2}-\frac{\partial^2 K}{\partial x_2^2}\Big)=\big(2(t^2+1)K_2-4ts^2K_3+s^4K_4\big)/(rr')^2. \tag{42}$$

When evaluating the non-zero quantities related to T_{l2} and T_{23}, we substitute D for K in Equations (35), (37)–(40) and E for K in Equations (37) and (38) and use Equations (7)–(9). Then

$$\frac{\partial^2 K}{\partial x_3 \partial x_1} = -\frac{s}{r^2} D_1 \tag{43}$$

$$\frac{\partial^3 K}{\partial y_1 \partial x_3 \partial x_1} = -(s^2 D_2 - t D_1)/(r^2 r') \tag{44}$$

$$\frac{\partial^3 K}{\partial y_3 \partial y_1 \partial x_1} = -(s^2 D_2 - t D_1)/\big(r(r')^2\big) \tag{45}$$

$$\frac{\partial^3 K}{\partial y_2 \partial x_3 \partial x_2} = D_1/(r^2 r') \tag{46}$$

$$\frac{\partial^3 K}{\partial y_3 \partial y_2 \partial x_2} = D_1/\big(r(r')^2\big) \tag{47}$$

$$\frac{\partial^2}{\partial y_3 \partial y_2}\Big(\frac{\partial^2 K}{\partial x_1 \partial x_2}\Big) = -s D_2/(rr')^2 \tag{48}$$

$$\frac{\partial^2}{\partial y_3 \partial y_1}\Big(\frac{\partial^2 K}{\partial x_1^2} - \frac{\partial^2 K}{\partial x_2^2}\Big) = -(s^3 D_3 - 2ts D_2)/(rr')^2 \tag{49}$$

$$\frac{\partial^4 K}{\partial y_3 \partial y_1 \partial x_3 \partial x_1} = (s^2 E_2 - t E_1)/(rr') \tag{50}$$

$$\frac{\partial^4 K}{\partial y_3 \partial y_2 \partial x_3 \partial x_2} = -E_1/(rr'). \tag{51}$$

The quantities (35)–(51) are then the basic isotropic quantities computed in the two local coordinate systems. Quantities involving gravity anomalies could have been obtained from functions similar to D and E, but where the degree-variances σ_j are multiplied with $(i-1)$ and $(i-1)^2$ respectively.

5. General covariance expressions

The general covariance expressions $\operatorname{cov}(L_1(T), L_2(T)) = K(L_1, L_2)$ are then obtained using Equations (35)–(51). We introduce of few more "shorthand" expressions

$$K_{ijkn} = \frac{\partial^4 K}{\partial x_i \partial x_j \partial y_k \partial y_n}, \qquad i,j,k,n = 1,\ 2,\ \text{or}\ 3$$

$$K_{\Delta kn} = \Big(\frac{\partial^2}{\partial x_1^2} - \frac{\partial^2}{\partial x_2^2}\Big)\frac{\partial^2 K}{\partial y_k \partial y_n},$$

and similarly for $K_{ij\Delta}$ and $K_{\Delta\Delta}$. Then

$$\begin{bmatrix} \text{cov}(T_\Delta, T_\Delta) & \text{cov}(T_\Delta, 2T_{12}) \\ \text{cov}(2T_{12}, T_\Delta) & \text{cov}(2T_{12}, 2T_{12}) \end{bmatrix} = R(180^\circ - 2\alpha) \begin{bmatrix} K_{\Delta\Delta} & 0 \\ 0 & 4K_{1212} \end{bmatrix} R(2\alpha' + 180^\circ)$$

$$\begin{bmatrix} \text{cov}(T_{13}, T_\Delta) & \text{cov}(T_{13}, 2T_{12}) \\ \text{cov}(T_{23}, T_\Delta) & \text{cov}(T_{23}, 2T_{12}) \end{bmatrix} = R(90^\circ - \alpha) \begin{bmatrix} K_{13\Delta} & 0 \\ 0 & 2K_{2312} \end{bmatrix} R(2\alpha' + 180^\circ)$$

$$\begin{bmatrix} \text{cov}(T_{13}, T_{13}) & \text{cov}(T_{13}, T_{23}) \\ \text{cov}(T_{23}, T_{13}) & \text{cov}(T_{23}, T_{23}) \end{bmatrix} = R(90^\circ - \alpha) \begin{bmatrix} K_{1313} & 0 \\ 0 & K_{2323} \end{bmatrix} R(\alpha' + 90^\circ).$$

Quantities such as $K_{\Delta\Delta}$ or K_{1212} are easily obtained using the subroutine `COVAX`, (for the degree-variance models described in [8]). However, the computation of $K_{13\Delta}$, K_{1313}, K_{2323}, and K_{2312} requires some small modifications. A modified version of `COVAX` is available on request from the authors.

6. Conclusion

We have here derived covariance expressions for torsion balance observations based on an isotropic covariance function (or reproducing kernel). It is planned to evaluate these functions using degree-variance models and compare these values with values obtained from empirical data.

References

[1] S. K. Jordan. Self-consistent statistical models for the gravity anomaly, vertical deflections and undulation of the geoid. *Journal of Geophysical Research*, 77, No. 20:3660–3670, 1972.

[2] W. Keller and S. Meier. Kovarianzfunktionen der 1. und 2. Ableitungen des Schwerepotentials in der Ebene. Technical report, Veröffentlichungen des Zentralinstituts für Physik der Erde, No. 60, Potsdam, 1980.

[3] Helmuth Moritz. Advanced least-squares methods. Technical report, Reports of the Department of Geodetic Science, 175, Columbus, Ohio, 1972.

[4] Ivan I. Mueller. Geodesy and the torsion balance. *Journal of the Surveying and Mapping Division, American Society of Civil Engineers*, 89, No. SU 3:123–155, 1963.

[5] R. A. Nash, Jr. and S. K. Jordan. Statistical geodesy—an engineering perspective. *Proceedings of the IEEE*, 66, No. 5:532–550, 1978.

[6] C. C. Tscherning. Representation of covariance functions related to the anomalous potential of the earth using reproducing kernels. Technical report, The Danish Geodetic Institute, Internal Report No. 3, Charlottenlund, 1972.

[7] C. C. Tscherning. Computation of second order derivatives of the normal potential based on the representation by a Legendre series. *manuscripta geodaetica*, 1:71–92, 1976.

[8] C. C. Tscherning. Covariance expressions for second and lower order derivatives of the anomalous potential. Technical report, Reports of the Department of Geodetic Science, 225, Columbus, Ohio, 1976.

25

Contribution to the Geometry of the Helmert Transformation

Abstract

Considering the fact that the determination of the Helmert transformation of one point set to another point set is a non-linear problem of adjustment, a geometrical theory for this problem is treated, and as a result of this theory a simple and numerically strong method for the computation of the parameters of the Helmert transformation is presented.

Introduction

In the q-dimensional Euclidian space $\mathbf{R}^q$ two ordered point sets are given, each consisting of n points. We suppose the point sets given by two $q \times n$-matrices X resp. Y, the column vectors of which are q coordinates of the points numbered 1, 2, ..., n in the first resp. second point set. We call the point sets the X-set and the Y-set, respectively.

The Helmert transformation is then a similarity transformation in $\mathbf{R}^q$, which applied to the X-set brings the points of this set to the 'best possible coincidence' with the points (with the same numbers) of the Y-set; that is when we use the designation 'Helmert transformation' instead of similarity transformation; we normally mean a certain similarity transformation characterized by some optimization condition, which should be specified, and we also think about the way in which the parameters of such a transformation have to be determined.

As in so many other problems in geodesy, the determination of the Helmert transformation is made using least-squares adjustment and like most such problems this problem is in reality a non-linear adjustment problem for $q > 2$, but unlike almost all other non-linear adjustment problems, that of the Helmert transformation admits a geometrical theory which in the opinion of the author is interesting and even beautiful, and—what is perhaps more important—it suggests a method for the computation of the parameters, which seems to be numerically well behaved.

The central problem in the adjustment is to find a similarity matrix, i.e., a matrix

$$A = rU,$$

where r is a non-negative number and U is an orthogonal matrix, i.e., a matrix such that

$$U^{\mathrm{T}}U = UU^{\mathrm{T}} = I.$$

In adjustment problems, the unknown quantity is normally a vector—here it is a vector and a matrix, a similarity matrix, the relevant entries of which naturally could be expressed as a vector, but it was obvious to try to look directly for the matrix itself, whereby the richer structure of matrices could be applied in the computations. We look upon the $q \times q$ matrices as constituting an algebra and at the same time a normed vector space. This idea is illustrated in Section 1 where it is applied to a linear problem.

In Section 2 we treat the problem of finding that similarity matrix which approximates a given square matrix best. It turns out that the hard core of the problem is the following: given a matrix N, find an orthogonal matrix U such that NU is symmetric, a problem which can be solved easily as soon as the singular value decomposition of N is found.

In Section 3 we try to use the insight gained in the introductory sections on the Helmert transformation, but this section is very short as the difficulties have already been overcome.

A few examples in Section 4 shall illustrate what happens in special situations when certain eigenvalues coincide or vanish, and in Section 5 we find a method of successive approximation which in most cases should be preferred to the method given in Section 2 based on the singular value decomposition.

In this paper, the best transformation is defined as that transformation (of a given type) for which the sum of the squares of the distances between corresponding points of the Y-set and the X-set after the transformation is as small as possible. Of course it is possible to make some variations of the method where these distances enter with different weights, etc.

1

A similarity transformation in $\mathbf{R}^q$ may be written as

$$z \to a + Az,$$

where a is a q-dimensional vector and A is a $q \times q$ matrix which may be written as the product of a positive number and an orthogonal matrix. In this section we shall drop this condition for A such that A is any $q \times q$ matrix and the transformation is then an affine transformation.

If η means the n-dimensional column vector with all its entries equal to 1, we may formulate the problem of this section as follows in matrix algebra:

Find the vector a and the matrix A such that the Euclidean norm of the matrix V, where

$$V = a\eta^{\mathrm{T}} + AX - Y,$$

is as small as possible.

The Euclidian norm of a matrix M, also called the Schur norm or the Frobenius norm, is the square root of the sum of the squares of its entries or

$$\|M\|^2 = \mathrm{tr}(M^{\mathrm{T}}M) = \mathrm{tr}(MM^{\mathrm{T}}).$$

To this norm corresponds a scalar product: for A and B being $q \times q$ matrices the scalar product is

$$\langle A, B\rangle = \mathrm{tr}(AB^{\mathrm{T}}) = \mathrm{tr}(A^{\mathrm{T}}B),$$

such that

$$\|M\|^2 = \langle M, M\rangle.$$

Now suppose that the origin of the coordinate system coincides with the center of gravity of the Y-set, then

$$Y\eta = 0.$$

Suppose, moreover, that we have already translated the X-set such that its center of gravity also coincides with the origin, then

$$X\eta = 0,$$

and now a is the remaining translation.

We must remember that the columns of X and Y from now on are the coordinates of the X- and the Y-sets with respect to the center of gravity of the respective sets!

We find

$$\begin{aligned}\|V\|^2 = \mathrm{tr}(VV^{\mathrm{T}}) &= \mathrm{tr}\big((a\eta^{\mathrm{T}} + AX - Y)(\eta a^{\mathrm{T}} + X^{\mathrm{T}}A^{\mathrm{T}} - Y^{\mathrm{T}})\big)\\ &= q\|a\|^2 + \mathrm{tr}\big((AX - Y)(X^{\mathrm{T}}A^{\mathrm{T}} - Y^{\mathrm{T}})\big).\end{aligned}$$

We have used here

$$\mathrm{tr}\, A^{\mathrm{T}} = \mathrm{tr}\, A$$

and

$$\mathrm{tr}(AB) = \mathrm{tr}(BA),$$

when these expressions have a meaning.

The two terms in this expression are non-negative, so the minimum of $\|V\|$ can only be attained for $a = 0$, i.e., we have found that the translation shall be that translation which brings the gravity center of the X-system over in the gravity center of the Y-system, and so we have reduced the problem to a homogeneous problem namely that of finding the matrix A which minimizes

$$\|V\|^2 = \mathrm{tr}\big((AX - Y)(X^{\mathrm{T}}A^{\mathrm{T}} - Y^{\mathrm{T}})\big).$$

An easy calculation gives

$$\|V\|^2 = \operatorname{tr}(XX^{\mathrm{T}}A^{\mathrm{T}}A) - \operatorname{tr}(YX^{\mathrm{T}}A^{\mathrm{T}}) - \operatorname{tr}(XY^{\mathrm{T}}A) + \operatorname{tr}(YY^{\mathrm{T}}).$$

By differentiation with respect to A we find:

$$\begin{aligned} d\|V\|^2 &= \operatorname{tr}(XX^{\mathrm{T}}A^{\mathrm{T}}dA) + \operatorname{tr}\big(XX^{\mathrm{T}}(dA^{\mathrm{T}})A\big) - \operatorname{tr}(YX^{\mathrm{T}}dA) - \operatorname{tr}(XY^{\mathrm{T}}dA) \\ &= \operatorname{tr}\big((XX^{\mathrm{T}}A^{\mathrm{T}} - XY^{\mathrm{T}})\,dA\big) + \operatorname{tr}\big((AXX^{\mathrm{T}} - YX^{\mathrm{T}})\,dA^{\mathrm{T}}\big) \\ &= 2\operatorname{tr}\big((AXX^{\mathrm{T}} - YX^{\mathrm{T}})\,dA^{\mathrm{T}}\big) = 2\langle AXX^{\mathrm{T}} - YX^{\mathrm{T}}, dA\rangle. \end{aligned}$$

A necessary condition for a minimum is that $d\|V\|^2$ is zero for all matrices dA, i.e.,

$$AXX^{\mathrm{T}} - YX^{\mathrm{T}} = 0,$$

and if XX^{T} is a regular matrix the solution is

$$A = YX^{\mathrm{T}}(XX^{\mathrm{T}})^{-1}.$$

We also find

$$\tfrac{1}{2}d^2\|V\|^2 = \operatorname{tr}\big((dA)XX^{\mathrm{T}}dA^{\mathrm{T}}\big).$$

If XX^{T} is regular, then XX^{T} is strictly positive definite and therefore this expression for $\frac{1}{2}d^2\|V\|^2$ is a strictly positive definite quadratic form in the entries of the matrix dA and therefore

$$A = YX^{\mathrm{T}}(XX^{\mathrm{T}})^{-1}$$

corresponds to a minimum.

It is easy to see that XX^{T} is regular if and only if the X-set consists of q linearly independent points and at least one point more which must not coincide with the gravity center of the first-mentioned q points. We shall not analyze what happens if this condition is not satisfied.

2

As a second preliminary exercise, consider the following problem:

Given a $q \times q$ matrix M, find the best approximating similarity matrix A, i.e., find a matrix $A = rU$, where $r \geq 0$ and U is a $q \times q$ orthogonal matrix: $U^{\mathrm{T}}U = 1$, such that $\|M - A\|$ is as small as possible, where $\|\cdot\|$ is the Euclidean norm in $\mathbf{R}^{q^2}$:

$$\|M - A\|^2 = \operatorname{tr}\big((M^{\mathrm{T}} - A^{\mathrm{T}})(M - A)\big) = \sum_{i,j=1,\ldots,q} (M_{ij} - A_{ij})^2.$$

As it does not seem unreasonable to solve our original problem by first finding the general matrix M solving the problem in Section 1 and then finding the best similarity matrix A approximating M, so it follows that this problem is related to our original problem.

Write

$$2E = \|M - A\|^2 = \operatorname{tr}(M^{\mathrm T}M) - 2\operatorname{tr}(M^{\mathrm T}A) + \operatorname{tr}(A^{\mathrm T}A)$$
$$= \operatorname{tr}(M^{\mathrm T}M) - 2\operatorname{tr}(M^{\mathrm T}A) + qr^2.$$

The $q \times q$ orthogonal matrices form a group, and near the unity matrix I, the unit of the group, the group may be parametrized by

$$U = \exp S,$$

where

$$\exp R = 1 + \frac{R}{1!} + \frac{R^2}{2!} + \frac{R^3}{3!} + \cdots + \frac{R^n}{n!} + \cdots$$

for any $q \times q$ matrix (this series converges in the norm $\|\cdot\|$ for every R), and where S is a skew-symmetric matrix:

$$S^{\mathrm T} = -S.$$

Therefore a neighborhood V of $U \in O(q)$ (the orthogonal group in q dimension) may be parameterized by

$$V = U \exp S,$$

and we have

$$2E = \operatorname{tr}(M^{\mathrm T}M) - 2r\operatorname{tr}(M^{\mathrm T}U\exp S) + qr^2,$$
$$dE = -r\operatorname{tr}(M^{\mathrm T}U\exp S\,dS) + \big(qr - \operatorname{tr}(M^{\mathrm T}U\exp S)\big)\,dr,$$
$$d^2E = -r\operatorname{tr}(M^{\mathrm T}U\exp S\,dS^2) - 2\operatorname{tr}(M^{\mathrm T}U\exp S\,dS)\,dr + q\,dr^2.$$

For $S = 0$, i.e., for the point U we get

$$dE = -r\operatorname{tr}(M^{\mathrm T}UdS) + \big(qr - \operatorname{tr}(M^{\mathrm T}U)\big)\,dr,$$
$$d^2E = -r\operatorname{tr}(M^{\mathrm T}UdS^2) - 2\operatorname{tr}(M^{\mathrm T}UdS)dr + qdr^2,$$

so $dE = 0$ if and only if

$$\langle M^{\mathrm T}U, dS^{\mathrm T}\rangle = 0$$

for every skew-symmetric matrix dS, and

$$r = \tfrac{1}{q}\operatorname{tr}(M^{\mathrm T}U).$$

The first of these conditions says that $M^{\mathrm T}U$ shall be orthogonal to all skew-symmetric matrices, but as the symmetric and the skew-symmetric matrices constitute orthogonal complementary subspaces of $\mathbf{R}^{q^2}$, the conditions may be expressed as

$$B = M^{\mathrm T}U \text{ is symmetric}$$

and

$$r = \tfrac{1}{q}\operatorname{tr} B \geq 0;$$

and if $dE = 0$ then

$$d^2E = -r\,\mathrm{tr}(BdS^2) + q\,dr^2 = r\,\mathrm{tr}(dS^{\mathrm{T}}BdS) + q\,dr^2.$$

This quadratic form in dr and $dS_{ij}, 1 \le i < j < q$, (the entries above the diagonal in dS) cannot be non-positive definite, so there can be no maximum for E. On the other hand if B is positive definite, then d^2E is also positive definite, and $A = rU$ is a minimum point.

We are now faced with the following problem:

Given a matrix M^{T}, find an orthogonal matrix, U, such that

$$B = M^{\mathrm{T}}U \text{ is symmetric.}$$

In order to find B and U we shall use the singular value decomposition of M^{T}: Any $q \times q$-matrix M^{T} can be decomposed as

$$M^{\mathrm{T}} = \Phi\Lambda\Psi^{\mathrm{T}},$$

where Φ and Ψ are orthogonal matrices and Λ is a diagonal matrix, cf. [1].

Evidently,

$$MM^{\mathrm{T}} = \Psi\Lambda^2\Psi^{\mathrm{T}}$$

and

$$M^{\mathrm{T}}M = \Phi\Lambda^2\Phi^{\mathrm{T}},$$

so the diagonal elements of Λ^2, i.e., the squares of the diagonal elements of Λ, are the eigenvalues of MM^{T} and of $M^{\mathrm{T}}M$, and the column vectors of Φ resp. Ψ are the corresponding eigenvectors of $M^{\mathrm{T}}M$ resp. MM^{T}. From this does not follow that they should be computed that way however, but it does follow that the diagonal elements λ_i, $i = 1, 2, \ldots, q$, of Λ are determined apart from their signs. Clearly we may—and we shall—suppose that they are non-negative and arranged in non-increasing order: $\lambda_q \ge \lambda_{q-1} \ge \cdots \ge \lambda_2 \ge \lambda_1 \ge 0$.

Let

$$\epsilon = \mathrm{diag}(\epsilon_q, \epsilon_{q-1}, \ldots, \epsilon_2, \epsilon_1)$$

be a diagonal matrix with the diagonal elements $\{\epsilon_q\}$ all of which are ± 1. This is an orthogonal and symmetric matrix! There are evidently 2^q different such matrices, and we have

$$M^{\mathrm{T}} = \Phi\epsilon\Lambda\epsilon\Psi^{\mathrm{T}} = \Phi(\epsilon\Lambda)(\Psi\epsilon)^{\mathrm{T}}.$$

Now define

$$B_\epsilon = \Phi(\epsilon\Lambda)\Phi^{\mathrm{T}},$$

and

$$U_\epsilon = \Psi\epsilon\Phi^{\mathrm{T}}.$$

Here U_ϵ is orthogonal as a product of three orthogonal matrices, B_ϵ is symmetric, and

$$B_\epsilon = \Phi(\epsilon\Lambda)(\Psi\epsilon)^{\mathrm{T}}(\Psi\epsilon)\Phi^{\mathrm{T}} = M^{\mathrm{T}}U_\epsilon,$$

and we have found 2^q solutions to our problem. All the U_ϵ are different and if all the $\{\lambda_i\}$ are also different and if $\lambda_i > 0$, then the B_ϵ are also different.

It is almost evident that there are no other solutions when the $\{\lambda_i\}$ are all different. The interested reader should be able to see what happens if this is not the case, at least after reading Section 4.

If $\lambda_1 > 0$, then B_ϵ is positive definite for $\epsilon = \{1, 1, \ldots, 1, 1\}$, and thus corresponds to a minimum point.

From

$$r = \tfrac{1}{q} \operatorname{tr} B \geq 0$$

follows that not all of the 2^q solutions (B_ϵ, U_ϵ) can be used for the solution of our minimum problem. In order to find out if some non-definite solutions B may really be relevant, we must investigate the quadratic form

$$\operatorname{tr}(dS^{\mathrm{T}} B\, dS)$$

occurring in the expression for d^2E. For simplicity, we write S instead of dS.

We have then

$$\operatorname{tr}(S^{\mathrm{T}} BS) = \operatorname{tr}(\Phi^{\mathrm{T}} S^{\mathrm{T}} BS\Phi) = \operatorname{tr}\big((\Phi^{\mathrm{T}} S^{\mathrm{T}} \Phi)(\Phi^{\mathrm{T}} B\Phi)(\Phi^{\mathrm{T}} S\Phi)\big) = \operatorname{tr}(\bar{S}^{\mathrm{T}} D\bar{S}),$$

where

$$\bar{S} = \Phi^{\mathrm{T}} S\Phi,$$

but

$$\bar{S}^{\mathrm{T}} = \Phi^{\mathrm{T}} S^{\mathrm{T}} \Phi = -\Phi^{\mathrm{T}} S\Phi = -\bar{S},$$

so $\bar{S}$ is also skew-symmetric, and the transformation

$$S \to \bar{S} = \Phi^{\mathrm{T}} S\Phi$$

is an automorphism of the set of $q \times q$ skew-symmetric matrices and therefore one of the two quadratic forms

$$\operatorname{tr}(S^{\mathrm{T}} BS) \qquad \text{and} \qquad \operatorname{tr}(\bar{S}^{\mathrm{T}} D\bar{S})$$

is definite if and only if the other is so, i.e., we may concentrate on the second of them.

We have

$$\operatorname{tr}(S^{\mathrm{T}} DS) = \operatorname{tr}(DSS^{\mathrm{T}}) = -\operatorname{tr}(DS^2).$$

Because D is diagonal, only the diagonal entries of S^2 play a role. S is parameterized by the $q(q-1)/2$ entries above the diagonal. Each diagonal entry in S^2 is minus the sum of the squares of $q-1$ different from those $q(q-1)/2$ entries, and each of these entries occurs exactly in two diagonal entries of S^2. The example $q = 4$ will make this clear:

$$\operatorname{tr}\left\{\begin{bmatrix} \lambda_1 & 0 & 0 & 0 \\ 0 & \lambda_2 & 0 & 0 \\ 0 & 0 & \lambda_3 & 0 \\ 0 & 0 & 0 & \lambda_4 \end{bmatrix} \begin{bmatrix} 0 & s_{12} & s_{13} & s_{14} \\ -s_{12} & 0 & s_{23} & s_{24} \\ -s_{13} & -s_{23} & 0 & s_{34} \\ -s_{14} & -s_{24} & -s_{34} & 0 \end{bmatrix}^2\right\}$$

$$= \operatorname{tr}\left\{ \begin{bmatrix} \lambda_1 & 0 & 0 & 0 \\ 0 & \lambda_2 & 0 & 0 \\ 0 & 0 & \lambda_3 & 0 \\ 0 & 0 & 0 & \lambda_4 \end{bmatrix} \right.$$

$$\left. \times \begin{bmatrix} -s_{12}^2-s_{23}^2-s_{14}^2 & -s_{13}s_{23}-s_{14}s_{24} & s_{12}s_{23}-s_{14}s_{34} & s_{12}s_{24}+s_{13}s_{34} \\ -s_{13}s_{23}-s_{14}s_{24} & -s_{12}^2-s_{23}^2-s_{34}^2 & -s_{12}s_{13}-s_{24}s_{34} & s_{12}s_{14}+s_{23}s_{34} \\ -s_{12}s_{23}-s_{14}s_{34} & -s_{12}s_{13}-s_{24}s_{34} & -s_{13}^2-s_{23}^2-s_{34}^2 & s_{13}s_{14}-s_{23}s_{24} \\ -s_{12}s_{24}+s_{13}s_{34} & -s_{12}s_{24}+s_{13}s_{34} & -s_{13}s_{14}-s_{23}s_{24} & s_{14}^2-s_{24}^2-s_{34}^2 \end{bmatrix} \right\}$$

$$= -(\lambda_1+\lambda_2)s_{12}^2-(\lambda_1+\lambda_3)s_{13}^2-(\lambda_1+\lambda_4)s_{14}^2 \\ \quad -(\lambda_2+\lambda_3)s_{23}^2-(\lambda_2+\lambda_4)s_{24}^2-(\lambda_3+\lambda_4)s_{34}^2.$$

It is not difficult to see that the general result is

$$-\operatorname{tr}(S^{\mathrm{T}}DS) = \sum_{1\le i<j\le q} (\lambda_i + \lambda_j)s_{ij}^2,$$

which evidently is strictly positive definite if and only if all the sums of pairs of $\{\lambda_i\}$ are positive, and this is the case 1) if all signs are $+$ and 2) if the sign for λ_1 is $-$ and all the others are $+$, so it seems that there will normally be two minima.

Given a point x in $\mathbf{R}^{q^2}$, a closed submanifold of $\mathbf{R}^{q^2}$ must contain at least one point y such that $\|x-y\|$ attains its minimum there, but the set of similarity matrices rU such that $r>0$ consists of two closed subsets: those for which the determinant of U is 1, and those for which it is -1. These two sets are differential manifolds with the common boundary consisting of the point 0. Therefore a point x in $\mathbf{R}^{q^2}$ will normally have a point y_+ and a point y_- on these two submanifolds, respectively, which minimizes the distance. The only difficulty comes from the point 0, but the only point x, for which y_+ and y_- coincide (in 0) is $x=0$, i.e., the zero matrix.

In Section 4 we shall see an example which illustrates what may happen if some of the eigenvalues of $M^{\mathrm{T}}M$ coincide.

By substitution we find the minimum values of $2E$:

$$\min 2E = \operatorname{tr}(M^{\mathrm{T}}M) - 2r\operatorname{tr} B + r^2 = \operatorname{tr}(M^{\mathrm{T}}M) - \frac{\operatorname{tr}^2 B}{q}.$$

If $\{\lambda_i\}$ are the non-negative square roots of the eigenvalues of $M^{\mathrm{T}}M$, this becomes

$$\sum_{i=1}^{q} \lambda_i^2 - \frac{1}{q}\left(\sum_{i=1}^{q}\lambda_i\right)^2$$

or

$$\sum_{i=1}^{q} \lambda_i^2 - \frac{1}{q}\left(\sum_{i=1}^{q}\lambda_i - 2\lambda_1\right)^2.$$

But which of these minima is attained on the manifold consisting of the proper similarities (those with $\det U = 1$)?

For the first of the minima we have for all the eigenvalues of B

$$\lambda_i \geq 0,$$

so $\det B \geq 0$, and for the second of them $\det B \leq 0$. But

$$B = M^{\mathrm{T}} U,$$
$$\det B = \det M \det U,$$

and therefore if M is regular then the first minimum value is less than the second one and corresponds to a proper similarity if and only if $\det M > 0$, and vice versa for the second minimum. If and only if $\det M = 0$, the two minimum values are equal.

In two dimensions, $q = 2$, everything simplifies radically. Every 2×2 matrix M may be decomposed according to

$$M = \begin{bmatrix} a & b \\ c & d \end{bmatrix} = \begin{bmatrix} \frac{a+d}{2} & -\frac{c-b}{2} \\ \frac{c-b}{2} & \frac{a+d}{2} \end{bmatrix} + \begin{bmatrix} \frac{a-d}{2} & \frac{c+b}{2} \\ \frac{c+b}{2} & -\frac{a-d}{2} \end{bmatrix} = A_1 + A_2.$$

Here we find

$$\langle A_1, A_2 \rangle = \operatorname{tr}(A_1 A_2^{\mathrm{T}}) = \tfrac{a^2-d^2}{4} + \tfrac{c^2-b^2}{4} - \tfrac{c^2-b^2}{4} - \tfrac{a^2-d^2}{4} = 0,$$

$$A_1 A_1^{\mathrm{T}} = \left(\frac{a+d}{2}\right)^2 + \left(\frac{b-c}{2}\right)^2 = \det A_1 = \tfrac{1}{4}\|M\|^2 + \tfrac{1}{2}\det M,$$

$$A_2 A_2^{\mathrm{T}} = \left(\frac{a-d}{2}\right)^2 + \left(\frac{b+c}{2}\right)^2 = -\det A_2 = \tfrac{1}{4}\|M\|^2 - \tfrac{1}{2}\det M.$$

That is, A_1 and A_2 are similarity matrices, and A_1 is orthogonal to A_2 in the Euclidean metric. We also see that A_1 is a proper similarity matrix ($\det A_1 \geq 0$), every such matrix may be written in this form, A_2 is an improper similarity matrix ($\det A_2 \leq 0$), and every such matrix may be written in this form.

Therefore we have for $q = 2$:

$$\min_{\substack{A \text{ is a proper} \\ \text{similarity matrix}}} \|M - A\| = \sqrt{\tfrac{1}{4}\|M\|^2 - \tfrac{1}{2}\det M} \qquad \text{for } A = \begin{bmatrix} \frac{a+d}{2} & \frac{b-c}{2} \\ \frac{c-b}{2} & \frac{a+d}{2} \end{bmatrix},$$

and

$$\min_{\substack{A \text{ is an improper} \\ \text{similarity matrix}}} \|M - A\| = \sqrt{\tfrac{1}{4}\|M\|^2 + \tfrac{1}{2}\det M} \qquad \text{for } A = \begin{bmatrix} \frac{a-d}{2} & \frac{b+c}{2} \\ \frac{b+c}{2} & \frac{d-a}{2} \end{bmatrix}.$$

For $q > 2$, neither the proper nor the improper similarity matrices form linear subspaces of $\mathbf{R}^{q^2}$, so in these cases the problem to find the best Helmert transformation is a non-linear one.

3

After this double preparation, we should be able to attack the problem directly with the hope of an easy victory.

The problem differs only from that solved in Section 1 therein that instead of minimizing

$$2E(A) = \mathrm{tr}(Y^{\mathrm{T}}Y) - 2\,\mathrm{tr}(XY^{\mathrm{T}}A) + \mathrm{tr}(XX^{\mathrm{T}}AA^{\mathrm{T}})$$

over all $q \times q$ matrices A, we shall minimize over all similarity matrices A. Here we have already homogenized the problem by using the gravity center of the Y-point set as coordinate origin and having translated the X-point set such that the gravity centers of the two point sets coincide.

Now

$$A = rU, \qquad U^{\mathrm{T}}U = I,$$

so

$$AA^{\mathrm{T}} = r^2;$$

$$2E' = \mathrm{tr}(YY^{\mathrm{T}}) - 2\,\mathrm{tr}(XY^{\mathrm{T}}A) + r^2\,\mathrm{tr}(XX^{\mathrm{T}}),$$

but in Section 2 we have solved the minimum problem for

$$2E = \mathrm{tr}(M^{\mathrm{T}}M) - 2\,\mathrm{tr}(M^{\mathrm{T}}A) + qr^2.$$

Here, if we put

$$M = \frac{q}{\mathrm{tr}(XX^{\mathrm{T}})}YX^{\mathrm{T}},$$

then

$$2E' = \left(\mathrm{tr}(YY^{\mathrm{T}}) - \frac{q}{\mathrm{tr}(XX^{\mathrm{T}})}\,\mathrm{tr}(Y^{\mathrm{T}}YX^{\mathrm{T}}X)\right) + \frac{\mathrm{tr}(XX^{\mathrm{T}})}{q}2E,$$

where the first term does not depend on A and so we have reduced the problem to that solved in Section 2 with the result

$$\min 2E' = \mathrm{tr}(YY^{\mathrm{T}}) - r^2\,\mathrm{tr}(XX^{\mathrm{T}}).$$

4

In order to achieve a better geometric understanding of the results of Section 2, I shall give a few illustrating examples here.

First let $q = 5$, and $M = I$, the 5×5 unit matrix.

We want to find orthogonal matrices, U, such that the distance $\|M - U\|$ is minimized, or, more generally, we will find those orthogonal matrices, U, such that

$$\frac{dE}{dU} = 0,$$

where

$$2E = \|M - U\|^2 = \operatorname{tr}(MM^{\mathrm{T}}) - 2\operatorname{tr}(MU^{\mathrm{T}}) + \operatorname{tr}(UU^{\mathrm{T}}) = 2(5 - \operatorname{tr} U),$$

and then find the geometrical significance of these solutions.

In this case, $M^{\mathrm{T}} = I$ and therefore U itself must be a symmetric matrix:

$$U^{\mathrm{T}} = U \qquad \text{and} \qquad B = U,$$

but

$$UU^{\mathrm{T}} = I,$$

so

$$U^2 = I.$$

The eigenvalues of an orthogonal matrix are (real or complex) numbers λ with $|\lambda| = 1$; from $U^2 = I$ follows

$$\lambda^2 = 1,$$

so the eigenvalues of U must be ± 1.

Such matrices U fall into six types:

$$U_0,\ U_1,\ U_2,\ U_3,\ U_4,\ U_5,$$

where the index is the number of negative eigenvalues -1, and so the distances d_i from M to U_i are

$$d_i = \sqrt{10 - 2(5 - 2i)} = 2\sqrt{i} \qquad \text{or} \qquad d = \{0, 2, 2\sqrt{2}, 2\sqrt{3}, 4, 2\sqrt{5}\}.$$

In these six cases the eigenvalues of the quadratic form $\operatorname{tr}(S^{\mathrm{T}}BS)$ occurring in the expression for d^2E are respectively

0	$2, 2, 2, 2, 2, 2, 2, 2, 2, 2$	strictly positive definite
2	$2, 2, 2, 2, 2, 2, 0, 0, 0, 0$	non-negative definite
$2\sqrt{2}$	$2, 2, 2, 0, 0, 0, 0, 0, 0, -2$	not definite
$2\sqrt{3}$	$2, 0, 0, 0, 0, 0, 0, -2, -2, -2$	not definite
4	$0, 0, 0, 0, -2, -2, -2, -2, -2, -2$	non-positive definite
$2\sqrt{5}$	$-2, -2, -2, -2, -2, -2, -2, -2, -2, -2$	strictly negative definite.

We see that 0 is the least distance from $M = I$ to any rotation matrix, that 2 is the least distance to any orthogonal matrix with determinant -1, that 4 is the greatest distance to any rotation matrix, and that $2\sqrt{5}$ is the greatest distance to any orthogonal matrix—here one with determinant -1; the remaining two cases do not correspond to local maxima or minima.

The matrices U_0 and U_5 are evidently uniquely determined

$$U_0 = \begin{bmatrix} 1 & & & & \\ & 1 & & & \\ & & 1 & & \\ & & & 1 & \\ & & & & 1 \end{bmatrix}, \qquad U_5 = \begin{bmatrix} -1 & & & & \\ & -1 & & & \\ & & -1 & & \\ & & & -1 & \\ & & & & -1 \end{bmatrix},$$

because for any orthogonal matrix W

$$W^{\mathrm{T}}U_0W = U_0, \qquad W^{\mathrm{T}}U_5W = U_5,$$

corresponding to the regularity of $\mathrm{tr}(S^{\mathrm{T}}BS)$ in these cases.

In the remaining cases this form has four or six vanishing eigenvalues, which seems to indicate a 4 resp. 6-dimensional manifold of solutions:

A matrix of type U_1 has four eigenvalues equal to 1, and one equal to -1; therefore it must represent a reflection with respect to a four-dimensional linear subspace of $\mathbf{R}^5$, and any such reflection corresponds to a matrix of type U_1. There are as many four-dimensional linear subspaces of $\mathbf{R}^5$ as there are directions in $\mathbf{R}^5$, i.e., four, and we see that the sphere with its center at M = unit matrix $\in \mathbf{R}^{5^2}$ and radius 2 touches the manifold of orthogonal matrices with determinant -1 in a four-sphere.

A matrix of type U_2 with three eigenvalues equal to 1 and two equal to -1 corresponds to a rotation which leaves all points of three-dimensional linear subspace invariant and reflects all the points in the complementary two-dimensional plane with respect to the origin. So the number of dimensions of manifold in $\mathbf{R}^{25}$ of matrices of type U_2 is the same as the dimension of the manifold of planes in $\mathbf{R}^5$, i.e., six, etc.

The reader may convince himself that the situation does not change drastically if we for M choose any orthogonal matrix instead of I.

Next we consider the case where $\det M = 0$, or more specifically where $M^{\mathrm{T}}M$ has one vanishing eigenvalue.

Then $\mathrm{tr}\, B$ will be invariant with respect to a change of sign for the vanishing eigenvalue, so two different similarity matrices will correspond to the same minimum value—one for a rotation, one for an improper similarity matrix. In this situation XY^{T} has one vanishing eigenvalue, and this means that at least one of the matrices X and Y is of rank $q-1$, i.e., at least one of the two point sets span an affine subspace of dimension $q-1$ in $\mathbf{R}^q$. It is clear that in this case the Helmert transformation is only determined apart from the orientation.

If $M^{\mathrm{T}}M$ has more than one vanishing eigenvalue, then at least one of the point sets span an affine subspace of dimension less than $q-1$ and so the determination of the transformation has an infinity of solutions, and correspondingly, $\mathrm{tr}(S^{\mathrm{T}}BS)$ has one or more vanishing eigenvalues.

5

After this short presentation of a theory for the Helmert transformation, we will turn to the question of the practical formulas for the computation of the parameters.

A given Helmert transformation may in infinitely many ways be represented as a parallel translation and a homogeneous similarity transformation around a point, the 'center' of the transformation, according to which point is chosen to be the center, and if the translation is chosen to be performed before or after the homogeneous similarity transformation. Here we shall suppose that the translation is performed first and that the translation vector is the vector from the center of gravity of the X-set to the center of gravity of the Y-set, such that the center of the similarity transformation becomes the common center of gravity for the two systems (after the translation) and that we use coordinates with this center as origin, i.e., exactly as we have described above. In this case, the coefficients of the translation will be mutually uncorrelated and uncorrelated with the entries of the similarity matrix. It should be almost evident that this choice is optimal from the numerical point of view.

We have seen in the foregoing sections how the similarity matrix, A, can be computed from the matrices YX^{T} and XX^{T}, but this calculation depends on the eigenvalue decomposition of $(YX^{\mathrm{T}})^{\mathrm{T}}YX^{\mathrm{T}}$, and it might be preferable to have a simpler method based on successive approximations, which—even if it does not converge in the general case—could be applied in normal geodetic and photogrammetric practice, where the two point sets are almost—i.e., apart from observation errors—similar.

As we have seen for two dimensions, the problem is not only linear but almost trivial, we may concentrate on three dimensions, $q = 3$.

First we have to choose a starting value, A_0, for our approximation process. The most obvious candidate would be

$$M = \frac{q}{\mathrm{tr}(XX^{\mathrm{T}})} YX^{\mathrm{T}},$$

but this matrix is not in general approximatively a similarity matrix; a much better choice is

$$A_0 = YX^{\mathrm{T}}(XX^{\mathrm{T}})^{-1},$$

in fact if we have exactly

$$Y = AX$$

for A a similarity matrix, then we would have $A_0 = A$, so under our assumptions A_0 is equal to a similarity matrix apart from a slight perturbation caused by observation errors. The price we have to pay for this better starting value is that it restricts the method to cases where the matrix XX^{T} is regular; we do not regard this to be a serious restriction, however.

First step: Given a matrix A_0 find one, A, in a neighborhood of A_0 and which is nearer to a similarity matrix than A_0 is:

Write

$$A = A_0(I + \epsilon),$$

then

$$A^{\mathrm{T}}A = (I + \epsilon^{\mathrm{T}})A_0^{\mathrm{T}}A_0(I + \epsilon) = aI,$$

where a shall be a scalar multiplied by the unit matrix.

This can, apart from second order expressions in the matrix ϵ, be written as

$$a(I - \epsilon - \epsilon^{\mathrm{T}}) = A_0^{\mathrm{T}}A_0.$$

Choose

$$a = \frac{\operatorname{tr}(A_0^{\mathrm{T}} A_0)}{q},$$

and

$$\epsilon = -\frac{1}{2}\frac{A_0^{\mathrm{T}} A_0 - a}{a},$$

then ϵ is symmetric and $\operatorname{tr}\epsilon = 0$ and $A = A_0(I + \epsilon)$ is apart from an error of second order (in ϵ), a similarity matrix.

We will iterate this step until we have a matrix, A, which is a similarity matrix within the calculation precision.

Second step: Now we have a similarity matrix, A, which is 'near to' A_0, and in many cases this may be sufficient, but here we wish to find that similarity matrix, A_2, which is the best possible in the metric we have chosen.

We know that A_2 should satisfy

$$M^{\mathrm{T}} A \text{ symmetric},$$

or

$$M^{\mathrm{T}} A - A^{\mathrm{T}} M = 0.$$

We shall find A_2 from A_1 by an iterative process along S, the manifold of similarity matrices.

We look for an approximation

$$A_2 = A_1(I + S),$$

where S is skew-symmetric, i.e., $I + S$ is an infinitesimal orthogonal matrix. We find

$$M^{\mathrm{T}} A_1(I + S) - (I - S)A_1^{\mathrm{T}} M = 0$$

or

$$(M^{\mathrm{T}} A_1)S + S(M^{\mathrm{T}} A_1)^{\mathrm{T}} = -M^{\mathrm{T}} A_1 + (M^{\mathrm{T}} A_1)^{\mathrm{T}},$$

or writing

$$C = M^{\mathrm{T}} A_1,$$

we have

$$CS + SC^{\mathrm{T}} = -C + C^{\mathrm{T}}.$$

By writing this in the matrix entries we find for $q = 3$ the following equation:

$$(\operatorname{tr} C - C^{\mathrm{T}}) \begin{bmatrix} s_{23} \\ s_{31} \\ s_{12} \end{bmatrix} = - \begin{bmatrix} c_{23} - c_{32} \\ c_{31} - c_{13} \\ c_{12} - c_{21} \end{bmatrix},$$

or

$$(\operatorname{tr}(M^{\mathrm{T}} A_1) - A_1^{\mathrm{T}} M) \begin{bmatrix} s_{23} \\ s_{31} \\ s_{12} \end{bmatrix} = - \begin{bmatrix} c_{23} - c_{32} \\ c_{31} - c_{13} \\ c_{12} - c_{21} \end{bmatrix}.$$

Now the manifold of similarity matrices is not linear, so by taking this linear step we may have left the manifold but we can get back into it by performing the first step again. By performing alternatively step 2 and step 1, we should 'normally' be able to arrive at a solution A_2, which in the precision of the calculation satisfies

1. A_2 is a similarity matrix;
2. $M^{\mathrm{T}}A_2$ is symmetric.

Then we only have to adjust the factor of multiplication putting

$$A = rA_2,$$

where

$$r = \frac{\operatorname{tr}(M^{\mathrm{T}}A_2)}{\operatorname{tr}(A_2^{\mathrm{T}}A_2)}.$$

It would perhaps be simpler in the approximation process to use

$$M^{\mathrm{T}} = XY^{\mathrm{T}}$$

and, accordingly

$$r = \frac{q}{\operatorname{tr}(XX^{\mathrm{T}})} \frac{\operatorname{tr}(XY^{\mathrm{T}}A_2)}{\operatorname{tr}(A_2^{\mathrm{T}}A_2)}.$$

This successive approximation method has been implemented on the computer of the Danish Geodetic Institute, and its precision seems to be optimal among approximation methods using only a priori data, i.e., data which do not use residuals computed in the foregoing steps.

A sharper result can be attained using the following idea:

First use the approximation method described above with the resulting similarity matrix A. Next compute the residual matrix

$$R = Y - AX.$$

Then

$$C = XY^{\mathrm{T}}A = XR^{\mathrm{T}}A + XX^{\mathrm{T}}A^{\mathrm{T}}A = XR^{\mathrm{T}}A + r^2XX^{\mathrm{T}}.$$

Here the second term is symmetric and the first one is relatively small; therefore in the equation

$$(\operatorname{tr} C - C^{\mathrm{T}}) \begin{bmatrix} s_{23} \\ s_{31} \\ s_{12} \end{bmatrix} = - \begin{bmatrix} c_{23} - c_{32} \\ c_{31} - c_{13} \\ c_{12} - c_{21} \end{bmatrix},$$

the right number can profitably be calculated from the first term, and the resulting matrix S may then be used as a last correction to A.

It should perhaps be mentioned that in our implementations of the method we always use the symmetrical left-hand matrix

$$\operatorname{tr} C - \tfrac{1}{2}(C + C^{\mathrm{T}})$$

instead of

$$\operatorname{tr}(C - C^{\mathrm{T}})$$

in order to have a simpler program. The value of A should be the same in the limit.

References

[1] G. H. Golub and C. Reinsch. Singular value decomposition and least squares solutions. *Numerische Mathematik*, 14:403–420, 1970.

26

Letter on a Problem in Collocation Theory

To the happy few.

Under Remark 4.1 in his Admont-lectures August 1986 F. Sansò mentions what I for years have looked upon as the main problem in collocation theory:

We know that the potential we want to find is an element of some Hilbert-space H_{k_0} (which I call H_0 in the following) but for technical reasons we use collocation in a space H_k (which I call H) based on a Bjerhammar sphere in which the potential almost certainly not is an element, so in the topology of H (which is stronger than that of H_0) our solution can not converge, even if we have information enough. The problem as I see it is not: does the collocation method (under certain circumstances) converge, but: in which topology does it converge.

After the summer school in Admont I saw that now I had to solve this problem. I believed it was a difficult problem, and therefore I was happy when I had constructed a rather complicated proof for the result I had found. But I found an error in the proof, and I had to cut in the stuff. After several frustrating iterations of this process, I was left with a small nucleus of one or two naive ideas. So fearing that the result should totally disappear between my fingers, I should like to communicate the rest of it to a few of my colleagues, who are supposed to be interested in this problem and to read a relevant mathematical text so as to see some of its implications to their field.

1

1. Let H_0 be a Hilbert space of harmonic functions based on a region $\Omega_0 \subset \mathbf{R}^3$. [This means:
   (a) all elements of H_0 are harmonic in Ω,
   (b) all functions which are harmonic in a neighbourhood of $\bar{\Omega}$ are elements of H_0.]

2. Let
$$\{ f_n \mid n = 1, 2, 3, \dots \}$$
be a sequence of bounded linear functionals on H_0.
3. The f_n are linearly independent (i.e. $\{ f_n \mid n = 1, 2, 3, \dots, N \}$ are linearly independent for every N).
4. The sequence $\{f_n\}$ is a fundamental subset of H^* (i.e. from $\phi \in H_0$ and $f_n\phi = 0$ for $n = 1, 2, 3, \dots$ follows $\phi = 0$).
5. Let
$$\{ P_n \mid n = 1, 2, 3, \dots \}$$
be a sequence of bounded linear operators
$$P_n : H_0 \to H_0,$$
such that for every n and every $\phi \in H_0$
$$f_i P_n \phi = f_i \phi \qquad \text{for } i = 1,\ 2,\ 3,\ \dots,\ n. \tag{1}$$

From (part of) these suppositions we shall now prove that P_n converges weakly to the identity operator for $n \to \infty$.

This means that *for all bounded linear functionals g on H_0 and all $\phi \in H_0$ we have*
$$\lim_{n \to \infty} g P_n \phi - g\phi = 0. \tag{2}$$
But from Lemma III.1.31 in [1] follows that it is enough to prove (2) for all g of a fundamental subset of H^* and all $\phi \in H$, so (2) follows from 4 and (1).

2

6. Let H be a Hilbert space of harmonic functions based on a region $\Omega \subset \mathbf{R}^3$ such that
7. $\Omega_0 \subset \Omega$;
8. $H \subset H_0$ (if $\Omega_0 \subset \Omega$ this follows from 7).

Then it follows from 6 that for $\phi \in H$, the restriction of ϕ to Ω_0 is an element of H_0.

We suppose that

9. if $\phi \in H$ and $\psi \in H$ have the same restriction to Ω_0, then $\phi = \psi$;
10. the set of all restrictions of elements of H to Ω_0 is dense in H_0. (Runge's theorem!)

Then it follows from 3 and 10 that for all N, $\{ f_n \mid n = 1, 2, 3, \dots, N \}$ are linearly independent also as linear functionals on H. Therefore we can for every n by using collocation in H for the n first f_i and with $f_i\phi$, $i = 1$, 2, 3, …, n, as 'observations' find P_n satisfying the conditions in 5.

3

If the sequence $\{P_n\}$ is found by collocation not in H but in H_0, then—naturally—the proof in Section 2 will still be valid, and it is well-known (but shall not be proved here) that in that case the convergence is even strong, but it does not seem to be so in the general case (?).

In a special case, however, we can prove that the convergence is strong:

11. Let H_1 be a Hilbert-space such that

$$H \subset H_1 \subset H_0,$$

and such that the injection I of H_1 in H_0 is compact (this will for instance be true if H_1 is a Hilbert-space based on a region Ω_1 and $\bar{\Omega}_0 \subset \Omega_1$).

12. $\phi \in H_1$ (and not only in H_0).

Then H_1 plays the role of H_0 and $\{P_n\}$ converges weakly in H_1 but as a compact operator 'transforms weak convergence into strong convergence,' we find that then the convergence in H_0 becomes strong:

$$P_n\phi \text{ converges weakly in } H_1$$

therefore

$$IP_n\phi \text{ converges strongly in } H_0.$$

References

[1] T. Kato. *Perturbation Theory for Linear Operators.* Springer, Berlin, 1966.

27

Approximation to The Earth Potential From Discrete Measurements

Introduction

Collocation was originally thought of as part of the mathematical foundation of physical geodesy, and the extent of the practical use of the technique of collocation has astonished me as much as it has disappointed me, that a solid and exact foundation of physical geodesy has advanced so little recently with—or (why not?) without—the theory of collocation.

My "Contribution" was written during few hectic and enthusiastic months in 1968 with all the weaknesses which follows—and perhaps some naive charm which also can follow from such a situation.

This paper written very slowly by an old sceptic man is meant as an essential complement to the "Contribution."

The main task of physical geodesy is from a reasonable set of observations at discrete points and along discrete orbits to calculate an approximation to the gravity potential. The classical approach is to interpolate the values of some linear functional of the potential to all points on a surface surrounding all the masses of the Earth and then apply the methods of boundary value problems, whereas the collocation method goes directly from the discrete observations without using interpolation.

The argumentations of the "Contribution" are such to induce the reader to believe, that for certain given types of observations it is possible for a given precision to choose such a set of those observations such that the potential can be calculated—according to the given precision—from the results of exact measurements of the observations—perhaps under certain conditions.

Several geodesists—among them myself—have felt the need of sufficient conditions and a specification of the calculation procedure in order to reach such a precision, and we have also seen attempts to solve this problem, popularly called "the convergence problem" for collocation, but—so far as I can see—without giving a satisfying answer.

Twelve years ago I wrote a paper to the Hotine symposium in Assisi 1978, where I postulated essentially the same result as in this paper, but, alas, in the proof there was a non-trivial lacuna, which I did not observe before I three years ago made a proof using another mathematical method.

In the following presentation of the proof I have tried to treat the mathematical hard-core of it rather generally in order to make the result easily adaptive for problems corresponding to other boundary value problems as those regarded here. A generalization to under-determined and/or over-determined b.v.p. should be almost a work of routine.

My best thanks to Else Udbye Poulsen for “technical assistance” and patience—during many years—and to Jørgen Eeg and Fernando Sansò for fruitful discussions and critique.

1. Smoothing

We have two open regions Ω_0 and Ω_1 in an Euclidean space, such that

$$\Omega_0 \subset \Omega_1.$$

Ω_1 may be that part of $\mathbf{R}^3$ outside a Bjerhammar sphere and Ω_0 the part outside the surface of the Earth.

If $H(\Omega_0)$ and $H(\Omega_1)$ are the sets of harmonic regular functions on Ω_0 and Ω_1, we have evidently

$$H(\Omega_1) \subset H(\Omega_0).$$

Now let H_0 and H_1 be acceptable Hilbert spaces of harmonic functions such that

$$H_0 \subset H(\Omega_0) \qquad \text{with norm} \quad \|\cdot\|_0$$

and

$$H_1 \subset H(\Omega_1) \qquad \text{with norm} \quad \|\cdot\|_1.$$

The definition of acceptable Hilbert spaces in an open set Ω is given in Section 2.

Supposing that the closure of Ω_0 is a subset of Ω_1 and that if Ω is any open region, such that the closure of Ω_0 (Ω_1) is contained in Ω then $H(\Omega) \subset H_0$ ($H(\Omega) \subset H_1$), it is evident that

$$H_1 \subset H_0.$$

It is clear that not all elements of H_0 are (restrictions of) elements of H_1, but we shall assume that elements of H_1 are dense in H_0 (Runge's theorem!).

In this set-up we shall study the problem of approximation of elements of H_0 by elements of H_1.

That H_1 is dense in H_0 means that for every $\phi \in H_0$ and every $r > 0$ there exists an element $\psi \in H_1$ such that

$$\|\phi - \psi\|_0 \leq r.$$

We can now define our approximation problem as follows:

Given $\phi \in H_0$ and $r > 0$, find elements $\psi \in H_1$ and $\delta \in H_0$ such that

$$\delta + \psi = \phi \tag{1}$$

$$\|\delta\|_0 \leq r, \tag{2}$$

and such that $\|\psi\|_1$ is as small as possible.

First we must prove that this problem has a solution and that the solution is unique.

The problem may be formulated as the problem of minimizing the norm $\|\psi\|_1$ under the side condition

$$\|\phi - \psi\|_0 \leq r,$$

therefore it seems reasonable to study the problem by the method of Legendre's multiplier, namely: find $\psi \in H_1$ such that

$$\|\delta\|_0^2 + \lambda\|\psi\|_1^2 \tag{3}$$

attains its minimum, where $\delta = \phi - \psi$. (It would have been more natural to place the factor λ before $\|\delta\|_0^2$ instead of before $\|\psi\|_1^2$, but it will be important in the following, that it is as in (3).)

For every $\lambda > 0$ the set of pairs

$$\{\, (\delta, \psi) \mid \delta \in H_0,\ \psi \in H_1 \,\}$$

with the norm

$$(\delta, \psi)\big(\|\delta\|_0^2 + \lambda\|\psi\|_1^2\big)^{1/2} \tag{4}$$

constitutes a Hilbert space $H_0 \oplus_\lambda H_1$.

Here it is important to observe that the equation

$$\delta + \psi = \phi$$

in this connection does not mean

$$(\delta, 0) + (0, \psi) = (\phi, 0).$$

(This equation has for a given ϕ only the solution $\delta = \phi$; $\psi = 0$!)

But (1) means

$$\delta + \mathcal{R}(\psi) = \phi,$$

where

$$\mathcal{R}(\psi) : H_1 \longmapsto H_0$$

is the restriction of ψ as a harmonic function on Ω_1 to Ω_0, such that for $x \in \Omega_0$,

$$\big(\mathcal{R}(\psi)\big)(x) = \psi(x).$$

Now $\mathcal{R}$ is a continuous operator from H_1 into H_0 (this will be proved in Section 2), therefore the mapping

$$(\delta, \psi) \longmapsto (\delta + \mathcal{R}\psi, 0)$$

is continuous and so the set

$$\delta + \mathcal{R}\psi = \phi \tag{5}$$

for ϕ given is a closed affine subspace of $H_0 \oplus_\lambda H_1$ parallel to the linear subspace

$$\delta + \mathcal{R}\psi = 0;$$

and from this follows that the orthogonal projection with respect to the norm (4) of the point $(0,0)$ into the subspace (5) exists, and this projection is evidently the unique solution of the minimum problem in Legendre's multiplier form; and we see that the expression

$$K(\lambda, \phi) = \lim_{\delta+\psi=\phi} \left(\|\delta\|_0^2 + \lambda\|\psi\|_1^2\right)^{1/2} \tag{6}$$

has a meaning for all $\phi \in H_0$ and all $\lambda > 0$.

(We shall again where we do not work in the space $H_0 \oplus_\lambda H_1$ use the same symbol for $\psi \in H_1$ and the restriction $\mathcal{R}\psi \in H_0$.)

In order to study the behaviour of $K(\lambda, \phi)$ as a function of λ for fixed values of ϕ, it is best to concentrate on the square of K:

$$f(\lambda) = \left(K(\lambda, \phi)\right)^2 = \min_{\delta+\psi=\phi} \left(\|\delta\|_0^2 + \lambda\|\psi\|_1^2\right), \qquad \lambda > 0.$$

Clearly f is positive, so the curve $y = f(\lambda)$ will remain in the first quadrant in the (λ, y) coordinate system.

Now let $\mu > 0$ be any value of λ, then

$$f(\mu) = \min_{\delta+\psi=\phi} \left(\|\delta\|_0^2 + \mu\|\psi\|_1^2\right) = \|\delta_\mu\|_0^2 + \lambda\|\psi_\mu\|_1^2$$

and

$$y_\lambda = \|\delta_\mu\|_0^2 + \lambda\|\psi_\mu\|_1^2 \tag{7}$$

is a straight line, such that

$$y_\lambda \geq f(\lambda) \qquad \text{for all } \lambda > 0.$$

(This follows from the fact that $\delta_\mu + \psi_\mu = \phi$ and the definition of $f(\lambda)$ as minimum.)

That is, (7) is the equation for a tangent to our curve and the curve must be concave (i.e. $y = -f(\lambda)$ is convex). The curve must also be increasing as its tangents at all points have positive inclination ($= \|\psi_\mu\|_1$). Clearly the tangent (7) meets the y-axis at the point $(0, \|\delta_\mu\|_0^2)$, and it is also clear from these results that

$$\|\delta_\lambda\|_0 \ \text{ is increasing}$$

and

$$\|\psi_\lambda\|_1 \text{ is decreasing,}$$

but this is not surprising!

As $f(\lambda)$ is concave it must also be continuous at least for $\lambda > 0$, but how is its behaviour when $\lambda \to \infty$ and when $\lambda \to 0$.

First it is evident that

$$f(\lambda) \leq \|\phi\|_0^2,$$

so f must as a bounded increasing function have a limit as $\lambda \to \infty$, and it is easy to prove that, in fact

$$\lim_{\lambda\to\infty} f(\lambda) = \|\phi\|_0^2.$$

Analogously as $f(\lambda) \geq 0$ and decreasing when λ decreases to 0, it must have a limit ≥ 0 for $\lambda \to 0$.

That

$$\lim_{\lambda\to 0} f(\lambda) = 0$$

follows from the fact that H_1 is dense in H_0: for given $\epsilon > 0$ and $\phi \in H_0$, we can choose $\psi \in H_1$ such that

$$\|\phi - \psi\|_0^2 < \tfrac{\epsilon^2}{2}$$

and $\lambda > 0$ such that

$$\lambda\|\psi\|_1^2 < \tfrac{\epsilon^2}{2},$$

then

$$\|\phi - \psi\|_0^2 + \lambda\|\phi\|_1^2 < \epsilon^2$$

and a fortiori

$$f(\lambda) < \epsilon^2;$$

i.e.

$$\lim_{\lambda\to 0} f(\lambda) = 0$$

(as evidently $f(0) = 0$, we have now proved that $f(\lambda)$ is continuous in the closed interval $\lambda \geq 0$).

Now we also have

$$\lim_{\lambda\to 0} \lambda\|\psi_\lambda\|_1^2 = 0 \tag{8}$$

and

$$\lim_{\lambda\to 0} \|\phi - \psi_\lambda\|_0^2 = \lim_{\lambda\to 0} \|\delta_\lambda\|_0^2 = 0;$$

it is for $\lambda \to 0$, ψ_λ converges in the space H_0 to ϕ even if ϕ is not in H_1 and ψ_λ therefore does not converge in H_1.

Besides (8) we shall later use

$$\lambda\|\psi_\lambda\|_1^2 \leq f(\lambda) \leq \|\phi\|_0^2. \tag{9}$$

2. A Lemma

The norm $\|\cdot\|$ for a Hilbert space (or more generally a Banach space) H is said to be an acceptable norm with respect to an open region Ω and the space H will be called acceptable if

1. for every open point set S which contains the closure of Ω every function which is harmonic in S is an element of H, i.e. the norm $\|\cdot\|$ is defined for all such functions,
2. for every point $x \in \Omega$ the linear functional evaluation at x is bounded, i.e. there exists a constant c_x such that for every $\phi \in H$

$$|\phi(x)| \leq c_x \|\phi\|.$$

It is almost trivial but shall nevertheless be proved here that:

In a Hilbert (or Banach) space with acceptable norm with respect to Ω, any sequence $\{\phi_i\}$ of elements, which converge strongly to an element ϕ, converges point-wise to ϕ in Ω:

For $x \in \Omega$

$$\big|\phi_i(x) - \phi(x)\big| \leq c_x \|\phi_i - \phi\|.$$

We shall now prove our main lemma, namely:

If Ω_0 and Ω_1 are open regions such that

$$\overline{\Omega}_0 \subset \Omega_1; \qquad \overline{\Omega}_0 \text{ is the closure of } \Omega_0$$

and $\|\cdot\|_i$ are acceptable norms for Hilbert (or Banach) spaces H_i with respect to

$$\Omega_i, \qquad i = 0, 1$$

then there exists a positive constant c such that for all $\phi \in H_1$,

$$\|\phi\|_0 \leq c\|\phi\|_1.$$

For the proof we shall use the not quite elementary "closed graph theorem," which is well-known in functional analysis, and says that

A closed linear operator

$$A\colon B_1 \mapsto B_0, \qquad B_1 \text{ and } B_0 \text{ are Banach spaces}$$

which is defined overall in B_1 is bounded.

We apply this theorem on the operator $\mathcal{R}$:

$$\mathcal{R}\colon H_1 \mapsto H_0\colon \mathcal{R}\phi(x) = \phi(x) \qquad \text{for } \phi \in H_1 \text{ and } x \in \Omega_0.$$

Clearly $\mathcal{R}$ is defined for all $\phi \in H_1$ and $\mathcal{R}$ is linear. It follows from the usual definition of closed-ness of linear operators, that $\mathcal{R}$ is closed if and only if for every sequence $\{ \phi_j \mid j = 1, 2, \dots \}$ of elements of H_1 which converges strongly in H_1 and for which the sequence

$$\{ \mathcal{R}\phi_j \mid j = 1, 2, \dots \}$$

converges strongly in H_0

$$\lim_{j\to\infty} \mathcal{R}\phi_j = \Big(\lim_{j\to\infty} \phi_j \Big), \qquad \text{where } (\cdot) \text{ denotes strong limits!}$$

All we need is to prove is that this condition is satisfied in our case, and also this is almost trivial:

Let

$$\Big(\lim_{j\to\infty} \phi_j \Big) = \phi \in H_1,$$

then from what we proved above

$$\lim_{j\to\infty} \phi_j(x) = \phi(x)$$

for all $x \in \Omega_1$, and therefore also for all $x \in \Omega_0$, that is

$$\lim_{j\to\infty} \mathcal{R}\phi_j(x) = \mathcal{R}\phi(x) \qquad \text{for all } x \in \Omega_0.$$

Then let

$$\Big(\lim_{j\to\infty} \mathcal{R}\phi_j \Big) = \psi \in H_0$$

then

$$\lim_{j\to\infty} \mathcal{R}\phi_j(x) = \psi(x) \qquad \text{for all } x \in \Omega_0$$

and therefore $\psi = \mathcal{R}\phi$, and the lemma is proved.

3. Smoothing collocation in the continuous case

From now on we suppose that the boundary of Ω_0 is a closed smooth surface and that

$$\mathcal{B}\colon H_1 \to L_2(\omega)$$

is a bounded linear operator defined on H_1 with functions f on ω, such that

$$\int_\omega f^2 \, d\omega$$

is finite. Further we suppose that the mapping $\mathcal{B}$ is injective, i.e. if for $\phi \in H_1$ we have $\mathcal{B}\phi = 0$, then $\phi = 0$.

In this case

$$\|\mathcal{B}\phi\|_{L_2} = \left(\int_\omega (\mathcal{B}\phi)^2 \, d\omega \right)^{1/2}$$

is a Hilbert norm and the set $\phi \in H_1$ with this norm constitutes a pre-Hilbert space.

Let H_B be the completion of H_1 with respect to this norm. Then H_B is a Hilbert space. At least in some case $\|\mathcal{B}\phi\|_{L_2}$ is an acceptable norm with respect to Ω_0. This is the case if $\mathcal{B}$ maps harmonic functions in H_1 to their values on ω, and also when Ω is the set of points outside the surface ω and

$$(\mathcal{B}\phi)_x = \left(\frac{\partial \phi}{\partial n}\right)_x \qquad \text{for all } x \in \omega,$$

and many other examples can be given.

In that case, i.e., if $\|\mathcal{B}\phi\|_{L_2}$ is acceptable with respect to Ω_0, we can use the theory of Section 1 putting

$$\|\phi\|_0 = \|\mathcal{B}\phi\|_{L_2},$$

but the problem we here have solved is in fact a continuous case of smoothing collocation (collocation with hybrid norm) and we can—if we use a space H_1, for which we know the reproducing kernel—use the same abstract formulas for the computation as in the discrete and finite case; the normal equations will in the continuous case be integral equations and not a finite set of linear algebraic equations. This is well-known but until now it has only been a belief, that it should be possible to achieve an arbitrarily good approximation to the potential by elements of H_1 in that way. Here we have a sufficient condition and a solid definition of such an approximation.

4. ...

In this section I want to prove that we—under the suppositions of the foregoing sections—for every wanted precision can find sufficiently many sufficiently well distributed points on ω and a form for smoothing collocation (in the discrete case) such that the calculated harmonic function in Ω_1 approximates the "true" potential according to the wanted precision in Ω_0.

It is on purpose that I do not use the word "observation" or "measurement," and it is for two reasons:

First the knowledge of the value of $\mathcal{B}\phi$ at the selected points must be exact. The influence of measuring errors is—as well as that of calculation errors—outside the scope of this paper; but it deserves a paper of its own!

Second I feel that in common language the sentence "the value of $\mathcal{B}\phi$ at the point x' presupposes that the functional" evaluating of $\mathcal{B}\phi$ at x' depends locally on ϕ, of its behaviour in a small neighbourhood of x etc. As such a locality seems to be inessential for the theory of this paper I can spare the reader—and myself—an unnecessary definition by not using the work observation.

The central quantity in this paper until now is $\|\cdot\|_0$, therefore it is reasonable to find an approximation to $\|\cdot\|_0$ from the values of $\mathcal{B}\phi$ at a finite number of points on ω—or more concretely—to look for an estimate for

$$\left| \|\phi\|_0^2 - \sum_i w_i \big(\mathcal{B}\phi(x_i)\big)^2 \right|$$

where ϕ is a function harmonic in Ω_0,

$$\{\, x_i \mid i = 1, 2, \ldots, N \,\}$$

are N points on Ω,

$$\{\, w_i \mid i = 1, 2, \ldots, N \,\}$$

are some positive weights to be applied on the squares of the evaluation of $\mathcal{B}\phi$ at those points.

But here we meet the difficulty that such an estimate does not exist for $\phi \in H_0$. This has to do with another fact, namely that evaluation in a point on ϕ of a function (in our case $\mathcal{B}\phi$) of L_2 is not a bounded functional.

The consequence of this seems to be that we cannot find the potential ϕ without supposing that it belongs to a space with a stronger topology than that of H_0. As a reasonable example I have selected a space, which I call B (it is a Banach space and not a Hilbert space).

It is defined as follows:

(a) First we define the spaces L_α for $0 < \alpha \leq 1$ of continuous functions on ω which satisfy the Lipschitz condition

$$|f(x) - f(y)| \leq kt^\alpha \qquad \text{for all } x, y \in \omega, \tag{10}$$

where t is the distance between the points x and y on ω. (Distances on ω are measured along (the shortest) geodesic connecting the points.)
Let k_α be the smallest value of k for which (10) is satisfied for a given α, then the norm $\|f\|_\alpha$, of f is defined by

$$\|f\|_\alpha^2 = \max |f|^2 + k_\alpha^2,$$

(As ω is compact and f continuous $\max |f|^2$ exists.)
Clearly with this norm L_α is a Banach space.

(b) The norm

$$\| \cdot \|_{B_\alpha} = \|\mathcal{B} \cdot \|_\alpha$$

is defined for all $\phi \in H_1$, and the set of $\phi \in H_1$ constitutes a pre-Banach space with norm $\| \cdot \|_{B_\alpha}$, its completion is the Banach space called B_α.

We want to prove that the norm $\| \cdot \|_{B_\alpha}$ is an acceptable one.

It is evident that point 1) of the definition in Section 2 is satisfied, so we have only to prove that 2) also is so.

Let $\phi \in B_\alpha$, then

$$\max_{x\in\omega} \big(\mathcal{B}\phi(x)\big)^2$$

exists and

$$\max_{x\in\omega} \big(\mathcal{B}\phi(x)\big)^2 \leq \|\phi\|_{B_\alpha}^2$$

but

$$\|\phi\|_0^2 = \int\limits_\omega (\mathcal{B}\phi)^2\, dx \leq \text{area}(\omega) \max_{x\in\omega} \big(\mathcal{B}\phi(x)\big)^2 \leq \|\phi\|_{B_\alpha}^2 \,\text{area}(\omega),$$

so if for every $x \in \Omega_0$

$$\frac{|\phi(x)|}{\|\phi\|_0}$$

is finite, so is

$$\frac{|\phi(x)|}{\|\phi\|_{B_\alpha}};$$

i.e. evaluation of ϕ at any point $x \in \Omega_0$ is also a bounded functional on B_α, and we have proved that B_α is an acceptable Banach space of harmonic functions in Ω_0.

From this fact and from the lemma of Section 2 then follows the existence of a constant K_α such that for every element $\psi \in H_1$

$$\|\psi\|_{B_\alpha} \leq K_\alpha \|\psi\|_1.$$

Here we shall use one more definition: By a centered tesselation of ω with radius $\rho > 0$, we shall mean a finite set of pairs

$$\{\,\omega_i,\, x_i\,\}, \qquad i = 1,\, 2,\, \ldots,\, N;$$

where

$$\omega_i \subset \omega, \qquad i = 1,\, 2,\, \ldots,\, N,$$

are disjoint open subsets of ω such that every point of ω is in the closure of (at least) one such subset ω_i, and the "centers"

$$x_i \in \omega_i, \qquad i = 1,\, 2,\, \ldots,\, N.$$

If

$$\rho_i = \sup_{x\in\omega_i} \text{dist}(x, x_i),$$

where $\text{dist}(x, x_i)$ means the distance between x and x_i, then the "radius" ρ is defined as

$$\rho = \max_{i=1,2,\ldots,N} \rho_i.$$

After these preparations we can prove the following estimate:

If

$$\{\, x_i \mid i = 1, 2, \ldots, N\,\}$$

are the centres of a centered tesselation on ω and the weights $\{\, w_i\,\}$ are defined as

$$w_i = \text{area}(\omega_i)$$

then for every $\phi \in B_\alpha$ we have

$$\left| \|\phi\|_0^2 - \sum_{i=1}^{N} w_i \big(\mathcal{B}\phi(x_i)\big)^2 \right| \leq \rho^\alpha \|\phi\|_{B_\alpha}^2 \,\text{area}(\omega). \tag{11}$$

Proof

$$\begin{aligned}
\left|\|\phi\|_0^2 - \sum_{i=1}^N w_i\big(\mathcal{B}\phi(x_i)\big)^2\right| &= \left|\sum_{i=1}^N\left(\int_{\omega_i}\big(\mathcal{B}\phi(x)\big)^2\,dx - \operatorname{area}(\omega_i)\big(\mathcal{B}\phi(x_i)\big)^2\right)\right| \\
&\le \sum_{i=1}^N\int_{\omega_i}\Big(\big(\mathcal{B}\phi(x)\big)^2 - \big(\mathcal{B}\phi(x_i)\big)^2\Big)\,dx \\
&= \sum_{i=1}^N\int_{\omega_i}\big(\mathcal{B}\phi(x)+\mathcal{B}\phi(x_i)\big)\big(\mathcal{B}\phi(x)-\mathcal{B}\phi(x_i)\big)\,dx \\
&\le \sum_{i=1}^N \operatorname{area}(\omega_i)\cdot 2\max\big|\mathcal{B}\phi(x)\big|K_\alpha\rho^\alpha \\
&= 2\operatorname{area}(\omega)\rho^\alpha\max|\mathcal{B}\phi|K_\alpha \\
&\le \operatorname{area}(\omega)\rho^\alpha\big(\max|\mathcal{B}\phi|^2+K_\alpha^2\big) = \operatorname{area}(\omega)\rho^\alpha\|\phi\|_{B_\alpha}^2.
\end{aligned}$$

For any function ϕ which is harmonic in an open set containing the closure of Ω_0 we shall define $\|\phi\|_N$ by

$$\|\phi\|_N^2 = \sum_{\omega_i} w_i\big(\mathcal{B}\phi(x_i)\big)^2$$

when the pairs

$$\{\,x_i,\, w_i\,\},\qquad i = 1,\, 2,\, \ldots,\, N, \tag{12}$$

are given. If the pairs (12) are related to a centered tesselation with radius ρ we shall write

$$\|\phi\|_\rho = \|\phi\|_N.$$

$\|\phi\|_N$ is not a norm, only a semi-norm, but it is easy to see that the approximation theory from the first sections works unchanged if it is applied to $\|\cdot\|_1$ and $\|\cdot\|_N$ instead of $\|\cdot\|_1$ and $\|\cdot\|_0$, and that it becomes the theory for smoothing discrete collocation.

The fact that $\|\phi\|_\rho$ for $\phi\in B_\alpha$ can be looked upon as an approximation to $\|\phi\|_0$—as follows from the theorem above—makes it tempting to try to prove that smoothing discrete collocation can be used as an approximation to smoothing continuous collocations; and this is the idea we shall take up in the following.

The central expression in Section 1

$$K(\lambda,\phi) = \lim_{\delta+\psi=\phi}\big(\|\delta\|_0^2 + \lambda\|\psi\|_1^2\big)^{1/2},$$

or, when minimizing harmonic functions δ_λ and ψ_λ are substituted:

$$K(\lambda,\phi) = \big(\|\delta_\lambda\|_0^2 + \lambda\|\psi_\lambda\|_1^2\big)^{1/2}. \tag{13}$$

We will write the expression corresponding to a centered tesselation with radius ρ:

$$K_\rho(\lambda,\phi) = \left(\|\delta_{\rho\lambda}\|_\rho^2 + \lambda\|\psi_{\delta\lambda}\|_1^2\right)^{1/2}. \tag{14}$$

Using (11):

$$\left|\|\delta\|_0^2 - \|\delta\|_\rho^2\right| = \left|\|\phi-\psi\|_0^2 - \|\phi-\psi\|_\rho^2\right| \le \text{area}(\omega)\rho^\alpha\|\phi-\psi\|_{B_\alpha}^2.$$

But as

$$\|\phi-\psi\|_{B_\alpha} \le \|\phi\|_{B_\alpha} + \|\psi\|_{B_\alpha}$$
$$\|\phi-\psi\|_{B_\alpha}^2 \le 2\left(\|\phi\|_{B_\alpha}^2 + \|\psi\|_{B_\alpha}^2\right)$$

and, as we saw above, that

$$\|\psi\|_{B_\alpha} \le K_\alpha\|\psi\|_1,$$

we have

$$\left|\|\delta\|_0^2 - \|\delta\|_\rho^2\right| \le 2\,\text{area}(\omega)\rho^\alpha\left(\|\phi\|_{B_\alpha}^2 + K_\alpha^2\|\psi\|_1^2\right) = \rho^\alpha(A + B\|\psi\|_1^2);$$

where

$$A = 2\,\text{area}(\omega)\|\phi\|_{B_\alpha}^2$$

and

$$B = 2\,\text{area}(\omega)K_\alpha^2.$$

We want to find for any given $\epsilon > 0$, λ_ϵ and ρ_ϵ such that

$$\|\phi - \psi_{\rho_\epsilon\lambda_\epsilon}\|_0^2 \le \epsilon. \tag{15}$$

This is accomplished in three steps:

1. Find λ_ϵ such that
$$\left(K(\lambda_\epsilon,\phi)\right)^2 \le \tfrac{\epsilon}{3}.$$
2. Find ρ_ϵ such that for any centered tesselation with radius ρ_ϵ
$$\left|\left(K_\rho(\lambda_\epsilon,\phi)\right)^2 - \left(K(\lambda_\epsilon,\phi)\right)^2\right| \le \tfrac{\epsilon}{3}.$$
3. $\left|\|\phi - \psi_{\rho_\epsilon\lambda_\epsilon}\|_0^2 - \|\phi - \psi_{\rho_\epsilon\lambda_\epsilon}\|_\rho^2\right| \le \tfrac{\epsilon}{3}.$

Point 1) is already proved in the first sections. When point 2) is proved it follows that

$$\left(K_\rho(\lambda_\epsilon,\phi)\right)^2 \le \tfrac{2\epsilon}{3}$$

and from this follows

$$\|\phi - \psi_{\rho_\epsilon\lambda_\epsilon}\|_\rho^2 \le \tfrac{2\epsilon}{3}. \tag{16}$$

Point 3) will be proved together with 2), and then from (16) and point 3) follows (15). So we have only to prove 2) and 3).

We have

$$\begin{aligned}\left(K(\lambda,\phi)\right)^2 &\le \|\delta_{\rho\lambda}\|_0^2 + \lambda\|\psi_{\rho\lambda}\|_1^2 = \|\delta_{\rho\lambda}\|_0^2 - \|\delta_{\rho\lambda}\|_\rho^2 + \|\delta_{\rho\lambda}\|_\rho^2 + \lambda\|\psi_{\rho\lambda}\|_1^2\\ &\le \left|\|\delta_{\rho\lambda}\|_0^2 - \|\delta_{\rho\lambda}\|_\rho^2\right| + \left(K_\rho(\lambda,\phi)\right)^2\end{aligned}$$

and

$$\left(K_\rho(\lambda,\phi)\right)^2 \leq \|\delta_\lambda\|_\rho^2 + \lambda\|\psi_\lambda\|_1^2 = \|\delta_\lambda\|_\rho^2 - \|\delta_\lambda\|_0^2 + \|\delta_\lambda\|_0^2 + \lambda\|\psi_\lambda\|_1^2$$
$$\leq \left|\|\delta_\lambda\|_0^2 - \|\delta_\lambda\|_\rho^2\right| + \left(K(\lambda,\phi)\right)^2$$

or

$$\left(K(\lambda,\phi)\right)^2 - \left(K_\rho(\lambda,\phi)\right)^2 \leq \left|\|\delta_{\rho\lambda}\|_0^2 - \|\delta_{\rho\lambda}\|_\rho^2\right| \leq \rho^\alpha\left(A + B\|\psi_{\rho\lambda}\|_1^2\right)$$
$$\left(K_\rho(\lambda,\phi)\right)^2 - \left(K(\lambda,\phi)\right)^2 \leq \left|\|\delta_\lambda\|_0^2 - \|\delta_\lambda\|_\rho^2\right| \leq \rho^\alpha\left(A + B\|\psi_\lambda\|_1^2\right)$$

that is

$$\left|\left(K(\lambda,\phi)\right)^2 - \left(K_\rho(\lambda,\phi)\right)^2\right| \leq \tfrac{\epsilon}{3}$$

if both

$$\rho^\alpha\left(A + B\|\psi_{\rho\lambda}\|_1^2\right) \leq \tfrac{\epsilon}{3} \tag{17}$$

and

$$\rho^\alpha\left(A + B\|\psi_\lambda\|_1^2\right) \leq \tfrac{\epsilon}{3}; \tag{18}$$

but in that case we also have

$$\left|\|\delta_{\rho\lambda}\|_0^2 - \|\delta_{\rho\lambda}\|_0^2\right| \leq \tfrac{\epsilon}{3},$$

so we only have to prove that given $\lambda > 0$ it is possible to chose ρ such that (17) and (18) are satisfied.

From (9)

$$\|\psi_\lambda\|_1^2 \leq \tfrac{1}{\lambda}\|\phi\|_0^2,$$

so (18) becomes

$$\rho^\alpha\left(A + \tfrac{1}{\lambda}B\|\phi\|_0^2\right) \leq \tfrac{\epsilon}{3},$$

and analogously (17) becomes

$$\rho^\alpha\left(A + \tfrac{1}{\lambda}B\|\phi\|_\rho^2\right) \leq \tfrac{\epsilon}{3}.$$

Now we have

$$\|\phi\|_0^2 \leq \|\phi\|_{B_\alpha}^2 \operatorname{area}(\omega),$$

and also

$$\|\phi\|_\rho^2 \leq \|\phi\|_{B_\alpha}^2 \operatorname{area}(\omega),$$

so defining

$$C = B\|\phi\|_{B_\alpha}^2 \operatorname{area}(\omega),$$

we see that the condition is

$$\rho^\alpha\left(A + \tfrac{1}{\lambda}C\right) \leq \tfrac{\epsilon}{3},$$

where A and C are independent of ρ and λ, and so we in fact can find ρ^α as postulated in points 2) and 3), as $\lambda > 0$ and $0 < \alpha \leq 1$; and we have thus proved that we can achieve any prescribed degree of approximation by using exact "observations" and smoothing collocation, but for the case $\lambda = 0$ i.e. non-smoothing collocation the method of proof breaks down.

Conclusion

I shall not try to draw the consequences of the result to which we have arrived, but a few remarks may be helpful.

1. I think it is necessary to use smoothing collocation in order to get convergence such that λ, the parameter of smoothing, becomes small when ρ, the radius of the tesselation becomes small. This has to do with the "Lelgemann-effect" (first pointed out by Lelgemann in his dissertation), namely the fact that the value of the functional (used by the collocation) on the potential solution of the collocation behaves rather exotically between the collocation points. It seems to be so that for non-smoothing collocation the B-norm is not necessarily bounded for $\rho \to 0$. By using the smoothing technique described here this effect is kept under control.
2. The weights of the observations are here determined such that the "density of weights" is constant. It is easy to see that it is enough that there exist two positive constants a and b such that for the density δ we have overall

$$a \leq \delta \leq b.$$

3. In the proof I have supposed that the potential ϕ, which we want to find, is a member of Banach space B such that for all $\psi \in B$, $\mathcal{B}(\psi)$ satisfies a Lipschitz condition on ω. I have recently found that continuity of $\mathcal{B}\psi$ on ω is sufficient, but the proof is a little more difficult. In any case it seems reasonable that $\mathcal{B}$ must be so that for all $\psi \in B$ the value of $\mathcal{B}(\psi)_\omega$ at all points of ω is a continuous functional on B, in fact I believe that this condition is necessary. And this is a thing which makes the treatment of real observations (observations with errors) more complicated, because unprecise observation results will not combine to continuous function on ω in the limit.

There are still many open questions in collocation theory, but I believe that with the ideas of this paper, which will be my last paper on collocation, it will be possible to create a solid foundation of physical geodesy. It is my hope that a few of my colleagues will understand these ideas, so I should be glad to answer questions—I can still write letters!

Bibliographical Notes

The "closed graph theorem" can be found in any book on functional analysis e.g. [2]. This theorem is essential for the proof, and it is what is needed to fill the lacuna in my Assisi paper [3].

Theoretically—but only theoretically—this paper can be read by readers without any knowledge of collocation, because I have used formulas or any other results from collocation theory. A good introduction to this is found in [4].

It is important to point out that I here have looked upon collocation under the analytical aspect. The fundamental treatise on collocation under the statistical aspect is [5]. I think the time has come where we should find out if collocation under the two different aspects is the same thing expressed in different languages.

In this paper I have used methods and symbols borrowed from the so-called K-method in the theory of interpolating spaces, see [1].

References

[1] J. Berg and J. Löfström. *Interpolating Spaces. An Introduction.* Springer, Berlin, 1976.

[2] J. L. Kelley and Isaac Namioka. *Linear Topological Spaces.* D. Van Nostrand, Princeton, 1963.

[3] Torben Krarup. A convergence problem in collocation theory. *Bollettino di Geodesia e Scienze Affini*, XL:225–240, 1981.

[4] Helmut Moritz. *Advanced Physical Geodesy.* Wichmann, Karlsruhe, 1980.

[5] F. Sansò. *Statistical Methods in Physical Geodesy*, volume 7 of *Lecture Notes in Earth Sciences.* Springer, Berlin, 1986.

28

An Old Procedure for Solving the Relative Orientation in Photogrammetry

Abstract

We describe the problem of relative orientation in terms of homogeneous coordinates concluding in a least squares problem in the observed image coordinates. The solution determines a rotational matrix for each image; these rotational matrices bring the images back to the normal position. The explicit formula for the rotational matrices is derived using properties of 'nearly' orthogonal matrices. The procedure is augmented by a special preliminary iteration step in order to cope with large rotations.

The method is described through a complete Pascal program.

1. Introduction

The problem of relative orientation is constantly subject to investigation. For some years we have been using a procedure which apparently is not well known, so we have decided to publish ideas originating from the early fifties when one of the authors finished his Master's thesis on photogrammetry. As so often before, the really good ideas emerge shortly afterwards. It shall be mentioned that ideas of J. Krames, [2], have influenced this paper, too.

The following section contains a description of the problem in terms of homogeneous coordinates. This is advantageous as we obtain a strict formulation of the subject matter. The rotational matrices which are necessary to bring back the images to the normal position are derived in Section 3. By means of a special procedure—a procedure we consider as the essential contribution of this paper—we try to orthogonalize these rotational matrices and thus obtain computational achievements not yet seen. In showing a commented version of the program, we want to introduce to photogrammetrists all details of the method and the features of a so-called zero iteration.

The procedure shall be looked upon as an efficient part of a larger program for solving the relative and the absolute orientation simultaneously.

2. The Problem in Terms of Homogeneous Coordinates

Let $\mathcal{U}$ and $\mathcal{V}$ be projective planes and let $\mathcal{W}$ be the projective space of dimension 8. In $\mathcal{U}$ and $\mathcal{V}$ we have chosen homogeneous coordinate systems in the orthodox way. Likewise, in $\mathcal{W}$ a homogeneous coordinate system is also applied with nine coordinates representing the entries w_{ij} of a 3×3 matrix. If W is such a matrix, then cW represents the same point in $\mathcal{W}$ provided $c \neq 0$.

Now we define a mapping of $\mathcal{U}$ and $\mathcal{V}$ onto $\mathcal{W}$:

$$w_{ij} = u_i v_j$$

or in matrix language

$$W = uv^{\mathrm{T}}$$

where $(u_1 : u_2 : u_3)$ and $(v_1 : v_2 : v_3)$ are coordinates for $u \in \mathcal{U}$ and $v \in \mathcal{V}$, respectively. It is not difficult to realize that this really is a mapping, i.e., to any points u and v there is a point $W \in \mathcal{W}$ independent of how the constant factors c_u and c_v are chosen. Because $c_W = c_u c_v$, then $c_W \neq 0$ if and only if $c_u \neq 0$ and $c_v \neq 0$.

The range of this mapping is denoted $\mathcal{P}$ and is called the product manifold of the two projective planes $\mathcal{U}$ and $\mathcal{V}$, cf. [1] or [4].

It is almost evident that the following statements are equivalent:

1. $W \in \mathcal{P}$;
2. The rank of the matrix W is 1;
3. Any two rows in the matrix W are proportional;
4. Any two columns in the matrix W are proportional;
5. All second order minors of W are 0;
6. $$\begin{vmatrix} w_{11} & w_{1s} \\ w_{r1} & w_{rs} \end{vmatrix} = 0 \qquad \text{for } \ r = 2,3 \ \text{ and } \ s = 2,3. \tag{1}$$

Accordingly, the range $\mathcal{P}$ consists of the set of points in $\mathcal{W}$ which satisfy the four independent (quadratic) equations (1). Not surprisingly, it is of dimension $8 - 4 = 4$. For each point of $\mathcal{P}$ there is one and only one pair of points $u \in \mathcal{U}$ and $v \in \mathcal{V}$ and vice versa.

The product of a non-singular matrix A and a matrix X has rank 1 if and only if X has rank 1. Consequently, the following holds true:

Let A and B be non-singular matrices. Then $AWB \in \mathcal{P}$ if and only if $W \in \mathcal{P}$. (2)

Any hyperplane in $\mathcal{W}$ can be described as

$$\langle H, W \rangle = \operatorname{tr}(H^{\mathrm{T}} W) = 0,$$

where H is a 3×3 matrix given apart from a constant factor (exactly as the coordinates of a point in $\mathcal{W}$). If we make the following particular choice Y for H

$$Y = \begin{bmatrix} 0 & 0 & 0 \\ 0 & 0 & 1 \\ 0 & -1 & 0 \end{bmatrix}, \tag{3}$$

we obtain a hyperplane $\mathcal{Y}$ determined by the equation

$$\langle Y, W \rangle = 0. \tag{4}$$

This hyperplane describes just the condition of coplanarity as shown in (21). Equation (21) is a more familiar way of writing this condition.

The intersection containing points common to $\mathcal{P}$ and $\mathcal{Y}$ we call $\hat{\mathcal{P}}$. It is a three-dimensional manifold, and $W \in \hat{\mathcal{P}}$ if and only if W satisfies the equations (1) and (4).

Since $\hat{\mathcal{P}}$ is of dimension 3 it is tempting to map $\hat{\mathcal{P}}$ onto a three-dimensional projective space $\boldsymbol{P}_3$. It certainly works well.

In $\boldsymbol{P}_3$ we denote the homogeneous coordinates $(x_0 : x_1 : x_2 : x_3)$. We try to solve for them so that

$$\begin{matrix} (x_1+x_0)(x_1-x_0) & : x_2(x_1-x_0) & : x_3(x_1-x_0) \\ (x_1+x_0)x_2 & : x_2x_2 & : x_3x_2 \\ (x_1+x_0)x_3 & : x_2x_3 & : x_3x_3 \end{matrix} \quad = \quad \begin{matrix} w_{11} : w_{12} : w_{13} \\ w_{21} : w_{22} : w_{23} \\ w_{31} : w_{32} : w_{33} \end{matrix}.$$

If $w_{33} \neq 0$, we choose x_3 so that $x_3^2 = kw_{33}$ with an appropriate value $k \neq 0$. Next we find

$$x_2 = \frac{w_{32}}{w_{33}} x_3,$$

or by use of (1),

$$x_2 = \frac{w_{23}}{w_{33}} x_3.$$

As all 2×2 minors in W vanish, we have

$$x_2^2 = kw_{22}.$$

Henceforth, $x_1 + x_0$ is determined as

$$x_1 + x_0 = \frac{w_{31}}{w_{32}} x_2$$

or equivalently as

$$x_1 + x_0 = \frac{w_{21}}{w_{22}} x_2,$$

as $\begin{vmatrix} w_{21} & w_{22} \\ w_{31} & w_{32} \end{vmatrix} = 0$, etc.

In this manner, we have determined a one-to-one mapping of $\hat{\mathcal{P}}$ onto the three-dimensional projective space $\boldsymbol{P}_3$.

We may proceed yet further. As long as we stay in a neighborhood of $\boldsymbol{P}_3$ with $x \neq 0$, we can normalize the coordinates: We put $x_0 = b/2$, i.e., we multiply all coordinates by $b/2x_0$ and then obtain the usual (model)coordinates in the three-dimensional affine space.

Before turning to the photogrammetric issues, we shall solve the following task:

For $W \in \mathcal{W}$ given, find the condition that

$$W + SW + WT \in \mathcal{Y} \tag{5}$$

where S and T are unknown skew-symmetric 3×3 matrices.

Therefore, we have

$$\langle Y, W + SW + WT \rangle = 0.$$

Hence,

$$0 = \langle Y, W \rangle + \langle Y, SW \rangle + \langle Y, WT \rangle = \langle Y, W \rangle + \langle S^{\mathrm{T}} Y, W \rangle + \langle Y T^{\mathrm{T}}, W \rangle$$

and

$$\langle Y + S^{\mathrm{T}} Y + Y T^{\mathrm{T}}, W \rangle = 0. \tag{6}$$

Let S be defined as

$$S = \begin{bmatrix} 0 & -s_3 & s_2 \\ s_3 & 0 & -s_1 \\ -s_2 & s_1 & 0 \end{bmatrix}, \tag{7}$$

and T in a similar manner whereupon we obtain

$$\begin{aligned} Y + S^{\mathrm{T}} Y + Y T^{\mathrm{T}} &= \begin{bmatrix} 0 & 0 & 0 \\ 0 & 0 & 1 \\ 0 & -1 & 0 \end{bmatrix} + \begin{bmatrix} 0 & s_3 & -s_2 \\ -s_3 & 0 & s_1 \\ s_2 & -s_1 & 0 \end{bmatrix} \begin{bmatrix} 0 & 0 & 0 \\ 0 & 0 & 1 \\ 0 & -1 & 0 \end{bmatrix} \\ &\quad + \begin{bmatrix} 0 & 0 & 0 \\ 0 & 0 & 1 \\ 0 & -1 & 0 \end{bmatrix} \begin{bmatrix} 0 & t_3 & -t_2 \\ -t_3 & 0 & t_1 \\ t_2 & -t_1 & 0 \end{bmatrix} \\ &= \begin{bmatrix} 0 & s_2 & s_3 \\ t_2 & -(s_1 + t_1) & 1 \\ t_3 & -1 & -(s_1 + t_1) \end{bmatrix}. \end{aligned}$$

So condition (6) can be expressed as

$$\langle Y + S^{\mathrm{T}} Y + Y T^{\mathrm{T}}, W \rangle = \left\langle \begin{bmatrix} 0 & s_2 & s_3 \\ t_2 & -(s_1 + t_1) & 1 \\ t_3 & -1 & -(s_1 + t_1) \end{bmatrix}, W \right\rangle = 0. \tag{8}$$

3. Determination of the Rotation Matrix

On the basis of the observations, we shall determine two orthogonal matrices: Φ, which rotates the right image back to the normal position, and Ψ^{T}, which rotates the left image back to the normal position (The matrix Φ shall not be confused with a rotational angle of the absolute orientation). The set of orthogonal matrices $\mathcal{O}$ is not a linear manifold, as $a\Phi + b\Psi$ normally is not an orthogonal matrix. So we embed $\mathcal{O}$ in the manifold $\mathcal{R}$ of 3×3 matrices. We shall work iteratively in $\mathcal{R}$, i.e., based on a preliminary guess at Φ we look for an improving matrix Φ' so that the improved guess is $\Phi'\Phi$.

The tangent manifold to $\mathcal{O}$ at Φ is

$$\Phi + d\Phi = \Phi(I + S),$$

where S is infinitesimal (A similar expression is valid for T: $\Psi + d\Psi = (I+T)\Psi$). There is no God-given correspondence between this tangent manifold $\mathcal{S}$ and the manifold of orthogonal matrices $\mathcal{O}$. The matrix $I + S$ is only orthogonal for S being infinitesimal.

Therefore, we look for a symmetric matrix R being a function of S so that $R(I + S)$ is orthogonal; that is

$$R(I + S)\big(R(I + S)\big)^{\mathrm{T}} = I$$

or

$$R(I + S)(I - S)R = I$$

or

$$R(I - S^2)R = I$$

or

$$R = (I - S^2)^{-1/2}.$$

Hence we introduce the following mapping

$$\Phi' = \operatorname{orth}(S) = \frac{I + S}{\sqrt{I - S^2}}, \tag{9}$$

which is defined through a corresponding power series at a neighborhood of $S = 0$.

The designation orth not only means that the result of the mapping is orthogonal, but also that $\Phi'\Phi$ is the orthogonal projection S of the tangent space at Φ onto $\mathcal{O}$ or in other words that

$$\operatorname{dist}(I + S, \Phi') = \min .$$

Of course, this assumes a metric being defined in $\mathcal{R}$. We choose the Euclidian metric defined through the Euclidian norm or the trace norm:

$$\|A\|^2 = \operatorname{tr}(A^{\mathrm{T}}A) = \sum_{r,s=1}^{3} a_{rs}^2, \qquad A = [a_{rs}]. \tag{10}$$

In this norm, the distance between A and $B \in \mathcal{R}$ is

$$\operatorname{dist}(A, B) = \|A - B\| = \sqrt{\|A\|^2 + \|B\|^2 - 2\langle A, B\rangle}. \tag{11}$$

Now we want to verify that

$$\operatorname{orth}(S^{\mathrm{T}}) = \operatorname{orth}(-S) = \big(\operatorname{orth}(S)\big)^{-1}. \tag{12}$$

Making a series expansion and investigating this leads to

$$\operatorname{orth}(S^{\mathrm{T}}) = \big(\operatorname{orth}(S)\big)^{\mathrm{T}}, \tag{13}$$

i.e., $(\Phi')^{\mathrm{T}} = (\Phi')^{-1}$, so Φ' is really orthogonal.

As a matter of form we also have to prove $\det \Phi' = 1$. Because Φ' is orthogonal, $\det \Phi'$ can only be $+1$ or -1. Next

$$\det\big(\operatorname{orth}(\Phi)\big) = \det 1 = 1;$$

so what remains is to prove that

$$\det\big(\operatorname{orth}(S)\big) = \det\big(\operatorname{orth}(\Phi)\big).$$

This follows from

$$f(t) = \operatorname{orth}(tS) \qquad \text{for } 0 \le t \le 1$$

varies continuously in this interval and so does $\det\big(\operatorname{orth}(tS)\big)$. Once it is $+1$, it can never become -1.

Finally, we have to prove that orth is an orthogonal projection,

$$\operatorname{orth}\colon \mathcal{R} \mapsto \mathcal{O}.$$

According to [3], this is equivalent to asking if $(I + S)(\Phi')^{\mathrm{T}}$ is a symmetric matrix. From (9) follows that this expression equals $\sqrt{I - S^2}$ which evidently is symmetric, and we have finished the proof.

After this we can turn to the computational aspects. Basically, we shall calculate $(I + S)/\sqrt{I - S^2}$, S being skew-symmetric.

The characteristic equation corresponding to the matrix S is

$$\det(S - \lambda I) = \lambda^3 + \sigma^2\lambda = \lambda(\lambda^2 + \sigma^2), \tag{14}$$

where $\sigma^2 = s_1^2 + s_2^2 + s_3^2 = \frac{1}{2}\operatorname{tr}(SS^{\mathrm{T}})$, cf. (7). As any matrix satisfies its own characteristic equation (Cayley-Hamilton theorem, see [6]), we get

$$S^3 + \sigma^2 S = 0 \tag{15}$$

or

$$S(S^2 + \sigma^2 I) = 0, \tag{16}$$

i.e.,

$$S^3 = -\sigma^2 S, \quad S^4 = -\sigma^2 S^2, \quad S^6 = \sigma^4 S^2, \quad S^8 = -\sigma^6 S^2, \quad \ldots .$$

For any polynomial in x^2

$$p(x^2) = a_0 + a_1 x^2 + a_2 x^4 + \cdots + a_n x^{2n}$$

the following is valid, if we substitute x with S:

$$\begin{aligned} p(S^2) &= a_0 I + a_1 S^2 + a_2 S^4 + \cdots + a_n S^{2n} \\ &= a_0 I + a_1 S^2 - a_2\sigma^2 S^2 + a_3\sigma^4 S^2 - \cdots \pm a_n \sigma^{2n-2} S^2 \end{aligned}$$

$$= a_0 I + \Big(a_1\sigma^2 - a_2\sigma^4 + a_3\sigma^6 - \cdots \pm a_n\sigma^{2n}\Big)\frac{S^2}{\sigma^2}$$

$$= p(0) - \frac{1}{\sigma^2}\Big(p(-\sigma^2) - I\Big)S^2.$$

The expression $\sqrt{I - S^2}$ is a continuous function for all σ, so the expression can in any finite interval be approximated uniformly by a series of polynomials

$$\frac{1}{\sqrt{I - S^2}} = I - \frac{1}{\sigma^2}\left[\frac{1}{\sqrt{1+\sigma^2}} - 1\right] S^2. \tag{17}$$

According to (9), this is to be multiplied by $I + S$. The product is written as

$$\frac{I+S}{\sqrt{I-S^2}} = (I+S)\Big[-\frac{1}{\sigma^2}\Big(\frac{1}{\sqrt{1+\sigma^2}} - 1\Big)(S^2 + \sigma^2 I) + \frac{1}{\sqrt{1+\sigma^2}} I\Big]. \tag{18}$$

By this, the expression is prepared for making use of (16); the result is

$$\frac{I+S}{\sqrt{I-S^2}} = I + \frac{1}{\sqrt{1+\sigma^2}} S + \frac{1}{\sigma^2}\Big(1 - \frac{1}{\sqrt{1+\sigma^2}}\Big) S^2. \tag{19}$$

The coefficient of S^2 is rewritten in the following brilliant way:

$$\frac{1}{\sigma^2}\Big(1 - \frac{1}{\sqrt{1+\sigma^2}}\Big) = \frac{1}{\sigma^2\sqrt{1+\sigma^2}}\big(\sqrt{1+\sigma^2} - 1\big) = \frac{1}{\sigma^2\sqrt{1+\sigma^2}}\,\frac{1+\sigma^2-1}{\sqrt{1+\sigma^2}+1}$$

$$= \frac{1}{1 + \sigma^2 + \sqrt{1+\sigma^2}}.$$

So the final version for "orthogonalization" of S is

$$\Phi' = \operatorname{orth}(S) = \frac{I+S}{\sqrt{I-S^2}} = I + \frac{1}{\sqrt{1+\sigma^2}} S + \frac{1}{1+\sigma^2+\sqrt{1+\sigma^2}} S^2 \tag{20}$$

with $\sigma^2 = \frac{1}{2}\operatorname{tr}(SS^{\mathrm{T}})$. This formula is implemented later on as a Pascal procedure.

4. The Adjustment and the Iterative Procedure

For the moment being, we suppose that $W \in \mathcal{W}$ is given. According to (2), $\Phi W \Psi$ is still in $\mathcal{W}$ when Φ and Ψ are non-singular matrices. We shall understand Φ and Ψ as rotational matrices so this condition is satisfied. The tangent manifold to $\mathcal{W}$ at Φ, Ψ is given as

$$\begin{aligned}\Phi W \Psi + d(\Phi W \Psi) &= \Phi W \Psi + d\Phi\, W\Psi + \Phi W\, d\Psi \\ &= \Phi W \Psi + \Phi S W \Psi + \Phi W T \Psi \\ &= \Phi(W + SW + WT)\Psi.\end{aligned}$$

Thereby the W-matrix has in the hyperplane $\mathcal{Y}$ been transformed to $W + SW + WT$ which is exactly condition (5). This again is equivalent to (8), and

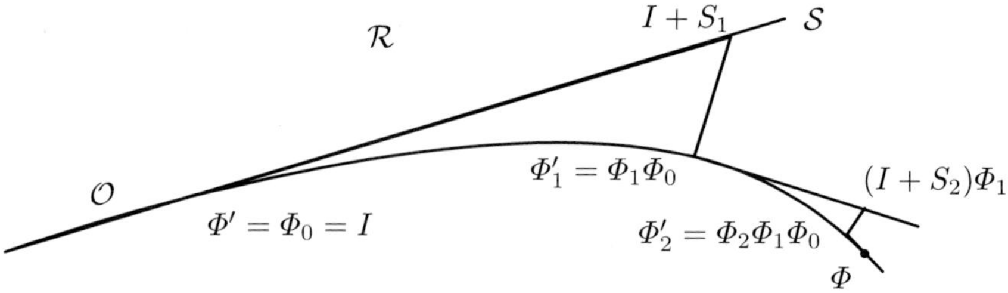

Fig. 28.1. The linear adjustment happens in the tangent space of the manifold $\mathcal{O}$ of orthogonal matrices at Φ_i'

we shall evaluate this expression into its explicit form. Recall that $\langle Y, W\rangle = \operatorname{tr}(Y^{\mathrm{T}}W) = 0$, and we obtain

$$(w_{22} + w_{33})(s_1 + t_1) - w_{12}s_2 - w_{13}s_3 - w_{21}t_2 - w_{31}t_3 = w_{23} - w_{32}. \quad (21)$$

This equation may be interpreted as an observational equation with unknowns $s_1 (= t_1)$, s_2, s_3, t_2, and t_3. There are as many observation equations as measured point coordinates. The coefficients of the unknowns are

$$2(w_{22} + w_{33}), \quad -w_{12}, \quad -w_{13}, \quad -w_{21}, \quad \text{and} \quad -w_{31},$$

with the right-hand side $w_{23} - w_{32}$. The unknowns are determined through a least-squares procedure with appropriate weights.

We repeat the definition of $W = uv^{\mathrm{T}}$. The measured point coordinates in the left image are designated (x', y') and in the right image (x'', y''). These coordinates and the camera constants are in the following way collected in vectors: $u = (x', y', c_1)$ and $v = (x'', y'', c_2)$. So we have

$$\begin{array}{lll} w_{11} = x'x'' & w_{12} = y'x'' & w_{13} = c_1x'' \\ w_{21} = x'y'' & w_{22} = y'y'' & w_{23} = c_1y'' \\ w_{31} = x'c_2 & w_{32} = y'c_2 & w_{33} = c_1c_2. \end{array}$$

The adjustment solution determines the two skew-symmetric matrices S and T. They are orthogonalized by means of (20) and we multiply to the left and to the right to obtain the updated rotational matrices:

$$\Phi' = \operatorname{orth}(S)\,\Phi; \qquad \Psi' = \Psi\operatorname{orth}(T). \quad (22)$$

Initially Φ and Ψ are put equal to the identity matrix. This procedure is repeated a few times till a given numerical accuracy is attained.

The two matrices Φ and Ψ rotate the coordinate systems symmetrically with respect to a vertical plane through the x-axis. This originally vertical plane is rotated a small amount in each iteration.

The product $\Phi_n\Phi_{n-1}\cdots\Phi_2\Phi_1$ of orthogonal matrices is not commutative. It is important that the preliminary orthogonal matrix $\Phi_{n-1}\cdots\Phi_2\Phi_1$ is multiplied to the left with Φ_n.

In the present context, we solve the problem of relative orientation; the problem of the non-vertical symmetry plane is part of the subsequent absolute orientation.

5. Practical Experiences with the Procedure

The method described has been used successfully for more than five years at the former Danish Geodetic Institute. The method has been tested both for speed and convergence.

The present method has been shown to be superior in speed compared to all other standard methods applied for relative orientation on the given computer. Often, the speed was multiplied by a factor of 2.

With reference to convergence, we have tested the method for all imaginable situations which occur in aerial and terrestrial photogrammetry. With less than 4 iterations—including the zero iteration—we have obtained results which we could have expected within the numerical accuracy of the computer.

In the following, we present a commented version of a newly written Pascal program. Many sophisticated considerations are to be found in the comments about finding elegant implementations on a computer.

```
program relative_orientation;
{$N+}
const
  umax = 6;  {umax .. maximum number of unknowns}
  vmax = 28;
type
  obs_line = 1..umax;
  tri_len = 1..vmax;
  vec  = array[tri_len] of double;
  vec1 = array[obs_line] of double;
  arrV = array[1..3] of double;
  arrM = array[1..3,1..3] of double;
  arrA = array[1..4] of real;
var
  bool: boolean;
  sign,iteration,g,h,i,j,k,nop,p,q,u,u1,v,w1: integer;
  base,c1,c2,rhs,sigma,x,y,z: real;
  M: vec;
  obs: vec1;
  filnavn: string;
  infile,outfile,w_file: text;
  DUM, PHI, PSI, PSIT, S, T, W, W0: arrM;
  sum,x1,x2: double;
  A: arrA;
  ans: char;
```

The function NLL is a normal equation solver using the method of Cholesky. The upper triangular part of the normal equations is packed in a one-dimensional array N; this happens by means of the OTN procedure. The function value is the a posteriori variance factor (of the unit weight) $\hat{\sigma}^2$. The integer s is the number of unknowns plus one, viz. the right-hand side. The boolean BS is true if you want the back substitution to take place.

```
function NLL (var N: vec; var s,fr,fc: integer; var BS: boolean): real;
var
  r,c,p1,I,Ir,Ic: integer;
  sum: double;
begin
  for r:=fr+1 to s do
  begin
    Ir:=(r*r-r) div 2; I:=Ir+fc;
    for c:=fc+1 to r do
    begin
      sum:=0; Ic:=(c*c-c) div 2;
      I:=I+1;
      for p1:=fc+1 to c-1 do
        sum:=sum+N[Ir+p1]*N[Ic+p1];
      N[I]:=N[I]-sum;
      if r<>c then N[I]:=N[I]/N[Ic+c];
    end;
    if r <> s then N[I]:=sqrt(N[I])
    else
    begin
      NLL:=N[I];
      if BS then
        for c:=s-1 downto 1 do
        begin
          Ir:=I-1; I:=I-1; Ic:=(c*c+c) div 2; N[I]:=N[I]/N[Ic];
          for p1:=c-1 downto 1 do
          begin
            Ir:=Ir-1; Ic:=Ic-1;
            N[Ir]:=N[Ir]-N[I]*N[Ic];
          end;
        end;
    end;
  end;
end;
```

The procedure OTN creates from the coefficients in the single observation equation B the contribution to the normal equations and puts it at the proper places in the array of the normal equations NL.

```
procedure OTN (var B: vec1; var NL: vec; s: integer);
var
  i,r,c: integer;
```

```
begin
  i:=0;
  for r:=1 to s do
    for c:=1 to r do
    begin
      i:=i+1; NL[i]:=NL[i]+B[r]*B[c];
    end
end;
```

Now follows an implementation of formula (20). Note that in this procedure `S` is the usual S plus the identity matrix I.

```
procedure orth (var S: arrM);
var
  V: arrV;
  a, b: double;
  i, j: integer;
begin
  V[1]:=-S[2,3];
  V[2]:= S[1,3];
  V[3]:=-S[1,2];
  a:=1;
  for i:=1 to 3 do
    a:=a+sqr(V[i]);
  b:=a;
  a:=sqrt(a);
  b:=b+a;
  for i:=1 to 3 do
    for j:=1 to 3 do
      S[i,j]:=S[i,j]/a+V[i]*V[j]/b;
end;
```

A simple procedure for transposing an $r \times s$ matrix, $B = A^{\mathrm{T}}$ is the following:

```
procedure transp (var A,B: arrM; r,s: integer);
begin
  for i:=1 to r do
    for j:=1 to s do
      B[i,j]:=A[j,i];
end;
```

A simple procedure for multiplication of two 3×3 matrices, $C = AB$ is the following:

```
procedure matrixmult (var A,B,C: arrM);
var
  j: integer;
  s: double;
begin
  for i:=1 to 3 do
```

```
    for k:=1 to 3 do
    begin
      s:=0;
      for j:=1 to 3 do
        s:=A[i,j]*B[j,k]+s;
      C[i,k]:=s;
    end;
end;
```

Initialization of the matrices A, Φ, and Ψ:

```
begin
  for i:=1 to 4 do
    A[i]:=0;
  for i:=1 to 3 do
    for j:=1 to 3 do
    begin
      PHI[i,j]:=0;
      PSI[i,j]:=0;
    end;
  for i:=1 to 3 do
  begin
    PHI[i,i]:=1;
    PSI[i,i]:=1;
  end;

  assign(infile,'rel.in');
  assign(outfile,'rel.out');
  assign(w_file,'w.in');
  rewrite(outfile);
  rewrite(w_file);
```

Opening of the output file is for the results and opening of the input file `rel.in`. The input file is expected to contain the following information:

```
6
150 150
 16.012   79.963   -73.930   78.706
 88.560   81.134    -5.252   78.184
 14.618   -0.231   -76.006    0.036
 86.140   -1.346    -7.706   -2.112
 13.362  -79.370   -79.122  -78.879
 82.240  -80.027    -9.887  -80.089
```

The first line contains the number of observations, i.e., number of measured point coordinates `nop`. The second line contains the left and right camera constants `c1, c2`, not necessarily equal. Then follows line-wise the observations x', y', x'', and y''. The present example is taken from [5], p. 179.

In practice the rotational angles s_i and t_i are not necessarily small. This may happen if the model is constituted of images from different flying lines; in order to handle this situation, we have introduced a zeroth iteration. During that iteration we rotate the left image so that the barycenter of measured points lies at the positive part of the x-axis and the right image so that the barycenter lies at the negative part of the x-axis. In doing so, we have determined the two rotational angles around the z-axes.

Once and for all the W matrix is computed for each point. The result is stored in the `w_file`.

```
writeln('Zero iteration?  [Y/N]: ');
readln(ans);
if UpCase(ans)= 'Y' then iteration:=-1 else iteration:=0;

reset(infile);
read(infile,nop);
read(infile,c1,c2);
c1:=-c1;  c2:=-c2;

for q:=1 to nop do
begin
  for i:=1 to 4 do
    read(infile,obs[i]);
  W[1,1]:=obs[1]*obs[3]; W[1,2]:=obs[2]*obs[3]; W[1,3]:=c1*obs[3];
  W[2,1]:=obs[1]*obs[4]; W[2,2]:=obs[2]*obs[4]; W[2,3]:=c1*obs[4];
  W[3,1]:=obs[1]*c2;     W[3,2]:=obs[2]*c2;     W[3,3]:=c1*c2;
  for i:=1 to 3 do
    writeln(w_file,W[i,1]:15:5,W[i,2]:15:5,W[i,3]:15:5);
end;
close(w_file);
```

The number of unknowns `u` is 5, and the coefficients of the normal equations `M` are initialized. The coordinate observations are read and the `W` matrix is computed.

```
rhs:=1;
u:=5;
u:=u+1; v:=(u*u+u) div 2; w1:=(u*u-u+2) div 2;

repeat
  reset(w_file);
  iteration:=iteration+1;
  writeln(outfile,'Iteration    ', iteration:2);
  writeln(outfile,'Number of points   ',nop);
  for q:=1 to v do
    M[q]:=0;
  for q:=1 to nop do
  begin
    for i:=1 to 3 do
      readln(w_file,W[i,1],W[i,2],W[i,3]);
    if iteration > 0 then
```

```
    begin
      matrixmult(W,PSI,WO);
      matrixmult(PHI,WO,W);
      obs[1]:=(W[2,2]+W[3,3])*2; obs[2]:=-W[1,2];
      obs[3]:=-W[1,3];           obs[4]:=-W[2,1];
      obs[5]:=-W[3,1];           obs[6]:=W[2,3]-W[3,2];
    end
    else
    begin
      A[1]:=A[1]+W[1,3]; A[2]:=A[2]-W[2,3];
      A[3]:=A[3]-W[3,1]; A[4]:=A[4]-W[3,2];
    end;
    if iteration > 0 then OTN(obs,M,u);
  end;
```

By this, the input of observations has finished (If the observations occur with varying weights all the terms `obs[x]` must be multiplied by the square root of the particular weight of that single observation). Then comes the proper adjustment procedure. It performs according to the description given in Section 4. The iteration is stopped after a specified number of passes or after the norm of the right-hand side `rhs` has reached a specified small (machine dependent) number.

```
  writeln(outfile,'The unknowns');
  g:=0; h:=0; bool:=true;
  if iteration > 0 then
  begin
    sigma:=sqrt(NLL(M,u,g,h,bool)/(nop-u+1));
    for q:=w1 to v-1 do
      writeln(outfile,M[q]:16:10);
    writeln(outfile,'Sigma   ',sigma:2:10);
    sum:=0;
    for i:=w1 to v-1 do
      sum:=sqr(M[i])+sum;
    rhs:=sqrt(sum);
    writeln(outfile,'Right hand side  ',rhs:2:10);
    S[1,1]:= 1;        S[1,2]:=-M[w1+2]; S[1,3]:= M[w1+1];
    S[2,1]:=-S[1,2];  S[2,2]:= 1;        S[2,3]:=-M[w1];
    S[3,1]:=-S[1,3];  S[3,2]:=-S[2,3];  S[3,3]:= 1;

    T[1,1]:= 1;        T[1,2]:=-M[w1+4]; T[1,3]:= M[w1+3];
    T[2,1]:=-T[1,2];  T[2,2]:= 1;        T[2,3]:=-M[w1];
    T[3,1]:=-T[1,3];  T[3,2]:=-T[2,3];  T[3,3]:= 1;
    writeln(outfile,'S  ');
    for i:=1 to 3 do
      writeln(outfile, S[i,1]:10:5, S[i,2]:10:5, S[i,3]:10:5);
    writeln(outfile);

    orth(S);
```

```
      writeln(outfile,'S-orth  ');
      for i:=1 to 3 do
        writeln(outfile, S[i,1]:10:5, S[i,2]:10:5, S[i,3]:10:5);
      writeln(outfile);
    end
    else
    begin
      sum:=0;
      for i:=1 to 4 do
        writeln(outfile,'A[',i:1,']/',nop:2, A[i]/nop:12:3);
      for i:=0 to 1 do
      begin
        sum:=sqrt(sqr(A[2*i+1])+sqr(A[2*i+2]));
        A[2*i+1]:=-A[2*i+1]/sum; A[2*i+2]:=-A[2*i+2]/sum;
      end;
      S[1,1]:=A[1]; S[1,2]:=-A[2]; S[1,3]:=0;
      S[2,1]:=A[2]; S[2,2]:= A[1]; S[2,3]:=0;
      S[3,1]:=0;    S[3,2]:=0;     S[3,3]:=1;
      T[1,1]:=A[3]; T[1,2]:=-A[4]; T[1,3]:=0;
      T[2,1]:=A[4]; T[2,2]:=A[3];  T[2,3]:=0;
      T[3,1]:=0;    T[3,2]:=0;     T[3,3]:=1;
    end;
    matrixmult(S,PHI,DUM);
    for i:=1 to 3 do
      for k:=1 to 3 do
        PHI[i,k]:=DUM[i,k];

    if iteration > 0 then orth(T);

    matrixmult(PSI,T,DUM);
    for i:=1 to 3 do
      for k:=1 to 3 do
        PSI[i,k]:=DUM[i,k];

    writeln(outfile,'PHI  ');
    for i:=1 to 3 do
      writeln(outfile, PHI[i,1]:10:5, PHI[i,2]:10:5, PHI[i,3]:10:5);
    writeln(outfile);
    writeln(outfile,'PSI  ');
    for i:=1 to 3 do
      writeln(outfile, PSI[i,1]:10:5, PSI[i,2]:10:5,PSI[i,3]:10:5);
    writeln(outfile);
  until (iteration > 8) or (rhs < 10e-10);
```

Now follows an optional part, the purpose of which is to balance the rotations around the x-axis. The model coordinates are balanced as well.

Finally, we output the results: Standard deviations of the weight unit and the single unknowns, the rotational matrices Φ, and Ψ, the model coordinates of the measured points and the right-hand side. The (dimensionless) model coordinates are normed according to $b = 1$. In case of the data from [5], our values for the coordinates x, y, z have to be multiplied by 90 mm (the base length) to be compared to the result given at loc. cit.

For the final calculation of model coordinates, it is necessary to rotate the right coordinate system back through the angles t_1, t_2, and t_3. The reasoning is as follows: If $u' = \Phi u$ and $v' = \Psi v$ then $W' = u'v'^{\mathrm{T}} = \Phi u(\Psi v)^{\mathrm{T}} = \Phi u v^{\mathrm{T}} \Psi^{\mathrm{T}} = \Phi W \Psi^{\mathrm{T}}$. Thus, the matrix Ψ must be transposed.

```
   A[1]:=PHI[3,3]+PSI[3,3];
   A[2]:=PHI[2,3]+PSI[3,2];
   A[3]:=sqrt(sqr(A[1])+sqr(A[2]));
   A[1]:=A[1]/A[3]; A[2]:=A[2]/A[3];

   for i:=1 to 3 do
     for j:=1 to 3 do
       S[i,j]:=0; T[i,j]:=0;
   S[1,1]:=1; T[1,1]:=1;
   S[2,2]:=A[1]; S[3,3]:=A[1]; T[2,2]:=A[1]; T[3,3]:=A[1];
   S[2,3]:=-A[2]; T[3,2]:=-A[2];
   S[3,2]:=A[2];  T[2,3]:=A[2];

   matrixmult(S,PHI,DUM);
   for i:=1 to 3 do
     for k:=1 to 3 do
       PHI[i,k]:=DUM[i,k];
   matrixmult(PSI,T,DUM);
   for i:=1 to 3 do
     for k:=1 to 3 do
       PSI[i,k]:=DUM[i,k];

  writeln(outfile);
  writeln(outfile,'Standard deviation of the weight unit ',sigma:6:6);
  for q:=0 to u-2 do
  begin
    for i:=w1 to v do
      M[i]:=0;
    M[q+w1]:=1;  u1:=u-1; bool:=false;
    writeln(outfile,'Standard deviation of unknown ',(q+1):2,
                 sqrt(-NLL(M,u,u1,q,bool))*sigma:12:6);
  end;

  transp(PSI,PSIT,3,3);

  writeln(outfile);
  writeln(outfile,'Final PHI');
  for i:=1 to 3 do
    writeln(outfile, PHI[i,1]:10:5, PHI[i,2]:10:5, PHI[i,3]:10:5);
```

```
  writeln(outfile);
  writeln(outfile,'Final PSI');
  for i:=1 to 3 do
    writeln(outfile, PSIT[i,1]:10:5, PSIT[i,2]:10:5, PSIT[i,3]:10:5);
  writeln(outfile);
  writeln(outfile,'Camera constants  ', c1:8:2, c2:8:2);
  writeln(outfile,'Number of points   ',nop:2);
  writeln(outfile,'Model coordinates [b=1]');
  write  (outfile,'          x              y              z');
  writeln(outfile,'       W[2,3]-W[3,2]');
  reset(w_file);
  for q:=1 to nop do
  begin
    for i:=1 to 3 do
      readln(w_file,W[i,1],W[i,2],W[i,3]);
    matrixmult(W,PSI,W0);
    matrixmult(PHI,W0,W);
    z:=-1;
    y:=z*(W[3,2]+W[2,3])/(2*W[3,3]);
    x1:=z*W[3,1]/W[3,3];
    x2:=z*W[1,3]/W[3,3];
    base:=x2-x1;
    x:=(x1+x2)/(2*base);
    y:=y/base;
    z:=z/base;
    writeln(outfile, x:15:5, y:15:5, z:15:5, W[2,3]-W[3,2]:15:5);
  end;

  close(infile);
  close(outfile);
end.
```

Test computations show that the method—including the zeroth iteration—works well for any rotation around the z-axis, and in these cases the procedure normally converges in at most four to five iterations. Under usual circumstances three to four iterations suffice with or without the zeroth iteration.

References

[1] W. V. D. Hodge and D. Pedoe. *Methods of Algebraic Geometry*, volume 2. Cambridge Mathematical Library, 1952.

[2] J. Krames. Über ein graphisches Verfahren zum gegenseitigen Einpassen von Luftaufnahmen. *Österreichische Zeitschrift für Vermessungswesen*, pages 13–29, 1949.

[3] Torben Krarup. Contribution to the geometry of the Helmert transformation. Technical report, 1985.

[4] Igor R. Safarevic. *Basic Algebraic Geometry*, volume 213 of *Grundlehren der mathematischen Wissenschaften*. Springer, Berlin, 1974.

[5] Kurt Schwidefsky and Friedrich Ackermann. *Photogrammetrie*. B. G. Teubner, Stuttgart, 1976.

[6] Gilbert Strang. *Linear Algebra and Its Applications*. Harcourt and Brace and Jovanovic, San Diego, third edition, 1988.

A

Bibliography for Torben Krarup

* (1956) *Geodetic Tables.* Geodætisk Instituts Skrifter 3. Række Bind XXIV, København, with E. Andersen and B. Svejgaard.

* (1956) *A Method for Matrix Multiplication, Matrix Inversion, and Problems of Adjustment by Punched Card Equipment.* Geodætisk Institut Meddelelse No. 31, København, with B. Svejgaard.

* (1960) Udjævning med koordinater. (Adjustment using coordinates, in Danish.) (Proceedings) Referat af det tredie nordiske geodætmøde i København, Maj 1959, 229–233. Nordiska Kommissionen för Geodesi, København.

* (1961) GIER—Logical Organization. Ingeniøren, International Edition, 5:3–8, with B. Svejgaard.

(1962) Linear equations. In: Chr. Gram (ed.), *Selected Numerical Methods*, Copenhagen.

(1965) The adjustment procedure in tensor form. Unpublished note.

* (1966) Datalogiske aspekter af geodætiske beregninger. Unpublished, typed according to the manuscript, Geodætisk Institut.

(1967) The Theory of Rounding Errors in the Adjustment by Elements of Geodetic Networks. Bulletin Géodésique, 84:117–122.

* (1967) Problemer i forbindelse med netudjævning i koordinater. (Proceedings) Det femte Nordiska Geodetmötet i Stockholm, April 1966, 137–142, Nordiska Kommissionen för Geodesi, Stockholm.

* An asterisk indicates that the title is not included in this volume.

* (1967) Data processing and automation in cartography. UN Inter-regional Seminar on Application of Cartography for Economic Development, Krogerup, Denmark.

* (1967) Molodenskij's Problem. XIV General Assembly IAG, Section V: Dynamic Geodesy.

* (1968) A Framework for Least Squares Determination of the Potential of the Earth, A preliminary report. Letter to IAG SSG 5.31.

(1969) *A Contribution to the Mathematical Foundation of Physical Geodesy.* Meddelelse No. 44, Geodætisk Institut, København.

(1970) A Remark on Approximation of T by Series in Spherical Harmonics. Proceedings Fourth Symposium on Mathematical Geodesy, Trieste, 28–30 May, 1969, 153–154, Bologna.

* (1970) The Method of Least Squares Collocation. Studia geophysica et geodaetica, 14:107–109.

* (1970) Letter (on Convergence problems) to SSG 5:31.

* (1970) On Determination of a "Reasonable" Distribution of Masses Corresponding to a Potential Harmonic Outside a Closed Surface. Unpublished note, Geodætisk Institut.

* (1971) Lectures read at the Technical University of Berlin, May 1971. I: Geodesy and Mathematics today. II: Introducing Integrated Geodesy. Unpublished, Geodætisk Institut.

* (1971) Letter to the members of SSG 5:31 of the IAG.

* (1971) Dynamisk Geodæsi. (Proceedings) Den Sjette Kongressen i Helsingfors, Maj 1970, 250–252, Nordiska Kommissionen för Geodesi, Helsinki.

* (1971) The Geodetical Background for Integrated Geodesy. The New Geodesy and the Geodetic Tradition. First part of unfinished manuscript, Geodætisk Institut.

(1972) On the Geometry of Adjustment. Zeitschrift für Vermessungswesen, 97:440–445.

(1973) Remarks on the Discussion Yesterday. Bollettino di Geodesia e Scienze Affini, XXXII:145–147.

(1973) Letters on Molodenskiy's Problem. Unpublished, Geodætisk Institut.

(1973) On the Spectrum of Geodetic Networks. Presented at Symposium on Computational Methods in Geometrical Geodesy, Oxford, with K. Borre.

(1974) Mathematical Geodesy. An invited lecture given at the VII meeting of the Nordic Geodetic Commission. (Proceedings) Referat af 7. nordiske geodætmøde, København, May 1974, pp. 175–180, Nordiska Kommissionen för Geodesi, København.

(1974) Foundation of a Theory of Elasticity for Geodetic Networks. (Proceedings) Referat af 7. nordiske geodætmøde, København, May 1974, pp. 181–212, Nordiska Kommissionen för Geodesi, København, with K. Borre.

(1975) Integrated Geodesy. Methoden und Verfahren der mathematischen Physik, 13:77–124, with J. Eeg.

* (1975) Last letter to the SSG No. 4:31. La formule de Stokes est-elle correcte? Qui, certainement. Geodætisk Institut.

(1975) On potential Theory. Methoden und Verfahren der mathematischen Physik, Bibliographisches Institut AG, Manheim Wien Zürich, 12:79–160.

(1976) La formule de Stokes est-elle correcte? Commentaires sur le papier de W. Baranov. Bulletin Géodésique, 50:251–252.

(1978) Some remarks about collocation. In: H.Moritz and H.Sunkel (ed.s): *Approximation methods in Geodesy*, 193–209, H. Wichmann Verlag, Karlsruhe.

* (1979) A few ideas on collocation loosely formulated in order to bring the discussion on norm, statistics, and all that out of a dead-end track. Unpublished note, Geodætisk Institut.

(1979) Apropos some recent papers by Willi Freeden, on a class of integral formulas in the mathematical geodesy. Unpublished note, Geodætisk Institut.

(1979) S-transformation or How to live without the generalized inverse—almost. Unpublished note, Geodætisk Institut.

(1980) Integrated Geodesy. Bollettino di Geodesia e Scienze Affini, XXXIX: 315–330.

* (1980) Götterdämmerung over least-squares adjustment. Proceedings 14th Congress of ISP, Hamburg, with K. Kubik and J. Juhl.

(1981) A Measure for Local Redundancy. A contribution to the reliability theory for geodetic networks. Unpublished note, Geodætisk Institut.

(1981) A Convergence Problem in Collocation Theory. Bollettino di Geodesia e Scienze Affini, XL:225–240.

(1982) Non-Linear Adjustment and Curvature. Unpublished note, Geodætisk Institut.

(1982) Mechanics of Adjustment. Presented at Peter Meissl Gedenk Seminar über mathematische Geodäsie, Graz.

(1982) Non-Linear Adjustment and Curvature. Feestbundel ter gelegenheid van de 65ste verjaardag van Professor Baarda, 1:146–159, Delft.

(1983) Angelica Returning or The Importance of a Title. In: K.P. Schwarz and G. Lachapelle (ed.s): *Geodesy in Transition*, 109–118, The University of Calgary, Division of Surveying Engineering, Publication 60002.

(1984) Evaluation of Isotropic Covariance Functions of Torsion Balance Observations. Bulletin Géodésique, 58:180–192, with C.C. Tscherning.

(1985) Contribution to the Geometry of the Helmert Transformation. Unpublished note.

(1987) Letter on a Problem in Collocation Theory. Unpublished note.

(1991) Approximation to the Earth Potential From Discrete Measurements. Proceedings of the Geodetic Day in Honor of Antonio Marussi. Rome, October 9th, 1989. Atti dei Convegni Lincei, 91:67–85.

(1998) An Old Procedure for Solving Relative Orientation in Photogrammetry. Unpublished note, with K. Borre and J. Olsen.

Printing: Krips bv, Meppel
Binding: Stürtz, Würzburg